Reconstructing Archaeological Sites

Reconstructing Archaeological Sites

Understanding the Geoarchaeological Matrix

Panagiotis (Takis) Karkanas
Malcolm H. Wiener Laboratory for Archaeological Science, American School of Classical Studies at Athens
Athens, Greece

Paul Goldberg
School of Earth and Environmental Sciences, University of Wollongong
Wollongong, Australia
and
Institute for Archaeological Sciences, University of Tübingen
Tübingen, Germany

This edition first published 2019
© 2019 John Wiley & Sons Ltd

The right of Panagiotis (Takis) Karkanas and Paul Goldberg to be identified as the authors of this work has been asserted in accordance with law.

Registered Offices
John Wiley & Sons, Inc., 111 River Street, Hoboken, NJ 07030, USA
John Wiley & Sons Ltd, The Atrium, Southern Gate, Chichester, West Sussex, PO19 8SQ, UK

Editorial Office
9600 Garsington Road, Oxford, OX4 2DQ, UK

For details of our global editorial offices, customer services, and more information about Wiley products visit us at www.wiley.com.

Wiley also publishes its books in a variety of electronic formats and by print-on-demand. Some content that appears in standard print versions of this book may not be available in other formats.

Library of Congress Cataloging-in-Publication data applied for

ISBN: 9781119016403

Cover Design: Wiley
Cover Images: Courtesy of Paul Goldberg and Panagiotis Karkanas; Magnifying glass by igoscience is licensed under CC BY 3.0

Set in 10/12pt Warnock by SPi Global, Pondicherry, India
Printed and bound in Singapore by Markono Print Media Pte Ltd

10 9 8 7 6 5 4 3 2 1

To Ofer Bar-Yosef, friend, colleague, and source of inspiration, and APJ for support

Contents

Preface

Why did we decide to write another geoarchaeology book? That is a good question in light of the fact that geoarchaeology is finally becoming a mature discipline. We are both geologists by training, close friends, and we communicate by Skype, telephone, and at conferences. Yet, ironically, we penecontemporaneously and independently realized that archaeological deposits have been overlooked as valuable and necessary resources of data in the archaeological record. We came to the conclusion as practicing geoarchaeologists – with perhaps too much field experience – that the ultimate context of archaeological objects is the surrounding sedimentary matrix and, as a consequence, the anthropogenic and geogenic depositional processes that form it. Such a discovery is not really new, but in searching through the literature we discovered that there is no systematic or in-depth treatment of the topic – the geoarchaeological matrix – and how it can be thoroughly squeezed to provide a wealth of latent information that can be used to reconstruct geological and human histories at archaeological sites.

This book focuses on basic theoretical and practical aspects of depositional processes in archaeological sites and discusses in depth the role of stratigraphy, which fundamentally is an intellectual construct that helps us envision how deposits are organized in time and space.

The book is constructed in such a way as to offer archaeologists and geoarchaeologists – whether senior level researchers or students at the beginning of their studies – an inventory of information and tools that they can use to recognize depositional systems, processes, and stratigraphic units that will enable them to realize and interpret the stratigraphy of a site in the field. We feel that this approach is a constructive way to further the dialogue between archaeologists and geologists.

In order to achieve this broad goal, we provide examples, along with illustrations and drawings, photographs and in-depth descriptions that will help the reader visualize depositional processes and how these build the stratigraphy of a site. In addition, we include some of the practical tools that we have used over a number of years of working closely with archaeologists in many types of sites and localities. In sum, we have tried to offer a holistic approach to the study of archaeological deposits, from the broad fundamental aspects to the details. We hope that we will convince the reader that *deposits are artefacts* on a par with other objects of the archaeological record. After all, they furnish the essential aspect of the sedimentary matrix that encircles archaeological objects and features from virtually any site anywhere on the globe.

Acknowledgments

The notions that we have developed here would not have been possible without the participation of our friends and colleagues in the field over a lot of time during our professional lifetimes. It is they who put us on to thinking of these issues, whether intentionally or not. They are too numerous to name individually, but a few stand out and are listed alphabetically here:

V. Aldeias, T. Arpin, G. Avery, O. Bar-Yosef, F. Bordes, C.B. Bousman, K.W. Butzer, M. Chazan, M. Chech, A. Belfer-Cohen, B. Byrd, E. Chambers, M.B. Collins, N.J. Conard, M.-A. Courty, M. Dabney, H.L. Dibble, N. Efstratiou, W.R. Farrand, N. Fedoroff, C.R. Ferring, N. Goren, N.A. Goring-Morris, K. Harvati, D.O. Henry, J.F. Hoffecker, V.T. Holliday, E. Hovers, G. Huckleberry, Z. Jacobs, A.J. Jelinek, S. and E. Kaplan, D. Killick, N. Kyparissi-Apostolika, S.L. Kuhn, H. Laville, T.E. Levy, R.I. Macphail, R.D. Mandel, C.E. Miller, C. Mallol, C.W. Marean, S.P. McPherron, L. Meignen, S.M. Mentzer, A. van de Moortel, M. Morley, L.C. Nordt, E. Panagopoulou, A. Pérez-Juez, W. Parkinson, K. Pavlopoulos, R.G. Roberts, C. Roos, A.M. Rosen, D. Sandgathe, H.P. Schwarcz, R. Shahack-Gross, L.A. Schepartz, S.C. Sherwood, J.D. Speth, J.K. Stein, M.L. Stiner, C.B. Stringer, L.A. Sullivan, A. Turq, G. Tsartsidou, B. Vandermeersch, S. Villeneuve, S. Weiner, I. Whitbread, R. White, J.C. Woodward, H.T. Wright, J.C. Wright, and A. Yair.

We owe special thanks to Jamie Woodward, Vance Holiday, Ruty Shahack-Gross, Steve Weiner, Rich Macphail, Bill Parkinson, and Amalia Pérez-Juez for their insightful and constructive suggestions on early versions of various chapters. They really helped a lot.

Over the years we have received institutional support that in one way or another (providing funding or the availability of intellectual or material resources) has ultimately made this work possible from (in no particular order) the Alexander von Humboldt Foundation, the American School of Classical Studies in Athens, Boston University, the Care Foundation, Ephoreia of Palaeoanthropology-Speleology (Ministry of Culture, Greece), Harokopio University (Greece), the Cave Research Foundation, the Geological Society of America, the Society for American Archaeology, the Archaeological Institute of America, Harvard University, Simon Fraser University, the French Foreign Ministry, the French CNRS, the Hebrew University of Jerusalem, the University of Michigan, the University of Texas at Austin, the University of Tübingen, the University of Wollongong, the US National Science Foundation, the Malcolm W. Wiener Foundation, the National Geographic Society, and the LSB Leakey Foundation.

Abbreviations

BA	Bronze Age
EBA	Early Bronze Age
ESA	Early Stone Age
LBA	Late Bronze Age
LP	Lower Palaeolithic
LSA	Later Stone Age
MBA	Middle Bronze Age
MP	Middle Palaeolithic
MSA	Middle Stone Age
PPL	plane-polarized light
UP	Upper Palaeolithic
XPL	crossed polarized light

I

Introduction: A Depositional Approach to the Study of Archaeological Excavations

Figure I.1 Profile photographs from (a) Medieval Corinth and (b) Palaeolithic Theopetra, Greece (from Karkanas and Goldberg, 2017b). Note the complexity in both despite the vastly different ages. Scale in (a) = 3 m and in (b) = 1 m.

Look at Figure I.1. It shows two photographs from archaeological sites in Greece that we have studied: Figure I.1a is a profile of Middle and Upper Palaeolithic layers from the site of Theopetra and Figure I.1b is from post Late Roman to recent deposits in the southern part of Ancient Corinth. Without knowing anything about the sites and without any geological training, an observer is immediately struck by the complexity of these stratigraphic sequences. How do we make sense of these deposits and how do we incorporate this information into our understanding of the significant archaeological findings from these two important localities? That is the subject of this book.

Theoretical Issues

It is probably the only consensus in archaeology that stratigraphy is 'the jugular vein of archaeological practice' (McAnany and Hodder, 2009a,b). Stratigraphy has many definitions and is defined here as the spatial and temporal arrangement of depositional units. Regardless of the wording, it provides the framework of reconstructing the history of a site. Stratigraphic units, layers, features, cuts, or strata – also called context or locus according to different archaeological schools of thought – are made of sediments that are the product of natural processes and anthropogenic activities, and are

Reconstructing Archaeological Sites: Understanding the Geoarchaeological Matrix, First Edition. Panagiotis (Takis) Karkanas and Paul Goldberg.
© 2019 John Wiley & Sons Ltd. Published 2019 by John Wiley & Sons Ltd.

deposited on the surface of the earth. Archaeological deposits – those that contain artefacts and anthropogenic products – are thus by their very nature forcibly part of the archaeological record. In order to interpret the archaeology of a site correctly, it is a prerequisite to understand how stratigraphy is built and how the strata are formed. This book is about how we recognize the processes and activities that produce the deposits of a site and how these are organized in time and space to form a stratigraphic sequence.

Our personal understanding of stratigraphy and archaeological deposits is based on a conceptual model, which is described by a few fundamental propositions:

- A site is a three-dimensional arrangement of artefact-bearing deposits, therefore the fundamental unit of a site is the deposit, not the artefact or the pattern of the artefacts.
- The deposits have accumulated by natural or anthropogenic processes or a combination of these.
- The fabrics of the sediment are indicative of the different processes involved in their formation.
- Traditional artefacts (pottery, flints, etc.) are fabric elements within archaeological deposits.
- Architectural features (e.g. walls, mosaics, etc.) may have their own typology and internal stratigraphy but the relationship between construction phases and the surrounding artefacts is mediated by the enclosing deposits.
- Natural deposits in a site may have a cultural meaning (e.g. sediments trapped within aqueducts).
- All elements that form a deposit should be treated as having equal importance with the traditional items of the archaeological record (architecture, pottery, lithics, objects, etc.) in the study of a site.
- Time resolution is essentially determined in the field by how finely we can recognize the vertical and horizontal extent of individual stratigraphic units, which are the proxies for activities and processes. Similarly, such units should be recognized and recorded only during excavation and not after.

For most prehistoric archaeologists, the above principles are known and generally accepted because prehistory traditionally evolved in parallel with geology. But this is not always the case, particularly for those who investigate the archaeology of historical periods, which traditionally treats archaeology as history, with a reliance on texts. Excavating a site by utilizing the above principles is not straightforward, as it demands knowledge of natural and anthropogenic sedimentary processes. However, most archaeology programs in academia do not include basic sedimentology courses. The outcome is that most archaeologists acquire knowledge of stratigraphy through practice in the field and detailed observations of what they see. The question that

underlies this reality is therefore can everyone 'see' all stratigraphic boundaries (or interfaces) in an excavation? Is it a matter of experience, knowledge, or talent?

In the following discussion we will show how stratigraphy is produced, and examine the fundamental elements and attributes that create it. At the same time, we will touch on common misconceptions about what constitutes a deposit, issues related to the nomenclature used, and the different approaches to investigating archaeological stratigraphy.

The Formation of Stratigraphy

Human earthen structures and anthropogenic sediments in general (floors, occupational surfaces, middens, fills, pits, mounds, etc.) have a materiality that, at an initial level, can be described and understood by using concepts and methods derived from the natural sciences, such as geology; in fact, no new terminology is required. The deposits of the site consist of materials that are overwhelmingly particles (clasts) of minerals and rocks. It is puzzling to assume that a unit can be described without referring to the fundamental attributes of sediments, mainly grain-size distribution, sorting, roundness, orientation, grading, colour, and ultimately the fabric produced by the organization of the attributes. All these attributes are not neutral and meaningless, or simply geometric features. The mineralogical content and chemistry are also fundamental properties of an archaeological deposit as they provide information about its source and post-depositional alteration.

Knowledge of the fundamental properties of human earthen constructions is a prerequisite for interpreting this aspect of site stratigraphy. On one hand, these properties serve to convey descriptive criteria to interested researchers, but more importantly they transmit information of the processes that produced them. In the realm of natural sedimentary basins, these descriptive terms imply certain depositional environments. For example, rounding is caused by abrasion during transport (e.g. streams) or by reworking by wave action, for example; transport does not fundamentally reduce the size but does selectively sort material (Folk, 1974). Each transporting medium produces certain types of sediments because the dynamics of transport and deposition are different.

In the realm of a site, human processes do not produce such attributes, which even in earthen constructions are mostly inherited from the original material at the natural source, but now organized in a different way through a 'human filter'. Anthropogenic activities admittedly produce new organizations, but the essential building blocks are the same. Human activities are no different in representing the elemental dynamics of transporting and depositing materials through actions such as trampling, digging, dumping, kneading and pugging, sweeping, brushing, discarding, and placing. Therefore, rounding can be produced by continuous trampling and scuffing, and brushing and sweeping may bring about sorting and lamination. Similarly, dumping may cause grading within the deposits, and pugging will produce orientation, whereas discarding may lead to clustering of grains (see Chapter 3). Some activities do produce new materials, such as burning (ashes) and other pyrotechnological activities (ceramics). Nevertheless, their accumulation and final deposition on the earth's surface is the product of the actions described above.

All the aforementioned attributes are the building blocks of every description of what is called a lithostratigraphic unit. These units are bounded by contacts, and contacts are the product of changes in the attributes, whether they be differences in composition or the way that they are organized. Sometimes, the contact itself records a change in the attributes, although they are rarely discerned as such in the field (further discussed in Chapter 4). However, all these attributes cannot be identified without knowing what to look for, and what they might mean. As is so often the case, without understanding how things can form – in this case, units – it is difficult and challenging to discern and describe stratigraphic entities. In addition, all material and earth science studies in an archaeological site use descriptions derived from physical sciences, mainly mineralogy, petrology, sedimentology, and soil science (pedology). During the last two decades geoarchaeological studies have made significant inroads toward archaeological interpretation of urban sites by providing microstratigraphic histories within rooms, and data on the life of buildings and the use of space (e.g. Matthews, 2005; Matthews et al., 1996; Shahack-Gross et al., 2005; Macphail and Crowther, 2007; Macphail et al., 2007; Milek and Roberts, 2013; Karkanas and van de Moortel, 2014). It will be confusing to have a separate and different system of description for archaeological deposits, and another one for geological ones.

Bear in mind that observation is not independent of theory. What we see is not objective because it is based on a conceptual model of organizing images (Hanson, 1958); in our case, sediment attributes and structures can be organized in a meaningful way, that is, stratigraphy, only if one has a conceptual model of what to look for. Therefore, it is not so much about interpretation of what we see but rather the stratigraphic organization of the ensemble of lines, surfaces, and features that are revealed in the field. Stratigraphic organization is not an object that can be simply described in terms of shape or colour; it is a concept gained after data are evaluated (Hanson, 1958). Stratigraphic theory also maintains that there is a hierarchy among the sedimentary features, which are decisive in correctly understanding a sequence.

Therefore, not every linear feature in the profile or surface on a plan view is a contact or interface. This is important because determining contacts implies other meaningful things rather than just a geometrical feature. It is the appreciation of this conceptual and holistic model of depositional processes and systems that is lacking from most archaeological approaches to stratigraphy. As a consequence, students of archaeology face a frustrating and perplexing task in trying to properly understand a stratigraphic sequence, simply because most do not have background knowledge of depositional and post-depositional processes.

The above discussion highlights that the weight should be shifted from constructing stratigraphy to acquiring knowledge of depositional processes that produced it. At this point, three issues should be discussed further:

a) Can anthropogenic depositional processes be described in the same way as natural (geogenic, pedogenic, biogenic) ones?
b) Except for disentangling stratigraphic relationships, what else is gained when someone succeeds in comprehending the depositional processes in an archaeological site?
c) Is it possible for archaeological education to provide the necessary background to understand such issues?

A. Describing Anthropogenic Depositional Processes

The first issue has been already mentioned above and is also treated extensively in Chapter 1. Here, suffice it to say that the majority of anthropogenic activities that produce deposits on a site are subject to broadly the same principles as natural depositional and post-depositional processes: mainly gravity and the dynamics of applied actions (e.g. brushing, sweeping, treading, dumping, kneading and pugging, and applying plasters), as well as chemical and physical transformations. The three-dimensional patterns of particles in anthropogenic deposits record the action of forces associated with human motions, if of course they are not disturbed later by natural processes. Note that we are talking about only the appearance of the final product and not about the intentions and the purpose of the construction or the cultural and social implications and symbolic inferences associated with the activities of making this product. Obviously, there are problems of equifinality because different activities can produce the same final product in terms of material content, fabric, and structure. But here we want to emphasize that by studying all archaeological deposits in the same way, we first eliminate the possibility of confusing products of natural processes with anthropogenic ones, and second, we restrict considerably the possibilities of interpreting the human activities related to this product. Interpretation of natural processes is more straightforward, as geological literature has much greater temporal depth that is built upon a set of universal sedimentary laws. Therefore, in difficult cases where the feature is not readily understood as human construction, it can be at least determined that this feature is not the product of a natural process.

The Nature of Anthropogenic Deposits

There is no need to consider anthropogenic processes equivalent to biological ones (Stein, 1987) in order to consider artefacts as part of the sediments. First, a deposit consists of materials that can be described using terms from physical sciences, and its properties can be analysed only by using methods from physical sciences. Second, all anthropogenic deposits can be seen as filling a three-dimensional space (the essence of stratigraphy), and the dynamics of material import (both human and 'natural') can be analysed in terms of rhythm/duration, intensity, and direction, or in terms of physical or chemical transformations.

Admittedly, most anthropogenic actions (e.g. sweeping, scooping, dumping, trampling, etc.) are not well studied from the sedimentological point of view – or from an ethnoarchaeological one for that matter. Consider, for example, earthworks. They have been studied from several perspectives, but we do not know what kind of sedimentary structures are produced by backfilling a trench using a spade, for example, although insights are beginning to emerge (Sherwood and Kidder, 2011; Karkanas et al., 2012). Moreover, some insights can be gained from sedimentological studies on dry grain flows in slopes where kinematic sieving results in vertical and slopeward sorting of particles such as in the filling of tombs (Karkanas et al., 2012), or from the investigation of engineering hydraulics as was shown by Sherwood and Kidder for Shiloh Mound in the USA (Sherwood and Kidder, 2011; Anderson et al., 2013). In general, anthropogenic depositional processes can be precisely described although they may be the product of an infinite human repertoire that is sometimes of enigmatic origin. A human structure, for example, though not readily seen in the field, can certainly be defined as such by using a combination of microscopic and chemical techniques (e.g. Terry et al., 2004). The same empirical links used to infer that a structure is a stone wall are also used to infer by microscopic analysis that something is a floor. The content and the organization of its components (fabric) are the key elements that allow us to decide if it is a structure and if

so, what kind. Even if the particular purpose for making this construction is not obvious – which is commonly the case, although much speculation will go into inferring this purpose – the association with other deposits, including the artefact content, guides us to the possible realistic interpretations.

There is quite a discussion in archaeology about the objectivity of a description, and if hard data are theory laden (Trigger, 2006). However, we have to continually remind ourselves that the final products that we observe are material deposits, which *are* the hard data for interpreting the archaeological record. Data are created within a theoretical framework of observation of phenomena, but this does not make them less objective. There are certain aspects of human activities that are constrained by deterministic values. There is real material patterning in archaeological sites that forms the basis of their interpretation (see discussion in Hodder (1992) and Boivin (2005)).

Consider a stone artefact. A first step would be to describe its material, and then the sequences of blows and the amount of force and directions of strikes needed to produce its shape. Ultimately, we would attempt to relate these aspects to a behavioural system, with the goal of connecting the movements that produce the flake with the intentions of the producer and finally incorporate them into a social or economic system after considering its context. In this book, we additionally stress that the context is provided by the patterning of the artefacts in relation to their sediment matrix, often – and wrongly – called 'dirt' or 'soil'. Indeed, the sediments and the artefacts together make for a certain patterning, which we believe is the basic raw patterning that is needed for interpretation. A constructed earthen floor, for example, has planar boundaries, a sharp upper contact, and a dense fabric consisting of parallel and horizontally oriented artefacts, and coarse natural particles embedded in a fine matrix. What we attempt to do here is to provide a certain approach of how archaeological sites should be studied. We surely do not reject other approaches, but we argue only that in the realm of the excavation in the field, this approach should be the leading and initial one.

B. Depositional Processes in an Archaeological Site

On this second issue related to the importance of depositional processes, the answer seems to be apparent. It appears that there is wide agreement that sediments and stratigraphy provide the ultimate context of the artefacts (Goldberg and Macphail, 2006; McAnany and Hodder, 2009a). Indeed, consideration of only the natural processes that are expected to occur in a modern city after its abandonment is enough to justify a depositional approach to their study. We are using this extreme example because normally in a modern city natural depositional processes are not visible before its destruction or abandonment, as they are buried by asphalt, runoff, and street sweepers, and refuse is carted off to the dump.

Nevertheless, natural processes may occur during the life history of a city, particularly in its open spaces such as yards, gardens, and plazas. In our hypothetical example, after abandonment the constructions will become part of the geomorphological terrain of the area. Houses will gradually collapse by weathering processes, and their materials will be redistributed by gravity, rainwash, runoff, and aeolian activity. These actions produce a series of mass wasting deposits and associated geomorphological configurations, such as stone avalanches, debris cones, debris flow, mudflows, slides, and overland flows (these processes are discussed in Chapter 2). Artefacts and materials of archaeological significance clearly will be affected by these processes, and it is these deposits and objects that future archaeologists will mostly dig.

Admittedly, there are some untouched, original parts of the constructions that will simply be buried by sediments. Erected constructions often have their own stratigraphy based on building and masonry typologies (e.g. the sequence of construction phases in a wall or a building based on the type of building material and the pattern in which the units are assembled, ashlar masonry vs. rubble masonry, etc.). Identification of archaeological phases in urban sites is mainly based on the detailed analysis of this masonry stratigraphy. However, the relationship between constructions or parts of the constructions is mediated by the surrounding sediments. In our examples, these sediments are the product of the natural processes, although they can be considered archaeological deposits in the sense that their content is anthropogenic.

In older sites, buildings are also buried within sediment that is the product of human activities. These include human-made earthen constructions such as mud floors and construction fills, as well as occupational deposits including hearths, food remains, and craft debris. In order to relate the construction phases with these deposits and their artefacts, we must first understand the stratigraphy of the deposits and correlate them with the building phases. Initially, this amounts to finding out those deposits, which abut against, or are cut by the different construction phases. However, it is the character of the deposit that will define whether they are floors or secondary fills, for example, and if their remains can be associated with a particular construction phase, its abandonment or destruction.

But then, why in some historical and urban sites are sediments not considered – or even avoided – as part of the archaeological record? Unfortunately, a conceptual idea underlying this position appears to prevent archaeologists from studying the sediments: often sediment is treated as something of no particular interest, and 'stuff' that the archaeologist has to get rid of in order to have a 'clean' and 'clear' picture of the artefacts and other objects and features (e.g. walls, plazas, 'feasting areas'; architecture), including their patterning. For the field archaeologist, sediments constitute 'noise' that has to be eliminated in order to reconstruct and understand the original arrangement of artefacts (thus often disparagingly termed 'dirt' rather than sediments) and extract the associated human behaviour (see, for example, Schiffer, 1987). So, even if we examine the sediments, we are studying them as a 'disturbing' secondary effect. However, by accepting this assertion – intentionally or not – we have built a false conceptual model for the depositional processes in a site. At the end, someone can rightly indicate cases where artefacts are not – or only partially – disturbed by natural processes; so why bother to do all these analyses? The artefacts are here, they can tell their story.

So what is an artefact? Only the material worked by humans? But are mud floors artefacts? Are dumping areas, pits, and earth-constructions artefacts? Is ash produced in a cooking hearth an artefact? We do not intend to enter directly into such a conversation, but we want to make the obvious point that *all anthropogenic deposits are of equal importance for understanding human behaviour in a site* unless someone wants to study only one aspect of a site, for example the artistic dimension. With minimal effort, it is obvious that ignoring the sediments as an anthropogenic record of behaviour – let alone the ultimate context for traditional artefacts – is the equivalent of deciding not to study, say, pottery from a site because you don't feel like it or are interested only in the lithics. All processes of discard produce deposits – albeit in a secondary context – that imply so much about the spatial use of a site and which carry social, symbolic, and ideological meanings related to the values of materials and space (e.g. Adams and Fladd, 2014).

Next, we should consider natural processes: do they disturb or modify the anthropogenic deposits? All natural sedimentary processes in a site are anchored by the actions of humans and reflect on their behaviour. Rainwash deposits accumulate in a street or a yard because the street or yard is an open space, constructed and maintained in a certain way to trap (or not) certain types of natural sediments. A mudbrick wall is eroded or collapsed, and associated breakdown products accumulate because the wall was made in a certain way and was left untended or abandoned. Constructions, maintenance, and use of space produce unique basins for natural sedimentation to occur. The same is true for the periodic or final abandonment of a site. Although natural processes will dominate, the types of deposit that will form are dictated by the type of constructions that reflect human decisions and actions. Thus, natural sediments in a site ultimately can be linked to a cultural meaning.

With the above assertion, we move on from considering artefact patterns as the fundamental contextual idea to the deposit (Stein, 2001). The deposit is the encoded relationship between sediments and the contained artefacts that provide the meaning of the archaeological record. However, we do not deny the dual nature of artefacts. They can be treated as part of the sediments in our conceptual model, but this does not restrict the possibility that other aspects can be studied on their own as well. A Roman cistern is a human construction that can be certainly studied by itself and in relation to others of the same kind. However, a Roman cistern is normally found inside deposits so all the above discussion is equally applicable to its study.

C. Historical Perspective: What we can learn from past and present views on operationalizing archaeological deposits and stratigraphy

Brown and Harris (1993) make a clear distinction between sites whose formation is the result of geological processes and those that are stratified by the activities of people. In the first case, they view geoarchaeology as playing a leading role in providing the interpretation of the stratification; however, in the second case, they treat geoarchaeology as one of many other specialties that can solve only some feature-specific problems. The theoretical issues that are implied from this approach explain why there are so few geoarchaeological studies of historical urban sites.

Geoarchaeology has been 'defined' or characterized in a number of ways over the years, depending on the problems or tools used, and the background of its practitioners (e.g. Butzer, 1982; Rapp and Hill, 2006; Renfrew, 1976; Waters, 1992). We follow a very broad view of the discipline as discussed in Goldberg and Macphail (2006), being simply those issues that occur at the interface between geology and archaeology. In any case, the development and expansion of geoarchaeology over the last two decades has been remarkable (Marriner, 2009; Butzer, 2009). Yet, most of the studies are related to what can be called landscape or regional geoarchaeology, focusing mainly on 'the combined study of archaeological and geomorphological records and the recognition of how natural and human-induced processes alter landscapes' (French, 2003). Butzer (2008), however, argues 'that geoarchaeology can and should be a great deal more

than the application of geoscience methods to archaeology'. More than four decades after Butzer (1973) and Renfrew (1976) introduced the term, geoarchaeology (regardless of its definition or characterization) that is practiced at the scale of the excavation is still woefully underutilized. This situation is particularly obvious in excavations of historical or urban sites where sediments are often not studied or removed to get at the finds, features, and especially the architecture. Even if there is an attempt to understand the deposits encountered in a site, it is usually done only to answer questions about feature-specific problems.

Many archaeologists, particularly non-prehistoric archaeologists, will be uncomfortable with this statement. Indeed, and somewhat ironically, the norm in Palaeolithic and Neolithic archaeology is to study the archaeological deposits using a variety of methods and concepts borrowed from the earth sciences. The eminent François Bordes was first a Quaternary geologist. Actually, Palaeolithic archaeology is probably the oldest branch of archaeology in which geology has always been an inseparable part. However, even in this case, we are of the opinion that many of the excavations are not carried out in the best possible way.

We do not mean to imply that the geoarchaeological studies are not of high quality, but that they are not fully integrated into the archaeological ones, particularly from the outset and the planning stages of a project. In particular, we want to stress the fact that because archaeologists do not typically have a geoarchaeological background, they do not always have full appreciation of the deposits they excavate. The specialist geoarchaeologists that come to the site at a particular time interval make some observations and collect some samples, and then return with reports that generally improve the interpretation of the site. On the other hand, in some cases we have heard complaints from archaeologists that geoarchaeological reports can be full of irrelevant information, not pertinent to the archaeological questions being asked. It is obvious that an interactive collaboration on a daily scale is lacking here.

Can this model be changed? Is it possible to have archaeologists having knowledge of natural or anthropogenic depositional processes, and special stratigraphic skills to interpret the spatial and temporal sequence of the deposits? Can we have geoarchaeologists that really get and understand the archaeological questions?

Before answering these questions it will be helpful to see what is happening in mainstream historical or urban archaeology in general. In architectural sites, the stratigraphy as exposed in profiles is often not representative of all deposits because of the size and complexity of these sites. So normally, the subsequent geoarchaeological study cannot reveal all aspects of the deposits and therefore offer a complete, detailed reconstruction. Harris (1989) in his treatise on the archaeological stratigraphy of urban sites justifiably accused geologists of leaving archaeologists unaided and without the adequate tools for describing and interpreting the stratigraphy. In our opinion he was right – not on his theoretical foundation but on an operational one – because there was a clear lack of describing the stratigraphic problem *and* in providing some *practical solutions*. The subsequent critique of the geoarchaeological community was generally in the right direction, but still it did not get to the point: how do we describe and interpret the complex stratigraphy of an urban site (Farrand, 1984; Stein, 1987; Collcutt, 1987)?

From a geoarchaeological perspective, the solution should take two pathways. Either a geoarchaeologist should be continually present on a site *or* archaeologists should obtain some geoarchaeological skills. Accepting only the first solution is just not practical. Geoarchaeologists, being specialists, have their own schedules, and most of the time they do not get much involved with the 'everyday' problems of an archaeological excavation. Moreover, there are far too few of them, especially in light of the number of excavations that take place each year around the globe. This is simply the reality.

The preceding should not be taken as a reproach to specialists. Nowadays, geoarchaeological studies on sites can be quite laborious, especially with the use of sophisticated techniques (Weiner, 2010). These efforts rightly focus on revealing details on the depositional history of a site or answering specific problems related to use of space, for example. However, trench or area supervisors seek someone that understands the essentials of sediments and depositional processes in order to help them understand the stratigraphy of their areas or the site as a whole. Even if we accept that Harris's approach is correct and there are no problems in his concept of archaeological stratigraphy, many of his descriptive terms (e.g. feature interfaces) nonetheless rely on interpretive skills of depositional and erosional processes (see Chapter 4).

The Harris Matrix is a simple system that graphically depicts the archaeological stratigraphy of deposition or truncation at a site on the basis of the physical relationships among strata. It has exerted an enormous impact on the development of archaeological stratigraphy. In several countries it has transformed not only recording practices but excavation practices too, especially in complex stratigraphic sequences of urban deposits. Viewing a stratum as the basic unit to be reconstructed, Harris revolutionized the way historical archaeologists were practicing excavation until that time. He was probably the first to appreciate the value of boundaries (interfaces) as stratigraphic entities on their own.

Nevertheless, two major problematic issues in Harris's approach are often overlooked (see, for example, the critiques of Barham (1995) and Warburton (2003): (1) the failure to explicitly define data and the physical criteria for recognizing interfaces and (2) the fact that post-depositional alterations can overprint visual features in the field that in essence produce a 'false' stratigraphy; colour is a noteworthy example of this. Harris favours the open excavation technique without leaving baulks and not necessarily drawing sections. However, as discussed in Chapter 4, such an approach often fails to recognize major unconformities (so-called 'feature interfaces') that do not have a clear vertical development, unless they are pits or similar features (often called cuts or negative contexts); similarly, major features that are not evident in plan view (e.g. 'stone lines') or sedimentary features such as grading, slides, slumps, and complex sequences of fills, will also be missed (Figure I.2). Furthermore, although application of the Harris Matrix appears to track well the relationships among the different stratigraphic units, in the end there is no clear picture of the three-dimensional internal structure and fabric of the

Figure I.2 Profile from 'Lower Town' at ancient Mycenae, Greece. The cobble lens shown partly at the left was recorded in a Harris Matrix during excavation as a feature 'inside' the surrounding sediments. However, its formation requires the existence of a surface demarcated at the level of its boundaries. The presence of this surface – as revealed by meticulous cleaning – was later confirmed by a crude stone line shown between the two arrows. Sedimentological reasoning led to identification of this subtle but important stratigraphic division. Excavation tools provide scale.

site because there is no hierarchical mapping of the features in three-dimensional space.

At this point, we should make clear that we do not have any particular problem with the application of the Harris Matrix in stratigraphic analysis. It definitely holds great advantages in recording complex stratigraphies in urban contexts, but if we apply it without knowledge of depositional processes we can end up with essentially an automated procedure for logging sequential events. The quality of our interpretations is only as good as the quality of the way it was inputted: garbage in, garbage out.

In further consideration of stratigraphy, it is surprising that there are different schools of approaching stratigraphy (for a review, see Warburton (2003)). In addition to the geological (which the authors favour) and Harris's approaches, there is also the architectural approach (discussed extensively in Chapter 4). It is tempting to assume that the reason for these different approaches does not rely on different conceptual ideas but on the practical difficulties in applying geological stratigraphy by archaeologists. In other words, it will be rather difficult for someone without the theoretical background in sedimentology and depositional processes to do a thoroughly accurate job of stratigraphic analysis. In conversations after hours during excavations we have commonly heard that doing stratigraphic analysis is just a matter of observation and everybody with a good eye can do the job. For example, Harris believes that units and their interfaces can be safely separated in the field without having any notion about the process that could have formed them. Unfortunately, this is a fundamental error, and sometimes even geologists believe that this is the case. Students in archaeological field schools are generally taught to draw 'exactly' what they see. It is evident with such an approach that the drawings convey little or no meaning about the dynamics of the depositional and post-depositional processes that produced a layer.

Geologists can usually construct a reasonably accurate stratigraphy, although they rely on just three apparent laws (e.g. superposition), which are implicit in their thinking because of their training. They are generally successful because they have knowledge of how depositional systems work: how sediments are deposited, which contacts are expected to be formed by deposition vs. those produced by erosion, and which contacts can occur that are not the product of artificial processes, such as the way the profile was scratched during cleaning, for example. They can often predict the continuation of a contact – or a deposit – even if it does not appear clear in the field; they can connect separate parts of the same contact by following and interpreting subtle features and changes in sedimentary attributes because of this understanding (Figure I.2). They can interpret geometrical arrangements and related distributions of materials, such as linear features (e.g. stone lines) or clustering and banding of objects or grains. Finally, they have the knowledge of how gravity and flows deposit materials. As a result, even though in some cases they do not know from field observations alone what processes formed a deposit, they are aware of most of the constraints and possibilities.

On the other hand, sedimentologists normally do not have a good idea of how human structures look in a site. They do not know how mud floors appear or what constitutes a construction fill and foundation layer. It is true that almost all people have an empirical knowledge of human-made structures, but these features are period- and sometimes even site-specific. In Chapter 3 we will provide some basic principles for recognizing anthropogenic sediments.

Note that up to now we have not used the term 'site-formation processes' but instead 'depositional processes' (including also erosional and post-depositional processes) because the first phrase as elaborated by Schiffer (1987) referred more to activities than to the final material products. Indeed, Barham (1995) pointed out that there is a lack of data to demonstrate how sedimentary structures may be diagnostic of different formational contexts as modelled by Schiffer. However, since it was introduced, 'site-formation processes' has taken on several meanings (Stein, 2001), with one of them actually referring to depositional processes (Goldberg and Macphail, 2006). Therefore, from now on the terms will be used interchangeably in the text.

While on the topic of conceptual approaches, we may bring into discussion another aspect of the standard description of textural properties of deposits as offered in several site manuals. Justifiably, Barham (1995) argues that the terminology and conceptualization used to describe deposits in archaeological sites – particularly in the mid-continental USA – is strongly viewed from the perspective of soils, namely, their formation (pedology) and nomenclature (e.g. Holliday (1990) and other references in that volume); other examples can also be found in the Old World (e.g. Kühn et al., 2010). Soils are definitely not synonymous with sediments, and among other things they are the product of alteration of sediments after their deposition.

By using a soil terminology to describe archaeological deposits, archaeologists and others are biased in their approach, as they commonly fail to recognize primary sedimentary structures or treat them equally with secondary alterations. For example, denoting an organic-rich Neolithic midden as 'an overthickened A-horizon' misses the point. Moreover, by denoting deposits in the field with terms such as '2Btb' implies that the layer/horizon is in fact a B-horizon. In other words, instead of describing a deposit, it is already given an interpretation before analyses have been undertaken to determine the validity of this description/interpretation. The same is true for the offhand use of the term 'ash' for any grey-coloured deposit found in a dig. Ash is a specific material that should be identified by microscopic techniques. In other words, soil field descriptions have a strong genetic flavour implied in the very terms that are used, and thus should be avoided in favour of a more objective descriptive terminology such as Unit 1, Unit 2, etc.

In conclusion, a clear distinction between depositional and post-depositional processes has to be made in order to understand the sequence of events that builds a site. In Chapter 2 we discuss in detail most of the post-depositional processes, including soil-forming processes, usually encountered in archaeological sites.

Admittedly, the above discussion concerns geoarchaeologists as well. For reasons that have to do with educational background, and sometimes too much specialization, there is not much appreciation and understanding of the stratigraphy and formation processes of a site in a holistic way. Studying a specific feature without integrating it with the totality of the site-formation processes – including those on the landscape scale – usually leads to a partial understanding of the feature. Nevertheless, in the best of circumstances, if the archaeologist has an appreciation of the site-formation processes, then geoarchaeologists working on specific questions can integrate their work safely and successfully with the rest of the archaeological studies of the site.

We have to admit that the geoarchaeological community has not succeeded in providing the necessary geoarchaeological knowledge to archaeologists. All geoarchaeological textbooks put in the same basket both landscape geoarchaeology and site geoarchaeology. Chapters dealing with natural depositional processes, for example, are mostly focused on high-energy or deepflow sedimentary processes (fluvial, coastal, and lacustrine deposits) or large-scale sedimentary processes (alluvial fans, glacial diamictons, etc.). Such geological environments are rarely encountered in a 'usual' archaeological site, except probably prior to excavation or with the rare, large-scale events such as landslides or floods, which would actually bury a site for a period of time, or remove it. Moreover, these processes are readily observed in the field, even by an untrained archaeologist, who at a later stage can ask the specialist to explain the significance of this deposit. The majority of the archaeological deposits in an excavation that are dug everyday are the product of small-scale mass wasting and overland flow.

Another issue is that the analysis of sediments in geoarchaeological textbooks focuses on the description of the evolution of processes, how and why such processes occur, and the presentation of the potentials of geoarchaeological studies. They do not explicitly convey how someone can practically recognize and differentiate them from other processes. It is tacitly implied that the readers have a basic knowledge of sediments and can transform a case study to practical field knowledge. However, both archaeologists and geoarchaeologists working on site-specific problems need to have a 'geoarchaeological guide' dedicated solely to sedimentary problems encountered during excavation. The same is true with geoarchaeological courses and seminars in archaeology departments. No doubt, an archaeologist should have an idea of the landscape dynamics and geomorphological processes, but the educational background that is needed in an excavation should be determined by the depositional processes encountered in an archaeological site and not in the whole Earth. These small-scale processes are hardly touched in general introductory courses of geoarchaeology.

Anyone who has worked in sites of different periods and cultures has faced the fact that urban and generally all constructed sites have a totally different depositional story to tell from those at a prehistoric site: natural sedimentary processes dominate at non-constructed sites, such as Palaeolithic ones. This is probably the reason why most geoarchaeological studies are conducted at these sites. Anthropogenic deposits are mainly composed of burnt and other organic remains (matting, bedding, and food residues) with very few basic constructions, if any (see Chapter 5). Nonetheless, the study of the microstratigraphy and microstructure of the deposits can reveal specific activities and use of space, such as *in situ* burning and dumping areas or activities related to cleaning and modifying living sectors; such studies can also reveal periods of stasis, when the site was left unoccupied. All these issues will be discussed and analysed in detail in Chapter 5.

The dichotomy between constructed and non-constructed sites is somewhat artificial, and certainly there is a lot of overlap. However, we believe that this distinction can be generally accepted because it serves the specific objectives of archaeological practice; it also gives special weight to the architectural sites that are the most neglected. Site-formation processes in urban and other architectural sites are much more complicated. Natural sedimentary features are not well expressed and are mainly of very small size, mostly submicroscopic. The geoarchaeological study of the archaeological deposits in such sites demands a range of expertise dictated by the infinite repertoire of anthropogenic settlement activities, an issue that even experienced geoarchaeologists face. In these sites post-excavation stratigraphy is of little help. An in-depth mastering of the depositional processes and stratigraphy is an essential prerequisite for interpreting correctly the archaeology of such sites. In Chapter 6 we attempt to describe the main depositional features encountered in architectural sites and give special attention to differentiating floor sequences from construction fills (levelling and foundation fills, often related to abandonment and destruction), occupational fills, and open vs. roofed areas. Such a differentiation is the backbone of any interpretation of the chronological and functional association between artefacts and the surrounding architecture.

Inevitably, a methodological chapter has its place in such a book. However, many other textbooks (e.g. Goldberg and Macphail, 2006; Weiner, 2010; Artioli, 2010) have addressed this subject very successfully and discuss in depth several methods that are suitable for studying archaeological and natural deposits. Although we believe it is not necessary to duplicate such efforts, we present in Chapter 7 a brief discussion of the most appropriate methods and the overall field strategies for documenting and sampling these deposits.

We have organized the book along two broad themes. The first theme, in Chapters 1 through 4, is focused on basic principles and includes discussions of the nature of natural and archaeological deposits, and the role of stratigraphy, a construct that helps us envision how they are organized in time and space. The second theme, in Chapters 5 and 6, takes an applied approach by looking at how these processes can be combined to lead us to interpretations about how sites form. This theme deals with inference, which incorporates the basic principles from the first levels and takes them to higher, more interpretative levels. So, in the later chapters we deal with specific concepts such as stables, living surfaces, or middens.

This book addresses fundamental issues related to archaeological deposits excavated in a site. It is built in such a way as to offer archaeologists a sufficient inventory of tools to recognize depositional units – and their modifications – and interpret the stratigraphy of a site in the field. However, we also utilize a complete analysis of the archaeological deposits that usually demands the expertise of a specialist and a range of sophisticated laboratory techniques. In this way we believe that we offer a holistic approach in studying a site from the fundamentals to the details. Thus, we envision that this book will be useful for both archaeologists and geoarchaeologists, beginning and advanced students, and professionals.

1

Principles of Site-formation or Depositional Processes

1.1 The Concept of the Deposit

In its broadest definition, an archaeological deposit is what encloses the archaeological finds and, as a result, the finds constitute an inseparable part of the deposit (cf. Stein, 1987). In other words, the archaeological deposit is the material that is excavated in order to 'reveal' the archaeology of a site. *Deposit* is a broader term than sediment: it constitutes more of a general and generic use. For example, 'glacial deposits' refers to a general class of sediments deposited in a glacial environment. However, archaeological sites contain only sedimentary deposits with very few exceptions like volcanic deposits. In the vast majority of cases, all archaeological objects were deposited penecontemporaneously (essentially at the same time as far as we can resolve) with the enclosing material; of course the degree of penecontemporaneity depends on temporal resolution, which is obviously greater for younger archaeological sites and less so for prehistoric ones dating to, say, one million years ago. Indeed, the separation of an archaeological object from its excavated matrix is artificial (and virtually impracticable to do). Understanding the deposits as part of the archaeological record has changed throughout time, disciplines, and regions. In many sites, excavated deposits are water sieved for recovering charcoal and macrobotanical remains, but this has not been always the case, and at many sites they do not even sieve the deposits to collect microarchaeological debris (e.g. lithic debitage, microfauna, beads) let alone botanical remains. To accept that the materials of archaeological interest are mostly deposited penecontemporaneously with their enclosing deposits is an essential tenet in interpreting formation processes.

This is not to deny that archaeological objects could not be deposited first and then covered with sediments, but we must realize that it is fruitless to treat objects and deposits separately from the excavation point of view and later in their interpretation phase. However, it should be noted that erected constructions (e.g. walls, bridges, dams, etc.) do not follow the above rule, and enclosing deposits are not always contemporaneous. Indeed, buildings are always filled with deposits after their construction, although some of them accumulated penecontemporaneously in the sense defined above (e.g. the first floor or occupational debris within the building). As we have already discussed in the Introduction, architectural features have an internal stratigraphy defined by the typology of masonry and other architectural features, and they have to be included within the depositional stratigraphy in order to be correlated with the surrounding archaeological objects.

To illustrate the above statements we will describe a rather rare example, but one that is clear and hopefully gets the point across. During a volcanic eruption the pottery left on the surface of a floor in a house was subsequently covered by volcanic tephra (the so-called Pompeii premise) (Schiffer, 1985; Binford, 1981). In order to expose the pottery we dig through the naturally deposited tephra. Obviously, the pottery was left undisturbed intact and covered by the tephra. The study of the pottery confirms that this is an assemblage left in place. Indeed, starting from a sedimentological (depositional) point of view, we might recognize that the material that covers the pottery is a natural free-fall volcanic tephra that covered the pottery left in the house without disturbing it. Free-fall tephra is volcanic material ejected into the air during an eruption, which then settles by gravity on the ground.

Although the above description appears obvious for most archaeological sites buried by volcanic tephra, we should also consider other possibilities in which, for example, the tephra is *not* a free-fall deposit. As Sigurdsson et al. (1982) describe very impressively, the AD 79 Vesuvius eruption produced individual sequences of different volcanic deposits in each of the cities of Pompeii, Herculaneum, and Oplontis. Each sequence points to different taphonomic histories of the archaeological material, which is far from the simplistic idea we have for this type of phenomenon; note that taphonomy is concerned with all processes happening to an organism after its death. In Pompeii, for example, all excellently

Reconstructing Archaeological Sites: Understanding the Geoarchaeological Matrix, First Edition. Panagiotis (Takis) Karkanas and Paul Goldberg.
© 2019 John Wiley & Sons Ltd. Published 2019 by John Wiley & Sons Ltd.

preserved human bodies are found inside a second volcanic deposit in the form of a pyroclastic surge, which stratigraphically lies above an initial free-fall tephra deposit. Pyroclastic surges and flows are fast-moving currents of superheated tephra, the former being more dilute than the latter and therefore they move faster and generate surges. The first thick free-fall tephra deposit was not responsible for the deaths in Pompeii, and in fact it forced the major part of the population to flee the city before the main disaster struck the following day. It produced only roof collapse in most houses. The second phase of the Vesuvius eruption first produced a very fast-moving hot and thin pyroclastic surge that probably killed by asphyxiation those who remained in the city; then a second stage, a hot, dense, and fine-grained pyroclastic flow, buried the human bodies. It is this last powdery material that preserved the fine details exhibited in the many human plaster casts that were obtained during the excavation of Pompeii (Sigurdsson et al., 1982). The details of the taphonomy of the archaeological materials could not have been revealed without the study of the burying volcanic deposits.

An even more complicated case involves Herculaneum, where ample evidence shows that tephra-rich mudflows (lahars) were also encountered in the final stages of the eruption. This type of deposit is often observed immediately after volcanic eruptions when eruption-induced rainstorms remobilize tephra and other material, and deposit them downhill as thick slurries. Although not reported in this case, there is a hypothetical possibility that archaeological material was deposited by such mudflows, which, for instance, resulted in the mixing of objects from a room and an adjacent yard. Depending on the local depositional processes in some cases, whole parts of the original archaeological assemblage could have been moved *en masse* by dense mudflows. Analysis of the pottery alone, for instance, would not reveal such subtle mixing processes.

The reason for presenting the above examples is to stress that it would be naïve to think that we can understand the patterning of the artefacts and features without understanding the sedimentary matrix that contains them. It is that simple. These sediments have accumulated by a certain process, and it is this process that explains the patterning of the artefacts. Even in cases where fragments of artefacts can be put back together, making one think that therefore they come from a primary context, it is the surrounding matrix that will lead us to confirm or reject this interpretation. There is no doubt that a separate analysis of the archaeological finds and the sediments would be mutually beneficial. However, within the framework of excavation both the finds and the deposits have to be examined together in order to interpret correctly the archaeological assemblage. In other words, it is the deposit as a whole that contains the archaeological finds, which will ultimately define the context.

The Pompeii example obviously is the exception. Most of the time, archaeological objects are affected by complex combinations of natural processes and therefore are contemporaneously deposited with the enclosing sediment or incorporated within the sediments by syn-depositional anthropogenic processes such as dumping, trampling, or construction of earthen structures (e.g. floors, walls, mounds). The archaeological objects, together with all other materials that are not of direct archaeological interest, are parts of the organization of the sediments. The way they have been incorporated inside the sediment is reflected in the fabric of the deposit, which is the three-dimensional arrangement of the constituents and their size, shape, and form (Stoops, 2003).

All these attributes inform us about the dynamics of deposition of the materials by natural forces. We have to consider that there are only a limited number of basic human actions that occur, such as laying down, dropping, compacting, and throwing materials. Therefore, compacting, kneading, and applying, sweeping and raking, trampling, dumping, backfilling, and levelling are fundamental actions that together with burning and animal husbandry activities form the majority of archaeological sediments (see Chapter 3). It is the combinations of these actions that lead us to interpretations of overall human activity.

Likewise, in the realm of the excavation, the possibility that the archaeological objects are buried at a later time by sedimentary deposits is meaningless, with very few exceptions. By definition, the archaeological finds are recovered by digging the deposits. If we accept that there is no way of interpreting correctly the archaeology of a site without understanding its stratigraphy (see Chapter 4) and context, then the smallest depositional unit recognized in the site is what defines the smallest informational unit, its time resolution, and its archaeological/anthropological significance (e.g. Goldberg et al., 2009a). When objects are referred to their excavated unit in order to define their stratigraphic position they cannot be treated separately from their matrix.

Nevertheless, in the case of a dumped accumulation of only bones (see section 5.2.1) or the construction of a wall, the archaeological objects can be recovered as a discrete continuous body, devoid of any sedimentary matrix. In this instance, the objects can have an identifiable discrete stratigraphic position on their own without being referenced to their enclosing sediment. Such features are usually walls and similar architectural constructions, and this is the reason why, during excavation, they are commonly treated as part of a different stratigraphy, that of the architectural or construction phases. In all other cases, archaeological objects are parts of the sedimentary deposits and at a first stage cannot be studied separately. Of course in a later stage individual archaeological objects can be studied on their own or combined into groups and assemblages, but their original context is defined in the excavation and in relation to the depositional unit with which it was associated.

In sum, archaeological objects cannot practically be separated from the deposits in which they are included, and in the majority of the cases they are part of the fabric of the deposits that characterize the way they accumulated in the site. In cases where the archaeological objects can be treated separately from their deposits, they already constitute a discrete field of archaeological study (e.g. architectural phases) and normally they are not excavated or they are removed at a much later time after being documented. Other rare exceptions are artefact lags (Figure 1.1), which are concentrations of archaeological objects that are left behind after their sedimentary matrix has been winnowed out by wind or rainwash in so-called 'deflated' deposits, for example (see section 2.7.1). In this case the assemblage is the direct product of a natural process of concentration, and therefore the objects by themselves make up the deposit. The same could be said when anthropogenic activities have led to a single concentration of objects in discrete, identifiable excavation bodies (e.g. shell middens) (Figure 1.2), and as a result constitute a single layer. Even with shells in middens, their orientations and fabrics can guide interpretations of how they accumulated or what happened to them afterwards. In all these cases, however, the fabric of the

Figure 1.2 Shell midden at Nelson Bay Cave, South Africa (LSA). Note the middle part of the section composed only of shell.

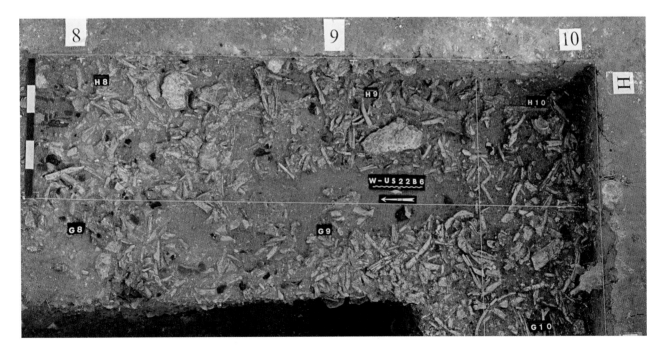

Figure 1.1 Dense concentration of bone that by itself makes up a layer. MP, Jonzac, France. Courtesy of Shannon McPherron.

assemblage has to be studied in addition to the typology because natural formation processes or certain anthropogenic processes impact the size distribution and arrangement of archaeological objects.

Another very compelling reason for studying archaeological deposits is to reconstruct the completeness of the archaeological record. Only by understanding the nature of the components in a deposit (minerals and organic matter) can we make an assessment of the possible absence of archaeological artefacts that were lost due to physical or chemical alteration (diagenesis – see section 2.7.2) (Karkanas et al., 2000; Weiner, 2010). The absence of certain materials (plant, bone, ash, shell, and others) is not always related to human choices, actions, and use of space. Post-depositional processes are often very aggressive and they can lead to disintegration and dissolution of several archaeological materials of interest. In Chapter 2 (section 2.7.2) we give a detailed overview of the processes that lead to the destruction of archaeological materials and how we study them.

1.2 Types of Archaeological Deposits

Archaeological deposits encompass all deposits found in an archaeological site, whether geogenic (natural) or anthropogenic. From a practical perspective, they can be separated into three broad categories: (a) those deposited by natural processes but without materials produced, modified, or re-organized by humans, (b) those deposited by natural processes but also containing anthropogenic materials, and (c) materials (natural or anthropogenic) deposited only by anthropogenic activities and processes. Natural sediments are materials deposited at the earth's surface and denote transport before deposition. Production of new materials (e.g. minerals) without transport is often part of soil formation (pedogenesis; see below), which follows incipient weathering stages in the formation of regolith (loose, heterogeneous superficial material covering solid rock) or saprolite (soft, thoroughly decomposed, and porous rock) (see relevant sections in section 2.7.3). Weathering produces or releases new materials, both physical grains as well as solutions rich in ions that can recombine to form new minerals elsewhere, below, or at the surface; these substances are transported and deposited by gravity, water, or air. Deposition can be physical or chemical and produce clastic or organic and chemical sediments (see Chapter 2 for details).

In the realm of an archaeological site, it is impractical to separate purely geogenic deposits from archaeological ones. Often, the majority of archaeological deposits are naturally deposited mixtures of geogenic and anthropogenic sediments. Geogenic deposits may provide important information about the nature of open areas in architectural sites where they are more 'connected' to geological systems and processes; they often can inform us about the prevailing environmental conditions in various types of sites. In addition, since the topography of the depositional surface influences the physical characteristics of the deposit, any previous anthropogenic activity or sediment would affect the deposition of subsequent natural deposits. Therefore, even post-abandonment geogenic deposits may reveal important aspects of the decay and destruction of a site or part of it, such as buildings and other features.

1.3 Anthropogenic Sediments

Human activities create a variety of anthropogenic sediments that include products, by-products, and refuse. Sediments can be organic and inorganic, and composed of materials used in a number of activities: the construction of dwellings, tool manufacture, remains from the processing and consumption of food and combustion, cloth, buttons, bottles, decoration materials, etc. We consider all these items to be sedimentary particles for a number of reasons:

1) Anthropogenic sediments can be integrated with naturally deposited sediments, as for example discarding artefacts in streets or yards or generally in open areas that are affected by natural sedimentary processes (water flow, wind, and gravity). In such cases, they behave as any other sediment and obey the natural laws of transport and deposition. Indeed, they are often the most important element of the fabric of naturally deposited archaeological sediments. Note that what are traditionally called 'natural disturbances' are included in this type of sediment, as, for instance, in the case where archaeological sediments have been redistributed by natural processes after the abandonment and destruction of a building or campsite (e.g. decay of a mudbrick wall or wind-eroded hearth) (Figure 1.3a,b).

2) They can be the result of transient reorganization of natural sediments by humans and are inseparable from the excavated deposits (e.g. mud floors, earthen mounds, backfilling of tombs). In cases of intentionally formed materials (e.g. mudbricks, mud floors), they almost always readily weather or decay back to 'natural' sediments. In addition, the original process of their construction is governed by the mechanics of the individual actions (the functional and technical aspects of an activity – see above) and therefore they can be interpreted by analysing their fabric and structure in a way similar to what one does for natural sediment (Figure 1.4).

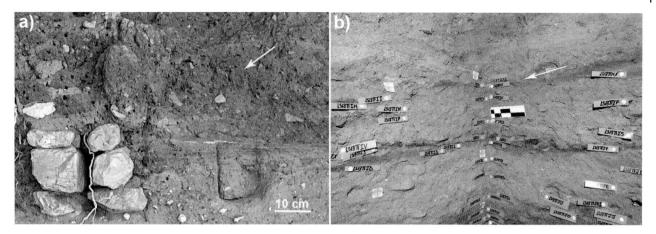

Figure 1.3 (a) Decay of the mudbrick wall at the upper left has produced an array of mudbrick fragments (arrow) floating in a fine-grained sedimentary matrix. EBA, Mitrou, Greece. (b) The ash and charred remains of the upper hearth (arrow) have been washed out, leaving only the red burnt substrate. Compare with the better-preserved hearth beneath it in the middle. MSA, Site PP5-6, South Africa. From Karkanas et al. (2015a).

Figure 1.4 Remnant of a constructed floor (F) fading into the darker surrounding sedimentary fill. LBA, Mitrou, Greece. Tag is 10 cm long.

3) They can be transformed into permanent artificial materials, such as pottery, brick, mortar, glass, and metals. However, as part of an anthropogenic deposit, their spatial distribution and orientation can be analysed using sedimentary techniques in order to provide information about the formation of the deposit that contains them (e.g. pottery distribution in an occupational deposit).

4) They can be materials that involve permanent chemical transformations, such as ashes. Ashes may be produced by both human and natural agents, and thus there is no reason for *not* treating them as sediments. Consider the somewhat hypothetical case of a natural ash accumulation resulting from a Holocene forest burned by a lightning strike. Such a build-up would readily be called a deposit by a geologist, and would be studied sedimentologically as any other chemical sediment (ashes are composed mainly of

calcium carbonate). In addition, ash produced at one location is easily distributed over a large area by wind or rainwash. Thus, for the same reason there is no real rationale for not considering it a deposit if it were to be found in an archaeological context, whether it would be from an intact fireplace, ash dump, or layer of burned goat dung.

Following the above reasoning, it is important to consider what constitutes a deposit: a standing mud wall is not archaeological sediment *sensu stricto*, but fallen parts of the wall *are* because they are elements within the excavated deposits. Strictly speaking and along the same lines, stone pavements or mosaic floors when intact are not sediments because they can be treated as continuous non-movable elements of anthropogenic construction (features) that can be clearly separated from the overlying and underlying deposits; they themselves are not commonly excavated. As such, they have their own internal stratigraphy that is based on architectural attributes and typology, and not on depositional stratigraphic principles.

At first glance, a constructed mud floor would not seem to be different from a mud wall or a mosaic floor. However, a mud floor is closely intermixed with non-constructed occupational debris and deposits (Figure 1.4). Thus, it is the product of a temporary resurfacing and reorganization of natural sediment in which parts of the floor are readily transformed into natural sediment. As a consequence, mud floors are excavated together with their occupational deposits and they cannot be practically removed as one single body from the overlying and underlying sediments, as in the case of mosaics, for example. Moreover, the internal stratigraphy of mud floors and their relation to the overall stratigraphy of the surrounding deposits can be analysed using depositional stratigraphic principles (see

Chapter 4). Mud floors can be studied typologically and architecturally, and thus form part of the architectural system of the site as well.

In summary, 'anthropogenic' sediment should be treated as natural sediment from a practical, methodological, and technical point of view. There is no doubt that there are 'grey areas' in this view. However, we believe that the above characterization is valid for the overwhelming majority of cases. We also wish to clarify that the reason we so painstakingly examine anthropogenic sediment is that terms like 'sediment' and 'deposit', which are derived from terminology used in the earth sciences, have been uncritically used in archaeology without having been analysed and defined (see below for discussion of the term 'deposit'). Consequently, such practices can lead to misconceptions and misunderstandings. We do believe that the misuse and lack of understanding of what deposits are and how they form has led archaeologists to criticize the application of earth science-based techniques and concepts in urban archaeological sites (see, for example, Harris, 1989).

1.4   Some Misconceptions of Site-formation and Depositional Processes

Schiffer (1972, 1987) was the first to champion, systematize, and formalize the concept of 'site-formation processes' and introduce a theoretical framework for analysing them. In his approach, past cultural systems related to the original behaviours associated with the artefacts (i.e. systemic context) are transformed by two kinds of formation processes, cultural and natural transformation processes (C- and N-transforms, respectively), and they are responsible for the formation of the archaeological context. Any object can participate in a behavioural system in several stages that include procurement, manufacture, use, maintenance, and discard and all these are included in C-transforms. However, in our minds it is only the last of any of those actions – whereby an object is ultimately deposited – that contributes to the building of a site. Among the different cultural transformation processes, the ones that lead to the final emplacement of an object in its depositional context participate in the formation of a site.

N-transforms are by definition transformation processes that disturb, alter, or obliterate the original patterns of the objects and obviously participate in the building of a site (see discussion in Shahack-Gross, 2017). However, as very nicely shown by Stein (2001), there is a misconception on the use of the term 'deposition' in the archaeological literature. Archaeologists often use 'depositional processes' to generally imply behavioural processes or cultural activities and their roles in creating the archaeological record. They

overlook the basic processes that led to the final emplacement of an object on the earth's surface.

We need to remind ourselves that excavation yields a body of sediment that includes, among others, objects that were once part of a behavioural system. In order to furnish an unbiased interpretation of the behaviour of the people that were using and discarding these objects, we must accept the idea that the basic raw data of this study are the excavated deposits that fill the three-dimensional space, particle by particle and object by object. The manner in which these objects and particles were laid down – one upon the other – should be the primary focus of the analysis of depositional processes.

Obviously, all activities related to the use of an artefact during its life history – except its final abandonment or discard – are not the ones that led to its final emplacement. They can be indirectly inferred – in the sense that they restrict the possibilities of the final emplacement – but they do not participate in this accumulation and cannot be revealed by the analysis of their depositional pattern alone. This *chaîne opératoire* should be the target of a 'meta-analysis' of the original data that are produced by the excavation of the deposits in a site, taking into account the characteristics of the object itself. For example, in order to understand the activities related to a tool, someone has also to consider the activities that are related to the procurement, manufacture, and maintenance of the tool. The analysis of the sediments that contained this tool will provide information on the relationship of this tool to the other archaeological objects. However, this relationship is always depositional in the sense that it is about the last human action of laying it down.

1.5   Soils and Post-Depositional Processes

A very broad definition of soil is the altered surface of the earth by atmospheric, physicochemical, and biogenic processes. According to Joffe (1949): 'The soil is a natural body of animal, mineral and organic constituents differentiated into horizons of variable depth which differ from the material below in morphology, physical make-up, chemical properties and composition, and biological characteristics.' Therefore, sediment and soil are discrete concepts and each term implies a set of materials and processes. In respect to sediments, soils are post-depositional creations. With time, sediment or rocks are altered by surface processes, their material reorganized into millimetre- to centimetre-sized soil aggregates (peds), and a vertical differentiation of the material is developed in the form of soil horizons (not 'layers' or 'depositional units'), which differ from each other in aspects such as colour, texture, composition, etc. (Figure 1.5a,b). Therefore, a more restricted definition would be that soil is the product of *in situ* alteration of sediments and rocks (and even previously developed soils) at the earth's surface for a considerable time so that the original sediment or rock structure has been modified, and a soil structure has developed.

It follows that soils have a strong built-in 'internal' temporal dimension, whether it spans years or millennia. Yet, from a purely pedological point of view, any sediment found at the earth's surface for a short period of time can be called an incipient soil. Indeed, the soil class 'Entisols' signifies those soils that have little or no evidence of the development of pedogenic horizons (Soil Survey Staff, 1999). For pedologists, plant growth is evidence enough that the unconsolidated materials are functioning as soils (Buol et al., 2003: 259). Thus, a one-year-old river flood sediment that is colonized by plants is considered soil by pedologists. Although this definition is probably valid for pursuing the particular interests of pedology, it is obviously not constructive for understanding depositional processes. For instance, data gleaned from Entisols generally do not provide more useful information on pedological-specific processes than can be provided by sedimentological studies.

From a sedimentological point of view, all transformations affecting the sediment after its deposition are lumped together under the term diagenesis (see section 2.7.2). These include mechanical disturbances produced by fauna and flora, loss of water, compaction, and chemical or biochemical transformations, such as mineral alterations, oxidation of organic matter, and precipitation of new material in voids (cementation). Since soil-forming processes also include many of the above processes there is an overlap between the two approaches.

Nevertheless, understanding an archaeological sequence is first and foremost about understanding the processes of deposition, which will reveal the original context of the deposit at the time of its formation. By treating almost all sediments as soils we are biasing our approach towards post-depositional processes, and indeed those that are related only to soil-forming processes. As such, this bias highlights the identification of soil horizons and not the substrate on which it has developed (cf. Holliday et al., 2007 vs. Velichko et al., 2009; Sedov et al., 2010). For example, it is of crucial importance to identify that a sequence is produced by slope processes and how they have originally affected the archaeological materials. Then, all post-depositional processes have to be taken into consideration to further elucidate the present pattern of the archaeological finds. Finally, if there are well-developed soils in the sequence, they can be studied from a pedological perspective in order to reveal information that cannot be revealed by treating them as sediments.

A well-developed soil can provide valuable information on climate, topography, and biogenic factors (Birkeland, 1999). The formation of a soil is environmentally controlled and, therefore, the conditions of moisture, temperature, seasonality, vegetation type, and organisms are decisive factors in soil development. However, probably the most important information is the establishment of how long it took for a particular soil to form, which is tightly linked to depositional stasis. A buried soil defines an old stable relief because, by definition, a soil is formed always on the surface and needs time to develop (Figure 1.6). In addition, in alluvial terrains the geoarchaeological mapping of surface soils with

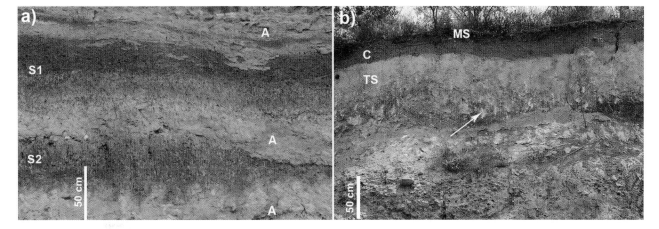

Figure 1.5 (a) Two main grey soils (S1 and S2) formed on parent material of alluvial sediments (A) that were deposited during periods of basin flooding. Note 'layering' of soils that is attributed to soil horizon formation during long periods of sedimentary stasis. Pleistocene, Aliakmon River, Grevena, Greece. (b) A lower soil (TS) characterized by the formation of white nodules and calcareous filaments (arrow) is truncated and overlain by brown colluvial sediment (C) with a clear and sharp erosional surface. A modern soil (MS) is developing on top of the colluvium producing a dark top humus-rich horizon. Ylike Lake area, Greece.

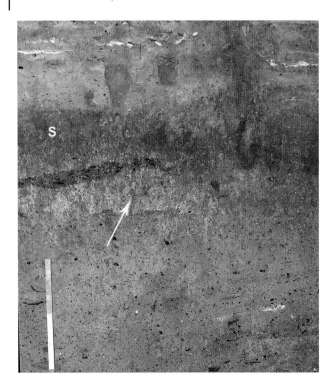

Figure 1.6 Dark soil formation (buried A-horizon (S)) on beige clay construction material at the Chalcolithic/BA boundary, Vésztő-Mágor, Körös area, Hungary. Note the numerous stripes attributed to root and faunal passages (some shown with arrow). The formation of the soil points to abandonment of the site during the end of the Chalcolithic.

different degrees of development have been effective tools in guiding the search for sites of different ages (e.g. one does not expect to find a 10,000-year-old PalaeoIndian site on a non-buried Entisol). Although such *soils* can be encountered in archaeological sites and landscapes, particularly during abandonment and pre-occupational periods, the overwhelming majority of the archaeological *deposits* are not soils, but sediments. The taking of 'soil' samples at archaeological sites – although seemingly a banal example – reflects a significant misconception in the distinction between these two fundamentally different classes of materials, although admittedly 'soil' sample is easier to say.

Obviously, the study of soils in the vicinity of archaeological sites for palaeoenvironmental reconstructions and for understanding human-induced soil disturbances is still an issue of pedology (Macphail, 1986; Macphail et al., 1987; Fedoroff et al., 1990; Holliday, 2004; Goldberg and Macphail, 2006). In the same vein, soils that normally make up the underlying substrate of a site can provide valuable information on the pre-occupation conditions of the landscape (Macphail, 1987; Sherwood and Kidder, 2011). Furthermore, soils encountered within an archaeological sequence should imply abandonment of the particular area, or an open, cultivated area, or one not in use (Figure 1.6).

In summary, all deposits in an archaeological site have to be treated as sediments. If soils are developed in the sequence they should be analysed as both sediment and soils. The pedological study should focus on aspects that cannot be revealed by sedimentological analysis alone.

Clearly post-depositional processes are those that act on a deposit after its accumulation and even after its burial by overlying deposits. Processes that are acting on a deposit as it accumulates on the earth's surface should normally not be included in post-depositional processes because they participate in the formation of this layer and eventually how we would observe it in the field. Therefore, they should be considered as syn-depositional processes and not post-depositional ones.

We do understand that figuring out all these processes may be considered a grey zone, but if we want to reconstruct a coherent and accurate history of deposition and human activity in a site, we should clearly separate the events that led to deposition – even if they were somehow destructive – and the events that affect the deposits after their final accumulation. Cases can arise when a layer is formed and later on trampling has modified and reorganized part of it. From a stratigraphic point of view, trampling effectively has produced a new stratum, since it produced new layer boundaries along with a new shape and form of the trampled deposits; the resulting unit normally will not fill the same space as the previous layer. Indeed, this process is not much different from the erosion and deposition of a fresh layer at the expense of a previously deposited one.

1.6 Recording Deposits and Site-formation Processes (Stratigraphy)

From the above, we have a clearer idea of what sedimentary deposits are and that archaeological sites are composed of them. So the question that arises is how do we 'manage' and organize them in a way that facilitates our understanding of site history. The answer is in stratigraphy, the way deposits are organized *both* in space and in time. Stratigraphy is the tool for understanding the chronological order of events and artefacts at a site, and it also provides the basis of the analytical units used to discuss the human activities at the site (Balme and Paterson, 2014). Thus, stratigraphy has two very fundamental dimensions: a spatial and a temporal one. Using stratigraphy, we are able to organize in chronological order archaeological features, artefacts, soils, and sediments. Surprisingly, however, for most archaeologists the purpose of stratigraphy is only time (Renfrew and Bahn, 2005). Indeed, the *Archaeological Site Manual* (MoLAS, 1994, section 1.2) specifies that '…any physical relationships between one context [≈ stratigraphic unit] and another are of no assistance in the study of site's stratification.'

Organization in space and time also implies connections and correlations among the different deposits. Indeed, the idea of stratigraphy as a simple layering of the 'soil' at a site (cf. Feder, 1997) is simplistic. From the very beginning of geological science, stratigraphy focused not only on the sequential order of 'regular' (stratified) formations, but also more generally on the spatial relations of rock masses of all kinds (Rudwick, 2008). In an archaeological site what we face are layers and features that can appear similar – or dissimilar – in different areas. As they build up on and next to each other through time they produce the stratigraphy of the site. Therefore, in order to extract from them a meaningful site history we have to correlate separate layers in space. This correlation might be established on time equivalence, which in turn can be based on absolute chronology, or by relative chronology, that is, equivalent stratigraphic position. Correlation also might be established through similarities in content and depositional characteristics. Nevertheless, archaeological content cannot be separated from the depositional characteristics because it is an integral part of the characterization of a deposit. If similar-looking deposits are not physically connected, their relative stratigraphic position and ultimately their chronology is what will define their contemporaneity. The building of stratigraphy and correlations of the different layers is based on some basic stratigraphic principles and notions that are discussed in detail in Chapter 4.

Since stratigraphy is a manifestation of the history of a site, it follows that there is no 'good' or 'bad' stratigraphy (Goldberg and Macphail, 2006). The idea that the 'stratigraphy of a site is not well preserved' is essentially wrong and is like saying that the history of a civilization is bad. A site might or might not preserve the entire history of Roman civilization, but its stratigraphy preserves the whole history of the site. In the same sense, a burrow does not 'disturb' the stratigraphy (it may disturb the deposits) because it is *part* of the stratigraphy.

On the other hand, our ability to read the stratigraphy is not connected to what the stratigraphy actually records. This ability is restricted both by theoretical misconceptions and by practical problems. The major theoretical misconception is based on the lack of understanding the fundamental aspects of a deposit as discussed above. For interpreting site history it is imperative to understand that the building blocks of stratigraphy are the deposits and how they have been formed.

We should also make clear that the theoretical aspects of stratigraphy are different from the operational ones. Indeed, there is so far no real theory in archaeological stratigraphy, except for constructing stratigraphy in the field (see, for example, Joukowsky, 1980; MoLAS, 1994; Roskams, 2000, 2001; O'Brien and Lyman, 2002). Since excavation is a destructive process it is inescapable that a way of recording and keeping-track of the stratigraphic units as they are excavated has to be established. One of these operational systems is the Harris Matrix (Harris, 1989; Harris et al., 1993), some aspects of which we have already discussed in the Introduction and will be further discussed in Chapter 4. Most commonly these works suggest a standardized description of archaeological stratigraphy and deposits. This description is often derived from soil description manuals and may also employ laboratory data such as carbonate, carbon, phosphorous, strontium, and silica content (e.g. Pietsch et al., 2013 and references therein). However, the superficial premise that these data (while useful) can define stratigraphic units and produce stratigraphic information is misdirected. Their reliance on soil nomenclature is problematic in the first place, and their major flaw lies in the same hidden premise that someone can do stratigraphy without understanding depositional processes.

As we have already discussed, without a conceptual model for organizing observations and data one cannot obtain anything meaningful (see Introduction). All these standardized descriptions are indeed helpful in keeping an objective track of the excavated sediments, but by no means can they lead us to interpret the stratigraphy by themselves. This situation is particularly evident when trying to interpret the stratigraphy of whole sectors or the site itself and not just a portion of a profile. Some archaeologists have already noted that standardized description of archaeological sediments leads to mediocrity (Adams, 2000), which could be interpreted as an illusionary belief that everything is under control.

From the above it is obvious that stratigraphy provides the context and the means for understanding the history of a site. So the question shifts to how can we unravel it. Ultimately we should be able to replay a three-dimensional movie of how the site was actually built. Not everyone can do this in their own minds, but at least when defining stratigraphy we have to envision some sort of a dynamic three-dimensional puzzle with complex interlocking pieces that can be put together in an appropriate order. Defining the pieces and their interlocking areas is crucial, and therefore we should equip ourselves with all theoretical and practical facets of depositional processes in an archaeological site before we present in more detail the analysis of site stratigraphy in Chapter 4.

2

Natural Sediments and Processes in Sites

2.1 Introduction

Natural formation processes contribute significantly to the accumulation of sediments in archaeological sites and to the preservation or destruction of archaeological patterns. Geogenic processes can add new material to the site by aeolian, colluvial, or water-flow processes but they may also redistribute previously deposited archaeological sediments; in addition, they can end up cementing deposits or precipitating calcium carbonate ($CaCO_3$), as in the case of springs or caves. Other deposits might include rainwashed lithics, pottery fragments, or particles derived from the gradual deterioration and collapse of a wall. Therefore, the content of a deposit does not determine the depositional process since it can equally affect anthropogenic or natural particles and objects. As a consequence, naturally deposited sediment can contain archaeological objects as well. Archaeological objects behave in a natural transporting medium (e.g. water, air, snow) in the same way as a geological clast (particle) (Gladfelter, 2001; Shackley, 1978) and therefore for the following discussion they are considered along the same lines.

Natural sediments in archaeological sites are predominately clastic and are the product of transport and deposition of mineral, rock, organic (bioclastic), or anthropogenic (human-made) particles (Table 2.1). The grain-size of clastic sediments ranges from submicroscopic particles (i.e. clay) up to larger sizes whose upper limit is dictated by the highest available energy in that natural system (e.g. a statue or a masonry block). Natural chemical sediments are infrequent in archaeological sites and they include all deposits on the surface or underground that have been precipitated from solutions. Sediments precipitated by biochemical processes – for example salt, gypsum, calcite and iron oxide encrustations, and soil accumulations – are also included in these types of sediments. Organic or carbonaceous sediments are mainly represented by the accumulation of large amounts of carbonized residues of plants. Exclusively organic matter deposits, such as peat, are infrequently found in archaeological sites, although peat and turf can be used in house constructions, as in Iceland (Bathurst et al., 2010; Milek, 2012); most often organic matter is a constituent of clastic sediments even if the particle size is (sub)microscopic.

All sediments are deposited by a variety of mediums, mainly air, water, and ice; therefore the principle physical agents are wind, water flow, glaciers, volcanoes, and gravity-induced mass flows. The latter, in particular, encompass all types of deposits that can be formed on slopes, such as falls, collapses, stone avalanches, and gravity flows whose water content alone cannot put them in motion, for example slides, mudflows, and debris flows. Obviously, the decay of an erected construction may produce such mass flows because gravity is the principal agent of destruction and movement of the eroded components. In the realm of an archaeological site, only wind, water, and gravity are the main contributors, and these are the focus of this chapter. High-energy water-flow sediments, such as river and lacustrine currents, and tide- and storm-induced currents are not frequently encountered in archaeological sites. These processes are covered by a large number of special sedimentology books (Reading, 1996; Reineck and Singh, 1980; Boggs, 2006). In this chapter, only small-scale wind, water-flow, and gravity-induced mass waste processes will be discussed.

All sediments are characterized by certain attributes, which include grain-size, sorting, grading, orientation, and the geometric relationships of their particles in three-dimensional space (i.e. sedimentary structures and fabric; see below). Each of these features can be attributed to a particular natural sedimentation process, and thus the depositional regime can be identified (Courty et al., 1989). Sedimentary structures are the direct manifestation of the medium of deposition and energy conditions prevalent at the time of deposition (Reineck and Singh, 1980: 8). They include all the bedding and surface features produced during sediment deposition. The most important ones are ripples, scour-and-fill structures, and parallel, cross, lenticular, graded, interlayered, and homogeneous bedding. It is the spectrum of sedimentary structures and

Reconstructing Archaeological Sites: Understanding the Geoarchaeological Matrix, First Edition. Panagiotis (Takis) Karkanas and Paul Goldberg.
© 2019 John Wiley & Sons Ltd. Published 2019 by John Wiley & Sons Ltd.

Table 2.1 Types of sediments and constituents, including terms for loose sediments and their lithified equivalents in parentheses (adapted from Goldberg and Macphail, 2006).

Clastic and bioclastic		Non-clastic	
Terrigenous and marine	**Volcaniclastic**	**Chemical** (many of these are consolidated)	**Biological and biochemical**
Cobbles, boulders, gravel (conglomerate)	lapilli, blocks, bombs	Carbonates: • typically marine (limestone) • terrestrial: travertine, flowstone, tufa	Plant fragments: peat (lignite, coal)
Sand (sandstone) Silt (siltstone) Clay (shale) Bioclastic: • coarse (e.g. coquina) • fine (chalk)	Ash (welded tuff)	Chlorides, sulfates (evaporites) Silicates (chert) Phosphates (phosphorite)	Bacteria (stromatolites) Diatoms (diatomite) Foraminifera (chalk)

Terrigenous clastic particles include mineral, rock and human-made material (pottery, bone, lithics, mortar, and other construction materials in general). Bioclastic limestones such as coquina that are composed of biologically precipitated shell fragments may be considered as both clastic and biochemical sediments.

their presence in certain combinations that are indicative of certain depositional environments and not the presence or absence of any one of them.

On the other hand, the texture of sediments (grain-size distribution, shape and roundness, orientation, and surface texture) is controlled mainly by the medium and mode of transport (bedload and suspended load; see below for details) and to a lesser degree by the depositing medium (Reineck and Singh, 1980: 8). They therefore provide better information on how the sediment was transported (e.g. wind transport) rather than where it settled (e.g. deposition in a pond).

Most types of sedimentary structures associated with textures coarser than silt are readily perceived with the naked eye. However, for observing finer textures and structures, microscopic techniques are needed. Archaeological deposits – in most cases – are difficult to study and interpret by field observations alone. They have a complex macrostructure, are mostly fine-grained, and commonly lack obvious macroscopic sedimentary structures. Thus, microscopic study is a necessary tool for characterizing archaeological deposits, in addition to field observations. Archaeological micromorphology concerns the study of undisturbed sediments, soils, and related materials at the microscopic level (Courty et al., 1989). Throughout this and the following chapter, analyses of the microscopic fabrics of sediments will be included in addition to field observations. However, micromorphological observations will be presented only for up-scaling the details of field analysis and not as self-standing observations. Micromorphological analysis requires special training. Nevertheless, presentation of some characteristic microscopic features gives an additional idea of how sedimentary processes work and the formation of fabrics and microstructures that eventually lead to macrostratigraphic

structures and fabrics. Indeed, microscopic observations should always be linked to the macroscopic structure and stratigraphy of the deposits because only in this way can the representation of the microscopic results be warranted; at the same time major changes in the depositional system can be recognized. Therefore, for a comprehensive study of sediments an iterative strategy has to be followed in which microstratigraphic and macrostratigraphic observations are continuously compared to each other.

2.2 Principles of the Transport and Deposition of Sediments

2.2.1 Physical Processes

The *fundamental* aspects of the processes that transport and deposit sediments are (a) their capacity to transport various grains sizes and (b) their pattern (turbulent or laminar flow; Table 2.2) and changeability (steady or fluctuating). In fluctuating processes, additional importance involves the periodicity or episodicity of the processes, as well as their catastrophic nature (Reading, 1996: 13). These essential features are mainly described by the mechanics of sediment grain movement and the physics of fluids. In this section only some general principles will be presented because it is beyond the scope of this book to consider these principles in detail. However, we are of the opinion that for better understanding of the field or microscopic appearance of the texture and fabric of the sediments it will be helpful to be able to visualize the process that produces these features by understanding some of their general properties. In addition, archaeologists would like to know what sedimentary processes and their resulting sedimentary structures mean in terms of their capacity to erode and move archaeological objects and generally whether the processes are catastrophic.

Water, air, and ice are the *fluids* that transport and deposit virtually all natural sediments (the effects of gravity are discussed below). These three media are characterized by different densities and viscosities. Density controls the buoyant forces acting on sedimentary particles and therefore the ability of the different media to move the grains. Water is more effective than air in moving grains because of its higher density, hence the ability of water to move large objects and particles. Experimental work by Boaz and Behrensmeyer (1976) has shown that density is the most useful variable in predicting whether a bone will move or not and what its velocity will be once it does move.

Viscosity describes the ability of the fluid to flow. Honey has higher viscosity than water. Water overloaded with sedimentary particles (e.g. mudflow) has a higher viscosity and a higher density than dilute water flows, so a higher force is needed for a mudflow to move, and this occurs on slopes where gravity is sufficiently large. In the same way, a mudflow 'freezes' when it meets the horizontal surface because gravity can no longer transport it. In contrast, a normal water flow that carries a small amount of sediment will easily move even on a horizontal surface and cannot stop moving abruptly. These differences are depicted in the fabric of the sediments, as is described below.

Turbulence can be described as random movements of parcels of fluid in contrast with laminar flow, which is movement in a straight stream (Figure 2.1). Understanding the nature of turbulence is of major importance for interpreting many sedimentary structures. Turbulence is the crucial mechanism for movement of sediment in suspension (Table 2.2). Rapid flow or low viscosity increases turbulence. Turbulence does not appear in high-viscous mudflows (overloaded with fine material) and as a result, features produced by suspension flow are not observed. The same is true with very slow flows like very shallow rainwashes (sheetwash).

The beginning of the grain movement of a sediment is determined by the velocity of flow and the grain-size of the sediments. When grains begin to move, erosion has taken place. Additional important parameters of grain entrainment by the flow are grain shape and position (orientation) of the grain, as well as turbulence of the flow. Coarse particles obviously need a higher fluid velocity to move, but this situation changes for very fine clay sediments. Below a diameter of approximately 0.1 mm, the energy needed for setting grains in motion increases with decreasing size. However, once in suspension clay particles remain in suspension longer than

Table 2.2 Types of processes and flow of clastic sediments (based on Bertran and Texier, 1999; Coussot and Meunier, 1996; see also Figure 2.17).

Sediment concentration	Low		Medium	High	
Fluid	Air	Water	Water + fines	Air	Water + fines
Process + motion type	Rockfall – debris fall (free-fall, failure)	Water flow		Grain flow (rapid motion)	Debris flow and mudflow (rapid motion = one phase flows)
		Streamflow (suspension and bedload = two phases flow)	Hyperconcentrated flow (suspension and bedload = two phases flow)	Debris avalanche (failure)	Solifluction (slow motion)
					Landslides (failure)

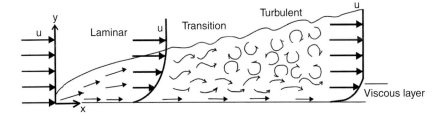

Figure 2.1 Turbulence and laminar flow, and their relation to fluid velocity (*u*) and viscosity. At low velocities or high viscosities, fluid flows smoothly in parallel layers, without lateral mixing, eddies, or swirls in the fluid. At high velocities and low viscosities, the fluid is characterized by irregular fluctuation, mixing, eddies, and vortices. Adapted from Incropera et al. (2007).

coarser particles such as sand or gravel. This part of the suspension is also called washload.

Depth of water and steepness decisively influence the velocity of the water and hence the entrainment of sedimentary particles (Blair and McPherson, 1994). In depths of water lower than a few tens of centimetres – sizes often encountered in archaeological sites – velocities are generally small and only certain bedforms can be produced, mainly small ripples and plane beds. The sedimentary particles that can be entrained are small and generally less than fine cobbles (see below; Table 2.3) even for moderate slopes of 5°. However, in steeper slopes (i.e. in the range of 20°) shallow flows can easily move cobbles and occasionally fine boulders, but such slopes are less frequent in architectural settlements.

Two major ways of movement are observed (Table 2.4): (a) suspension load in which sediment does not come into contact with the bed and is supported by fluid turbulence – in practice, only the finer material of silt and clay grade moves in this way – and (b) bedload or traction transport, which describes the movement of grains in intermittent or continuous contact with the bed by saltation (jumping), rolling, or creep. These latter processes are associated with the development of bedforms (traction carpets) on the sediment surface and involve only sand and gravel sizes (e.g. sand ripples, gravel lags, etc.) (Table 2.4). Another type of grain movement is grain flow. Grains do not move when they are at rest. Movement is achieved when the angle of rest (angle of repose) is exceeded due to the action of gravity on the

Table 2.3 Common grain-size scales used in geology (based on Wentworth, 1922).

Wentworth class	Size range
Boulder	>256 mm
Cobble	64–256 mm
Pebble	4–64 mm
Granule	2–4 mm
Very coarse sand	1–2 mm
Coarse sand	0.5–1 mm
Medium sand	250–500 μm
Fine sand	125–250 μm
Very fine sand	63–125 μm
Coarse silt	31–63 μm
Medium silt	16–31 μm
Fine silt	8–16 μm
Very fine silt	4–8 μm
Clay	<4 μm

Note that the size limits of categories in geology and soil science are not the same (Goldberg and Macphail, 2006).

Table 2.4 Types of sediment transportation and sedimentary structures produced.

Sediment transportation	Mechanism	Grain sizes	Sedimentary structures
Suspension load	Turbulence	Silt, clay, fine sand	Mainly plane bed and laminations
Bedload	Saltation, rolling, creep	Sand, gravel	Sand ripple, sand dunes and anti-dunes, plane bed (lamination), gravel lag, cross bedding

slope. In this case the grains in the layer start colliding vigorously with one another as soon as they begin to slip, and the grains starts moving downwards. This is often observed in slopes where sediment starts to move and then avalanches to the foot of the slope.

2.2.2 Sediment Properties

The entrainment, transport, and deposition of sediment particles by any agent are controlled by the physical properties of the grains. These include mainly grain-size, density (specific gravity), and shape (form and roundness). In sedimentology, grain-size, shape, and arrangement (fabric) are collectively described under the term 'texture'.

Grain-size

Most sedimentologists have adopted the Udden–Wentworth grade scale or a modified version of it (Table 2.3). The grain-size of a clastic sediment is a measure of the energy of the depositional environment. Coarser sediments are found in high-energy environments and therefore gravels are observed in rivers beds and along coasts, for instance. However, such coarse sediments are also found in slope deposits because gravity is a very efficient and powerful agent of transport. Therefore, gravel-sized deposits can be found in archaeological sites since slopes are always produced in the life history of a site due to substrate topographic irregularities but also during the erection of buildings and any kind of structures; during their abandonment gravity flow processes dominate their destruction. Occasionally, of course, river, stream, or beach gravel deposits may be encountered, but these are not the main focus of this book, as explained in the Introduction. In the realm of an archaeological site the majority of archaeological particles are generally coarse grained, particularly when the original size of the object is large. At the scale of a site, transport distances are too small to abrade and reduce considerably the size of these objects.

Sediment particles often consist of an aggregate of materials held together in a clay matrix. As already mentioned above, clay particles are glued together by strong cohesive forces and, thus, they are difficult to erode. These aggregates are transient features and normally do not persist over prolonged transport as they break down to their individual constituents. At a small scale and under conditions of low-energy transport, sediment and soil aggregates are frequently observed. Conventional grain-size analysis such as sieving (see below) fails to record this feature.

Grain-size distribution is defined by a set of mathematical parameters based on the physical and instrumental methods used in the analysis. Conventional methods for measuring grain-size include direct measurement with a calibre for gravel-sized particles, sieving for sandy sediment, and sedimentation in a water column for finer sediment (Syvitski, 2007). The latter is based on the fall time of particles in a settling tube. There are also various instrumental methods such as the laser-diffraction size analyser, which is based on the diffraction of light by different particle sizes. Grain-size data are plotted in histograms, cumulative curves, or frequency curves. The mathematical parameters of median, mean, standard deviation, and occasionally skewness and kurtosis summarize the results. Nevertheless, in the field it is very helpful to have an approximate estimation of these properties based on observations and some simple tests (Table 2.5a, 2.5b).

Table 2.5a Field estimation of grain-sizes in soils (modified from Colorado State University Extension, 2017. Estimating Soil Texture, CMG GardenNotes #214. Colorado State University, Fort Collins, CO).

1. **Feel test:** Moist soil is rubbed between the fingers
   - Sand feels gritty
   - Silt feels smooth
   - Clays feel sticky
2. **Ball squeeze test:** A moistened ball of soil is squeezed in the hand
   - Coarse texture soils (sandy) break with slight pressure
   - Medium texture soils (silty sands and sandy silts) stay together but change shape easily
   - Fine texture soils (clayey silty clays) resist breaking
3. **Ribbon test:** A moistened ball of soil is squeezed out between thumb and fingers
   - Ribbons less than 2.5 cm
     - Feels gritty = coarse texture (*sandy*) soil
     - Not gritty = medium texture soil high in *silt*
   - Ribbons 2.5–5 cm
     - Feels gritty = medium texture soil
     - Not gritty feeling = fine texture soil
   - Ribbons greater than 5 cm = fine texture (*clayey*) soil

Mean size is a strong descriptive property, but there is no clear correlation between mean size and the depositional environment. Grain-size depends on the energy of the local environment and not so much on the distance of transport. Nonetheless, all grain-size parameters depend on the size of the available particles in the surrounding environment; thus occupational deposits with large clasts are often destruction layers or packed demolition material.

Sorting refers to the distribution of sediment sizes around a common mode and is expressed as the standard deviation of sediment sizes around the mean. Lower values imply better sorting and thus more selective deposition or winnowing. A well-sorted sediment is one that has a very narrow range of size of grains. However, care has to be taken to determine whether someone uses this property for describing the whole deposit or a size fraction of it. For instance, a layer composed only of sand is a perfectly sorted sediment, but a sandy layer may be unsorted because it refers to the spread of size classes inside the sand grain range. In modern environments dune sands of aeolian origin show the best sorting; similarly, coastal sands are better sorted than fluvial ones. Sorting, therefore, appears to relate to the hydrodynamic characteristics of the environment, such as waves, current velocity and fluctuations, and turbulence.

In archaeological sites, the property of sorting is a valuable descriptive term. The majority of sediments in a site are unsorted because in most cases there are no steady or strong currents to produce enough sorting other than wind (Figure 2.2a). If the sorting is not related to a high-energy environment such a stream or river that crosses the site and the sediment is found in an occupational layer then the chance that this represents the result of anthropogenic activity is rather high. Construction, maintenance, and burning, for instance, are processes that can produce selective grain-size distributions. Ash, for example, is a well-sorted deposit because ash crystals have a narrow size range. Often, grey, fine sediments represent deposits rich in ashes (Figure 2.2b). Layers of fine-grained sediment conspicuously devoid of very large clasts and of significant lateral continuity may represent constructed surfaces.

Skewness, a measure of the symmetry of the size distribution, and kurtosis, a measure of the peakedness of the distribution, can be of value for comparing different layers and features. However, the interpretation of grain-size analysis depends heavily on the sampling procedure. A layer may be the result of several processes and not a single activity, and a depositional unit might consist of several sublayers with extensive lateral variation. Detailed field observations on the texture of sediment are nonetheless very valuable in grouping units of similar appearance and probably of the same origin. Laboratory grain-size

Table 2.5b Characteristics of textural classes in the field (modified from USDA, 1987. Soil Mechanics Level I, Module 3, USDA Textural Soil Classification Study Guide).

Texture	Dry	Wet				
		Stickiness	Plasticity	Ribboning	Grittiness	Smoothness
Sand	Loose	Non-sticky	Non-plastic	None	Very gritty	–
Sandy silt/silty sand	Soft to slightly hard	Non-sticky to slightly sticky	Non-plastic to slightly plastic	Slight to medium	Gritty	–
Sandy clay	Hard to very hard	Very sticky	Very plastic	High	Gritty	–
Sand/silt/clay	Slightly hard to soft	Slightly sticky to non-sticky	Slightly plastic to non-plastic	Slight to none	Somewhat gritty	Somewhat smooth
Silt	Soft to slightly hard	Non-sticky	Non-plastic	Slight	None	Very smooth
Silty clay	Very hard	Very sticky	Very plastic	High	None	Very smooth
Clay	Very hard to extremely hard	Very sticky	Very plastic	High	None	Smooth to very smooth

Figure 2.2 (a) Typical unsorted fills in an archaeological site. LBA, Mitrou, Greece. The layer above the level of the white line is less sorted than below and contains gravel-sized stone and sherds in a random distribution inside a finer matrix. The lower layer is better sorted but is still an unsorted mixture of fine gravel, sand, silt, and clay. Scale is 10 cm. (b) Layer 2 is a typical example of very well-sorted ash, composed mainly of silt-sized phytoliths and ash crystals, hence its homogeneity across the layer. Layer 1 is highly unsorted packing fill with bimodal distribution: large stone blocks floating in a fine matrix. Scale is 30 cm. BA, Kolona, Aegina Island, Greece.

analysis quantitates these observations, but it has to be used as corroborative information and not as sole the basis of any interpretation. This caveat is particularly true for anthropogenic deposits, which are mixtures of natural and 'cultural' particles. Only contextual and rather qualitative approaches such as micromorphology can provide a secure understanding of complex depositional and post-depositional processes that then can be further elaborated by quantitative approaches.

Grain Density

Density (or specific gravity, i.e. the ratio of the density of the material to that of a standard material, usually water) influences the settling and entrainment of the grains, so that lighter grains, such as charcoal, organic matter,

feathers, and often bone, can be easily separated by the other rock and mineral sedimentary particles during relatively high-energy water or wind flow and deposited as different layers. The same can happen when they are left to settle in standing water conditions: the lighter fraction will settle last (Figure 2.3).

Shape

The form and roundness of particles depend upon the conditions of transport; this is especially true for form, which is controlled by the composition and fabric of the original rock source and type of material. This statement is particularly true for archaeological materials such as pottery, which tends to break into platy clasts. Schists and generally rocks with fissility (tendency to

Figure 2.3 A lake shore deposit containing the remains of a conflagration event. Neolithic, Dispilio, Kastoria, Greece. The grey lacustrine sand unit in the lower part of the image contains large charcoal fragments, and some gravel and dissolved burnt construction material. The dark layer at the top (arrow) consists of fine-grained burnt materials and represents the lighter weight fraction that was last to settle in the water.

Figure 2.4 (a) Silhouette chart used to characterize the sphericity and roundness of particles. (b) Main classes of particle shapes. Adapted from Krumbein and Sloss (1963).

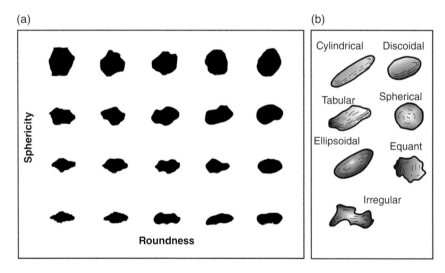

split along surfaces parallel to bedding) tend to produce tabular and elongated sediment grains whereas massive rocks tend to produce spherical grains. Form is a characteristic attribute of gravel-sized particles whereas the form and roundness of sand grains are not so helpful in determining the depositional environment; however, better rounded and spherical grains indicate a mature (i.e. long transport) sediment. Typical forms of gravel particles are oblate (disk), bladed, prolate (roller), and equant (spherical) (Figure 2.4). As a general rule, flat gravels are observed along sea coasts and rod-shaped gravels in rivers.

Roundness is the measure of the curvature of the corners of the grain. The roundness of gravels is independent of their shape and the better-rounded gravels indicate prolonged mechanical wear and tear and long transport. However, it has to be stressed that form is different from roundness so that for example, a spherical object can be angular as well. Rounded archaeological particles can also be formed by prolonged trampling and ploughing. However, human activities cannot produce ellipsoidal and probably discoidal forms (except when intentionally manufactured) but continuous breakage tends to produce more equant forms.

Bone form comprises a special category of its own by showing large variations in shape. As such, it is a variable that is difficult to quantify, although a shape index based on the ratio of the maximum length to maximum width has been suggested to indicate the degree to which an assemblage has been water sorted (Hill and Walker, 1972). Several bones have a higher surface area (e.g. upright scapula, a complete skull) and therefore are more transportable (Boaz and Behrensmeyer, 1976; Coard, 1999). It is also important to note that many bone fragments are splinters of microscopic size and cannot be analysed and measured without a microscope.

2.2.3   Fabric

Non-spherical objects have an orientation with respect to horizontal and vertical planes. Preferred orientation is observed when a considerable number of objects in a layer possess a special orientation. The most useful fabric elements are particles that show maximal dimensional inequalities because they are more readily aligned to the direction of flow. The orientation of a grain is described by the direction of the orientation of the long axis in relation to North and the inclination with reference to the horizontal. Data are plotted on a fabric diagram or statistically processed (vector analysis) (Bertran and Texier, 1999; Bernatchez, 2010).

As a general rule, fluvial gravels are mostly oriented transverse to flow and they also dip upstream (imbrication, which refers to particles lying at an angle to the bedding surface in a shingled arrangement) (Figure 2.5). Mass flows show only slight tendencies of orientation, mostly towards the flow, and are poorly or not at all imbricated (Figure 2.6). Archaeological occupational deposits do not show a preferred internal orientation, and trampling does not produce significant orientation (see Chapter 3). Human activities do not produce imbrication except when dumping material on a slope, which may result in week imbrication (Karkanas et al., 2012).

2.2.4   Sedimentary Structures

Sedimentary structures refer to the three-dimensional features produced by the deposition of sediments. The far most important is stratification, which can be defined simply as layering brought about by deposition. The term 'layering' is generally used for any arrangement of sediments in bodies with approximately planar-tabular or lenticular shape. Several other geometric features are produced during or immediately

Figure 2.6   Mass flow consisting of angular, platy gravel-sized limestone floating in a finer matrix. Note that gravels dip in the direction of the mass flow. MP, Combe Capelle, France.

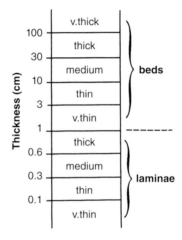

Figure 2.7   Classification of bed thickness.

after deposition of sediments. Sedimentary structures are considered the most valuable features for interpreting depositional environments.

In natural environments beds are bodies of sediments that possess a bottom and top surface and are characterized by composition and fabric, which distinguishes them from the bodies of sediments above and below. Their thickness is very small in relation to their lateral extent. These bodies are formed under generally steady physical conditions and sediment supply (Collinson and Thompson, 1989). A layer thinner than about 1 cm is termed a lamina (Figure 2.7). In archaeological sites the terms 'layer', 'stratum', or 'stratigraphic unit' are often used instead. In sedimentology 'layer' and 'stratum' are descriptive terms often used when definite differentiation between bed and laminae is difficult (Reineck and Singh, 1980: 97). In archaeological sites, it is justifiable to use the term 'layer' since strata are often thin, not

Figure 2.5   Imbricated, well-rounded, ellipsoidal, and discoidal fluvial gravels are shown in the middle of the image. Pleistocene, Aliakmon River, Kozani area, Greece.

traceable for long distances, and not of constant thickness (for an extensive discussion see Stein, 1990). In this section some major principles of the geometry of natural bedding will be presented. For an additional discussion of the nature of bedding in archaeological sites see Chapter 4.

The geometry of a bed or layer is defined by its thickness, the lateral continuity of its thickness, the orientation of the upper and lower contacts, and their manner of termination. Thus, a layer can be tabular, lenticular, wedge-like, etc. (Figure 2.8). The measurement of the inclination of a layer is also important. Depending on the characteristics of two layers, contacts can be gradational – they 'grade' into each other – or be sharp,

where clear changes in characteristics can be seen (Figure 2.9a,b). Vertical variations in grain-size or composition within each layer are often observed. Consequently, layers can be homogeneous or heterogeneous, rhythmic or gradational (Figure 2.10a–d). Layers are generally recognized by differences in grain-size, colour and shade, orientation of particles, and distribution of individual components. The latter is the most neglected and difficult to comprehend, where individual components in layers, like pebbles, pottery pieces, charcoal, and bone, have a certain distribution pattern in the space of the layer. Often this distribution produces certain geometries, like lines, clusters, bands and tails; the components may be parallel or non-parallel, planar,

Figure 2.8 Different geometries of layers in a sequence of burnt remains in the MP Kebara Cave, Israel. T = tabular; L = lentoid; W = wedge-like. A large elliptical burrow is also shown (B).

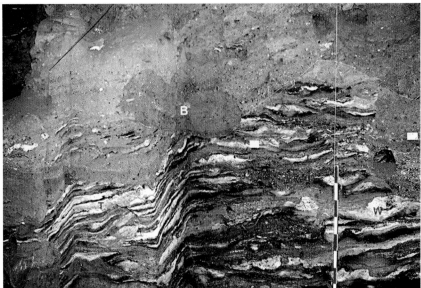

Figure 2.9 Examples of contacts between lithostratigraphic units. (a) The upper dark layer (1) is characterized by a sharp lower contact and a relatively diffuse upper one. All other contacts below, between layers 3, 4, and 5, are diffuse. Note also the contact of the channel fill is sharp and almost vertical (arrow). Such a contact might be the result of slide or collapse of the walls of the channel. Note hardhat for scale. MSA, Site PP5-6, South Africa. (b) A gradational contact through approximately 15 cm between layers 1 and 2 (scale at left is 30 cm long). The transitional zone was produced by gradual mixing of occupational debris with the underlying natural substrate. Classical, Molyvoti, Thrace, Greece.

Figure 2.10 Bedding characteristics. (a) Example of homogenous layers that can be confidently followed for several meters. Such formations are typical of lacustrine, fluvial, or lagoonal deposits. In this case, beach sands (BS) are overlain by backshore flood muds (FM) and aeolian sands (AS) with occasional evidence of soil formation as well (SS). Scale is 10 cm. Archaic – Classical, Delta Phaleron, Greece. (b) Heterogeneous construction fill at the base of a new building. Note fragments of reddish clay construction materials floating in the fill (an example with arrow). Scale is 30 cm. Neolithic, Koutroulou Magoula, Greece. (c) The bedded sequence at the lower part represents sheetwash deposits consisting of rhythmically alternating light sandy thick layers and darker, organic-rich thin laminae. MSA, Site PP9c, South Africa. (d) Reverse grading (upwards from fine to coarse) of gravels in a delta fan. Pliocene, Achlia, Crete, Greece.

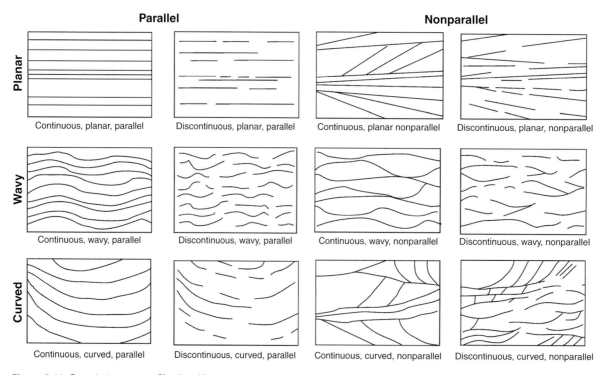

Figure 2.11 Descriptive terms of bed and laminae continuity, shape, and geometry (adapted from Campbell, 1967).

Figure 2.12 Bedding. (a) Torrential, mass flow deposits showing crude parallel arrangement of subrounded sherds and subangular stones. Note the formation of clusters (arrow) and stringers of clasts. Hellenistic-Roman, Messini, Greece. (b) Subparallel orientation of sherds, stones, and some bone (arrow). Such vague stratification of coarse particles shown in both images is mostly observed in mass flows. Angularity of stones implies short transportation. Pottery is softer, hence it rounds more easily. Scale is 30 cm. EBA, Palamari, Skyros Island, Greece.

curved or wavy, and continuous or discontinuous (Figures 2.11 and 2.12a,b). Random distributions are also often observed.

One type of distribution concerns grain-size parameters like sorting and grading, with the latter produced by vertical changes of grain-size. Grading can be normal (upwards from coarse to fine) and reversed (upwards from fine to coarse) (Figure 2.13a–c). Two types of normal grading are possible. In one type, successive increments of finer material are added so that there are no fines in the lower part of the graded bed (Figure 2.13d). In the other, the fines are distributed throughout the bed but each added increment has one less coarse grade, that is, it is finer than the one below it (see Figure 2.51a) (Reineck and Singh, 1980: 118). The first type is produced by currents that gradually decrease in velocity and therefore lose their capacity to transport coarser grains. The second type is produced from suspensions where all sizes are carried and settled gradually. Often high concentration flows result in sedimentation of the second type. The same is true for reverse grading, which is typical for some mud flows and grain flows.

One special type of bedding is cross bedding. A cross bed is defined as a single sedimentation unit consisting of internal laminae inclined to the main depositional surface. Cross beds in most cases are the result of the migration of current and wave ripples. They may also be produced in fluvial channel fills or on the inclined faces of beaches and river bars. Therefore, in the majority of archaeological sites cross bedding may be most often encountered as microscopic features in small water flow

features such as small gullies, canals, and channels (sewers, aqueducts, etc.) (Figure 2.14).

Other types of bedding are mainly the result of rhythmic alternation of depositional characteristics. In several cases a bed is made up of not one but two different types of layers that alternate repeatedly. Rhythmic or alternating bedding is a general term for describing bedding types composed of alternating sand and mud. They are mostly the result of alternating periods of current activity and quiescence. During current activity sand is supplied and deposited, and during quiet periods mud settles. Alternatively, in a generally quiet environment, such as a pond or a ditch or any hydraulic construction where mud is deposited, occasional rainwash may introduce coarser additions in the deposit.

Several other types of sedimentary structures are also observed, such as sole marks, bioturbation marks and burrows, deformation features, desiccations, and nodules and concretions. Except for sole marks, which are scouring features produced by the action of water movement on the depositional surface, all other sedimentary structures are post-depositional features and are discussed in section 2.7. Sole marks typical of turbulent current flows in deep-sea deposits are generally not encountered in the low-energy water flows of archaeological sites, but erosional features are often produced. Flowing water on a soft substrate may erode a gully or a channel (Figure 2.15a,b). Channels run for many meters and may be symmetrical or asymmetrical in form and possess U- or V-shaped profiles. However, very low energy water flow in relatively consolidated sediment

Figure 2.13 Grading. (a) Beds with normal grading (arrows) in fluvial gravelly deposits. Pleistocene, Aliakmon River, Kozani area, Greece. (b) Lateral grading in archaeological fill (arrow). Note crude downward coarsening of freshly broken tiles and stones, most likely the result of dry grain flow (see section 2.3.2). Fabric is open-work with little interstitial material. Late Roman, Corinth, Greece. (c) Reverse grading (arrow) at the head plug of a debris flow (see section 2.3.4). MSA, PP5-6, South Africa. (d) Photomicrograph of normal grading where all particles becoming finer upwards. MSA, Die Kelders, South Africa.

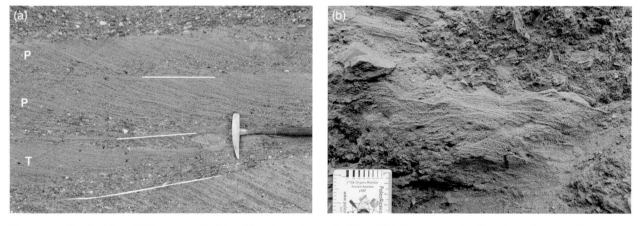

Figure 2.14 Fine bedding. (a) Planar cross bedding (P) and trough cross bedding (T) in fluvial deposits. Note the inclination of laminae in each bed. Normal grading from fine graved to sand is observed in all beds. Pleistocene, Kozani area, Aliakmon River, Greece. (b) Remnants of ripple bedded sands attached to the walls of a small gully. MSA, Site PP5-6, South Africa.

may produce very irregular, often denticulate, profiles with even reversed curvatures (see also section 2.7.1). In general, natural channels have a sinuous development that forms alternating meandering and straight reaches. Artificial channels and ditches (Figure 2.16) are normally straight, and their pattern follows constructions of the site. Otherwise, they can be very similar to natural erosional channels.

Figure 2.15 Fills in depressions. (a) Cross-section profile of a gully fill cutting through archaeological sequence at Site PP5-6, MSA, South Africa. Both straight and wavy forms are observed along the cut. Note that the base of the fill consists of massive material directly derived from the collapse of occupational layers of the eroded channel walls. The upper part of the fill is bedded and it is attributed to water flow (see section 2.4.2). Scale is 1 m. (b) An artificial channel filled with post-abandonment dark destruction debris. Note the continuation of the same dark fill along the sides of the channel. The brownish sediment at the base of the channel is part of the substrate that has been excavated to show more clearly the walls of the channel. Scale is 30 cm. Hellenistic, Molivoti, Thrace, Greece.

2.2.5 Some Remarks on the Interpretation of Textures, Fabrics, and Sedimentary Structures

Most layers in archaeological sites appear heterogeneous and structureless as they are a mixture of different sizes and components. However, careful observation may reveal distribution patterns not obvious at first glance. Mass flows, for instance, may produce massive sediments but subtle clustering, linear arrangement, crude orientation, and tiling up of coarse clasts may be observed occasionally. Small variations in sorting and overall grading, preservation of thin sandy laminae or clast tails can be useful in determining whether a relatively massive deposit is the result of anthropogenic dumping, for instance, or a mass flow. Although random distribution in practice implies that coarse elements float inside a finer matrix, in some cases this distribution might appear quite patterned in the sense that the coarse components have quite similar distances from each other (Figure 2.12a).

Nonetheless, the overall geometry and inclination of a layer and the nature of its bounding surfaces (contacts) are also decisive criteria for interpreting the formation of a layer. This is the most difficult issue during excavation, since most of the time the full geometry of a sedimentary body cannot be revealed at once. It is therefore of crucial importance to group deposits with the same appearance whereby revealing information on one deposit can be used to interpret the others as well.

The same holds with the case of the depositional substrate. Pre-existing depositional or erosional surfaces dictate sedimentation processes because their geometry

Figure 2.16 Spanish Civil War trench. Madrid, Spain. Its walls are carefully prepared, straight, and smooth. Note that the fill tends to sag in the middle and form festoons. Courtesy of Amalia Pérez-Juez.

and inclination provide the boundary conditions of the intensity and dissipation of the energy of the flow. Thus, illuminating and understanding the geometry of the substrate allows us to restrict the range of sedimentary processes that can occur above. Obviously, an inclined surface is prone to abrupt and high energy flows, such as mass flows with erosional capabilities and a catastrophic impact.

Sedimentary structures are not only the result of natural processes. Anthropogenic processes can result in certain types of stratification, such as in the case of pit fills, road constructions, and floor sequences (see Chapters 3 and 6). However, there are certain types of natural sedimentary structures and fabrics that cannot be produced by human activities. In general, all structures that result from deposition by currents and suspension (e.g. graded layers of well-sorted sediment, cross bedding, and imbrication of rounded and sorted pebbles) cannot be produced by human activities. Similarly, rhythmically bedded, well-sorted sediments are also difficult to make by human processes; however, types of gravel roads can mimic in some ways naturally deposited sediments (see section 6.4.4 for comparison).

Furthermore, the geomorphological and petrological conditions of the surrounding area of a site dictate the type of sediments and depositional processes that can occur. Therefore, knowledge of the petrology of the substrate and the geomorphology of the catchment area will guide any possible interpretations, helping to determine which types of sediments cannot be transported and deposited in the site by natural processes. The catchment area of the site dictates the petrology of the sediments but also the type and maturity of sediments deposited by natural processes. It is therefore imperative to record and

understand the sources and paths of sedimentary transport inside the archaeological site. In this light, the use of geological maps is therefore a must in understanding the geological nature of the materials at a site.

The architecture of sedimentary sequences – that is, the overall geometry of deposits and their spatial relationships – is very informative about the prevailing processes during their formation. Gravity flow processes will be associated with sloping sequences, and ponded sediments with flat lowland sites. Stacking of horizontal, tabular, and laterally persistent bodies implies low-energy depositional processes. Channels, lenticular bodies, and wedging out boundaries should be associated with frequent erosional events.

2.3 Mass Movement in Sites

Mass movement (or mass wasting) is the downslope movement of soil and rock material under the influence of gravity. Water or ice can indirectly trigger the movement by reducing the strength of the slope materials and lubricating them. A general term used for describing slope waste material is colluvium (Blikra and Nemec, 1998). Talus, scree, slope waste, and hillslope deposits are terms also used in the geomorphological literature.

Several mass movement processes are distinguished mainly by the type of movement (falls, slides, or flows), material (bedrock, or soil and sediment), and water content (rock and debris fall, rock and debris slide, solifluction, mudflow and debris flow, dry and wet grain flow). With increasing water content mass wasting processes are transformed to ones of fluvial transport (Figure 2.17).

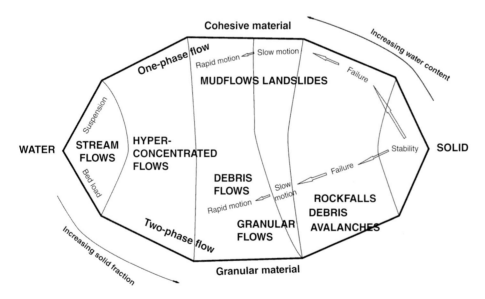

Figure 2.17 Classification of mass movements as a function of solid fraction, water, and material type. Adapted from Coussot and Meunier (1996).

Each of these processes has different impacts on the formation of archaeological sites and the context of the archaeological materials. Although not common, occupation on steep slopes – including coarse debris cones – is not unknown. Such occupations are short term (Bettinger, 1994), but both long-term camps (Sullivan and Sassoon, 1987) and more permanent constructions (John and Kočar, 2009) also occur. Moreover, mass movement processes have often been involved in substantial reworking of archaeological sites and thus analysis of their integrity requires a thorough understanding of site formation processes (e.g. Zilhão et al., 2006). Nevertheless, the presence of sloping deposits does not rule out the existence of *in situ* and non-reworked archaeological remains. Non-depositional periods can be associated with surface stabilization, which over prolonged periods can lead to soil formation

that may be also related to climatic changes or reforestation (Hinchliffe, 1999; Kittel, 2014). During these periods of exposure, occupation can readily occur but renewed slope deposition can erode the previous occupation depending on the particular type of conditions.

2.3.1 Slides and Slumps

Definition, Processes, Occurrences

Slides and slumps are relatively rapid downslope movements of a sediment or soil mass along a shallow and wide failure plane (Figure 2.18); slumps have curved failure planes and involve rotational movement of soil mass (Figures 2.18 and 2.19). In slides and slumps, movement may be concentrated along internal shear planes in addition to the basal one. Slides require a minimum slope gradient, which may vary according to the strength of the

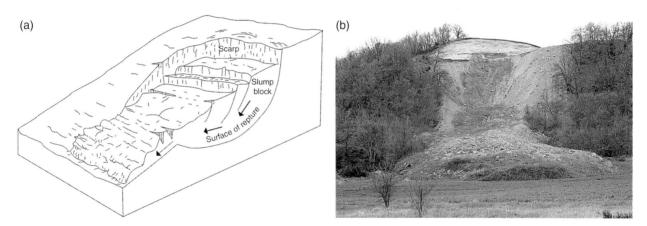

Figure 2.18 (a) Geomorphology of a landslide in cross-section. Adapted from USGS, http://pubs.usgs.gov/fs/2004/3072/. (b) Slide along a fluvial terrace. Aliakmon River, Grevena area, Greece. Note the lobes of material accumulated at the toe of the slide.

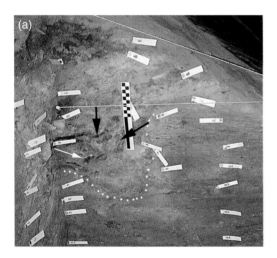

Figure 2.19 (a) A small slump at Site PP13B. MSA, South Africa. Note the subsidence of the main dark occupational lens and distortion and dragging down of some others (arrows). The channel-like diffuse boundaries of the slumping (dotted line) delineates the movement of material towards the observer. (b) Longitudinal section of the same slump showing dark occupational lenses that are fragmented, folded, and pinch out. Scale: excavated square is 25 cm across. From Karkanas and Goldberg (2010a).

substrate. Slide planes occur along a boundary between the soil and the rock substrate or between sediment materials with contrasting clay and sand content, and thus water permeability. Slope failure can be triggered by the removal of material from the toe of the slope, but most often it may occur after extreme rainfall or snow melt. Additional triggering factors are ground-acceleration during an earthquake and loss of vegetation cover due to human activity and grazing.

To our knowledge documented landslides inside archaeological sequences have been rarely reported except post-abandonment and destruction phases (Gleeson and Grosso, 1976; Macphail et al., 2016), although small-scale slides may occur in the slopes of a tell during its life history, as well as in earthen mounts and similar earthen constructions or in sites with large topographic differences. The slip plane will follow layers with contrasting competence such as aeolian sand bodies, organic-rich layers (e.g. dung or turf), or constructed floor surfaces, for example. If a side of the archaeological sequence has been undercut by a stream or waves, then sliding may occur in the remaining standing part of the sequence.

Recognition

Field Slides in archaeological sequences can be very difficult to recognize because often the sliding mass undergoes relatively small modification so that the initial structure of the material may be only partly observed in the final deposit. Often, large pieces of intact sediment are separated by macroscopic fractures. When disintegration of the mass is weak, the main diagnostic criterion is often the upslope tilting of the bedding in the different blocks (Figure 2.18a). Small-scale slides may be recognized as dipping bodies of sediment delineated by relatively clear boundaries from the surrounding sediment often approaching the shape of an open gully. If the unaffected, non-sliding part of the deposit contains characteristic features and layers, then those within the sliding body might be present as inclined features that thicken and thin, and locally pinch out across the profile as a result of the lateral spreading of the material (Karkanas and Goldberg, 2010a, 2013) (Figure 2.19).

The surface expression of slides may have been 'healed' or masked by deflation and erosion, and will appear to follow the general inclination of the slope. However, it is still possible to observe its internal structure, which should have been affected and disturbed as described above. Downslope, slides may be transformed to flows since deformation and water content frequently increase close to the base of the slope. Hence, features recognized in debris and mudflows may locally occur in slides as well.

Microstructure Diagnostic microscopic features of sliding are very few. Often subangular to smoothed aggregates are observed separated by fractures which may be referred to as a 'brecciated' microstructure (Bertran and Texier, 1999; Bertran, 1993; Karkanas and Goldberg, 2013) (Figure 2.20). However, the overall geometry of these fractures has to be associated with dilation and shearing movement, that is, movement of the aggregate blocks away from each other. Some areas show stretched and sometimes segmented clay-rich intercalations; clayey deposits can be aligned along sliding surfaces, producing striated b-fabrics (Figure 2.21). Rotational features similar to those in debris flows (see section 2.3.4) have also been reported in slides (Bertran and Texier, 1999) but normally are expected in more water-rich flows (debris flows, etc.). Nevertheless, all these plastic deformation features imply elevated water and clay content.

Orientation Analysis

According to Bertran and Texier (1999) and Bertran (1993) clast fabric in earth slides ranges from random to weakly planar. Bernatchez (2010) reports clast fabric data from an area of a small slump in the Middle Stone Age site PP13B, South Africa. The study confirms a planar arrangement of objects conforming to the slope of the cave in this area and some realigning in the downslope direction.

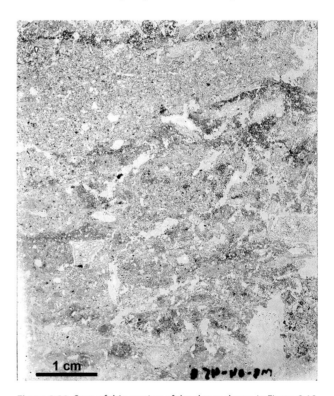

Figure 2.20 Scan of thin section of the slump shown in Figure 2.19. Dark occupational lenses are distorted and fragmented in place, producing loose aggregated sediment as a result of lateral spreading of material.

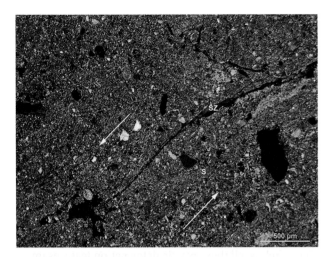

Figure 2.21 Shear zone (SZ) produced during sliding. Note the arrangement of finer particles along the shear zone (arrows show sense of movement). Reorganization of material due to deformation has produced sorted bands (S). XPL, MP, Denisova Cave, Siberia, Russia.

Figure 2.22 Pile of free fall accumulation of limestone blocks due to the collapse of the entrance area of a cave. Holocene, Site PP9 Cave, South Africa.

Effects on Archaeological Materials

As a consequence of the sliding process as described above, the original position of the archaeological objects will be affected. However, often whole blocks of sediment have been moved *en masse* so the overall integrity and context might be locally preserved. On the other hand, there may be instances (for example basal slide and internal slide zones, and fractured zones), where the original pattern can be totally destroyed. It has to be stressed that although slides may disturb the original position of archaeological objects they produce new stratigraphic bodies with new boundaries and therefore they cannot be considered post-depositional disturbances at least from the stratigraphic point of view (see section 1.5).

2.3.2 Rock and Debris Falls, and Avalanches and Grain Flows

Definition, Processes, Occurrences

Rock falls from cliffs, rock faces, or steep sediment sections involve the removal of any kind of rocky or soft mass from blocks to granules. They may also include free-falling, or downslope travel of rock over a very steep slope. Such free falls along cliffs often form sheets of debris, known as talus slope or scree. Falls below chutes or gullies form cones of debris, known as talus cones (Selby, 1993). Similar accumulations of debris cones are formed below the brow, along the dripline of a cave (Butzer, 1981). These are the products of the collapse of the entrance face of the cave and the free fall of material eroded from the area above the cave. Within caves, free-fall accumulations by collapse of the roof and walls are common (Figure 2.22) (see also section 5.2.1, on caves).

Rock debris avalanches are granular mass movements that originate in rocky or granular mass ruptures. Within debris avalanches the role of water, if there is any, remains minor. All cases of fall and debris avalanches involve loosening of the rock mass by physical and chemical weathering processes such as water erosion, freeze-thaw activity, large diurnal temperature differences, and salt precipitation in pores and along fractures.

In archaeological sites other than caves, falls are normally restricted to areas of the site close to cliffs and steep slopes or in eroding sections of the site carried out by waves and streams. Indeed, most of the falls occur in areas where the slopes are undercut by other processes. However, talus processes are likely to occur in tell slopes and similar mound-like archaeological sites where inclination is in excess of 35–30°, the angle of repose. In archaeological caves and some deep rockshelters the regular supply of breakdown material and the formation of debris cones at the entrance lead to the formation of talus forms towards the back of the cave (Bertran et al., 2008, 2010; Karkanas et al., 2008). In some rockshelters, on the other side, the major supply may come from the back wall and often a coarse talus is observed dipping to the entrance area (Donahue and Adovasio, 1990).

A special case of fall of major significance in archaeological sites is the collapse of construction materials from buildings and other constructions. During the effective life of a structure, material decayed from the walls forms a narrow and thin apron directly on the ground, with a slope away from the wall. These miniature talus slope forms usually consist of fine-grained sediment derived from the disintegration of mudbricks, plaster, or mortar from the walls (Macphail, 1994). The slope debris is rapidly enlarged during abandonment and particularly when the roof is removed and the wall starts collapsing

(McIntosh, 1977; Friesem et al., 2014a). Free-fall characterizes the collapse of masonry or dry-stone walls and large cones of collapsed mudbrick or building stones and construction debris accumulate at the base of the wall remnants (Figure 2.23).

The processes that describe the majority of fabrics of materials in talus forms are free-fall, grain flow, and slush avalanche or slide. Primary free fall occurs only in cliff and rock faces slopes in excess of 80°. Characteristics of free rock-fall and debris fall deposits are normal grading, lack of distinct depositional units (non-stratified), and a rather disorderly clast fabric (Nemec and Kazanci, 1999; Sanders et al., 2009).

Figure 2.23 Free-fall of masonry blocks (2). Note that the biggest blocks are found close to wall 1. Note water bottle for scale. LBA, 'Lower Town' of Mycenae.

In free falls, individual particles bounce and roll downwards by gravity until they reach a zone where they form a surface of particles of similar size (Figure 2.24). Large particles will roll over smaller ones and consequently will dominate lower slopes. Small particles will be trapped in the spaces between large ones and therefore will gradually dominate the upper slopes. Therefore, some degree of sorting will be formed, with the small particles accumulating at the crest and the larger particles with their greater energy reaching the foot of the slope. Consequently, change in grain-size will be observed along a dipping layer (Blikra and Nemec, 1998). Obviously, such clear patterns will be observed when clasts are rock fragments or consolidated sediment aggregates. Loose soil and sediment aggregates tend to disintegrate when they fall, and can only move for short distances downslope because their kinematic energy will be exhausted as they break into smaller pieces. Nevertheless, some small-scale and microscopic sorted fabrics may be preserved.

Dry grain flows are characterized by ubiquitous grain collisions and downslope movement of large masses of rock clasts forming elongated lobes. The dilatant behaviour of the mass as the grains bounce against each other leads to downward movement of finer clasts at the base of the flow while the coarse clasts become concentrated at the surface, a phenomenon called kinematic sieving. As a consequence, reverse grading will be observed in layers dominated by dry flow processes with coarser particles found at the top (Blikra and Nemec, 1998). In addition, strong imbrication and preferred orientations

Figure 2.24 Fallen masonry blocks from wall 1 have been distributed downslope (arrow) according to their size. A debris flow has picked up smaller masonry elements and moved them downslope. Note that stones in the overlying debris flow (2) are floating in a finer matrix with a random distribution but they do follow the direction of the flow. Scale is 1 m. EBA, Palamari, Skyros Island, Greece.

parallel to the slope are usually related to grain flow processes (Figure 2.25) (van Steijn et al., 1995; Bertran and Texier, 1999; Nemec and Kazanci, 1999; Bertran et al., 2008).

'Slush' flows in the form of dry rock debris slides are dry flows that redistribute debris from the head of a talus slope when the stability of the slope fails due to gravitational forces. This type of slide is different from grain flow processes because it characterizes a streaming type of flow whereas grain flow is a collisional flow (Ventra et al., 2013). This process leads to overthickened deposits in the lower slope and concave forms. The fabric of these deposits is most often chaotic.

Figure 2.25 Alternating beds of relatively sorted clast-supported fine gravels and cobbles (with arrows) probably as a result of kinematic sieving during dry grain flow. MP, Qesem Cave, Israel. From Karkanas et al. (2007).

Recognition

Field Dry slope processes of sediments dominated by rock particles produce slope deposits in the form of cones or sheets that are generally stratified and mostly coarse grained. Layers have tabular and lenticular forms often with sharp erosional contacts. Clasts are generally angular, open framework to clast-supported, and show lateral sorting or even reverse grading (Figures 2.23–2.25).

Because the distance of transport is small, talus deposits are often coarse grained and can include very coarse gravel and even boulder sizes. All rock clasts or durable particles (pottery, bone, lithics) are usually angular, although some show fresh fractured surfaces (Figures 2.13b and 2.26). A mixture of fresh and variably weathered clasts is often observed because of incremental accumulation and not mass deposition (Ventra et al., 2013). Open-work fabrics with few small interstitial clasts are characteristic of talus deposits in which rock fragments of similar size are in contact with each other (Figure 2.26). However, infiltration of fine-grained material can produce a closed clast-supported fabric where all of the voids are filled. This is usually the result of washing of fines into the voids of the original open-work. Note that in cases of matrix-supported fabrics where large clasts float in a fine-grained matrix the deposits may have originated as a debris flow or solifluction (see sections 2.3.3 and 2.3.4). In cases where the dominant particles consist of fine-grained sediment or soft sediment and soil aggregates, most of the characteristic features will be quickly lost by rainwash processes. Although such deposits may have been originally deposited by free-fall or debris avalanches, they are described by mud or debris flow processes or slides and slumps (see sections 2.3.1 and 2.3.4).

Figure 2.26 Talus (T) of whole and fresh fractured pottery tiles presumably formed from roof collapse. Some interstitial material has infiltrated in between the tiles. Roman, Ephesus, Turkey.

Piles of collapsed stone walls and associated masonry material will be characterized by open-work fabrics or clast-supported fabrics if disintegrated plaster and mortar is deposited (Figure 2.26) and later infiltrates between the stones together with other fine-grained washed in sediment. If the stones are of equal sizes, the fabric will be chaotic, but stones will be generally inclined and will follow the dip of the pile. A similar picture will be formed when mudbrick walls collapse. However, if the mudbrick piles are not quickly buried by other deposits, rainwash will soon disintegrate the mudbricks and produce a 'melted' deposit with pieces of mudbrick floating in a muddy matrix (see section 2.3 on water-flow processes). In cases where different sizes are involved, such as in dry-stone walls, some kind of lateral sorting might be observed. However, the size of such piles is normally small and consequently, sorting and grading will be rather ill developed, if present at all.

Microstructure At a microscopic scale, relatively fine-grained free-fall deposits show an open-work fabric with clasts having random orientations (Figure 2.27); unmodified free-fall roofspall accumulations are composed of angular, unsorted clasts. Dry grain flow processes are responsible for microscopic inverse grading, something that is often recorded in aeolian dune deposits (see section 2.5).

Orientation Analysis

Free-fall deposits tend to have random orientations along the sloping plane (Bertran et al., 1997). Shape plays an additional role in distribution along the slope: isometric particles (spherical shapes) more often tend to reach

Figure 2.27 Photomicrograph of free-fall roofspall showing open-work angular quartzite clasts having random orientations (XPL). Note apparent similarity with field photograph of Figure 2.26. From Karkanas et al. (2015a).

the lower parts of the slope than platy clasts. At the same time, platy clasts often slide before coming to a halt and therefore attain an orientation parallel to the local slope. Clasts with shapes that roll (e.g. ellipsoidal) can also more easily reach the lower parts of the slope, resulting in orientations perpendicular to the slope direction. In grain-fall deposits, on the other hand, clast orientation is parallel to the slope direction. Imbrication is strongly developed and often clasts dip upslope.

Analysis of the downslope movement of archaeological objects (bone, pottery, and lithics) on steep slopes shows a typical pattern where heavier and denser objects are moved further downslope than lighter and less dense ones; therefore bone tends to be concentrated in the upper parts of the slope, whereas lithic artefacts move down to the lower parts. Pottery shows a relatively intermediate behaviour (Rick, 1976).

Effects on Archaeological Materials

Slope sequences characterized by dry fall and avalanche processes are highly unstable and are prone to continuous remobilization, and as a consequence archaeological objects will be reworked. Older objects are located deeper and towards the more distal parts of the slope deposits. However, slope sequences are often stabilized during periods of reduced sediment inputs, soil formation, and plant colonization. Free falls and all the relevant dry slope processes are not particularly erosive except in cases of large-scale catastrophic falls and debris avalanches. Therefore, it is not strange that talus cones may preserve *in situ* archaeological remains on stabilized horizons and protected niches.

2.3.3 Solifluction

Definition, Processes, Occurrences

Slow periglacial soil movements, collectively designated 'solifluction', consist of frost creep, gelifluction, and frost heave (van Steijn et al., 1995; Bertran et al., 1995; Lenoble et al., 2008). Frost creep is produced by ice formation and results first in lifting of particles by the growing ice and then their slow downslope movement. Frost heave also results in the ejection of large stones to the surface. A common feature of periglacial environments is the sorting of stones in the upper decimetres of soils, leading to sorted polygons, sorted stripes, or stone-banked solifluction lobes (Bertran et al., 2010). Gelifluction involves relatively rapid localized mass displacement due to thawing of the ice during warm and humid periods. Solifluction operates widely on slopes as low as 2° where sediment undergoes freeze-thaw alternations, producing smooth slopes that have small-scale surface forms such as lobes, sheets, steps, and stripes; they may also lack any of these specific features (Figure 2.28). Today, solifluction

Figure 2.28 (a) Solifluction lobe (2) gouging into red fluvial sediment (1). Note flowing of coarse clasts inside the finer matrix. MP, La Ferrassie, France. (b) Folding (F) and faulting (arrow) due to solifluction at Kostenki (MP), Russia. 'K' refers to one of several krotovina. Courtesy of Vance T. Holiday.

operates in arctic environments but during the Pleistocene glacial periods solifluction deposits can be found in mid-latitude regions that experienced periglacial climates; therefore solifluction is widely reported in Palaeolithic sites of these regions (Lenoble and Bertran, 2004; Lenoble et al., 2008; Macphail, 1999; Bertran et al., 2010), but also in Palaeoeskimo sites (Todisco and Bhiry, 2008a,b). Solifluction deposits are also known from Palaeolithic rockshelters and caves (Bertran et al., 2008; Lenoble and Bertran, 2004).

Recognition

Field All types of sediments are prone to solifluction: fine-grained sediment as well as matrix and clast supported and open-work coarse-grained sediment. Poorly stratified but also very well stratified deposits are often encountered in the form of sheets, lobes, and plugs, and therefore planar and lenticular layers will appear in an excavation section. Since most coarse clasts are the product of shattering by frost action, they are usually angular with fresh surfaces and also tend to have platy or bladed shapes (Figure 2.29).

Natural pavement formation by frost heave and creep on solifluted sheets often produces open-work coarse-grained layers. These can be later filled with infiltrated fine-grained sediment and appear as 'clast-supported' deposits. Open-work layers often show vertical grading that can be normal or reverse. Repeated movement and burying of the previously formed pavements produces stone banks, which are well sorted but also show lateral grading (Figure 2.30). In fine-grained sediment, redoximorphic (reduction and oxidation) features are often observed in the form of mottles and dark brown to black iron and manganese aggregations and nodules as a result of intermittent waterlogging during thawing seasons. Cryoturbation features, frost cracks, and ice lenses

Figure 2.29 Thick layer of angular platy frost shattered clasts from Pech de l'Azé II, France (MP). Trowel on top of the layer for scale.

Figure 2.30 Image of stone concentration from MP, La Ferrassie, France. Such concentration of stones are attributed to frost sorting and heaving. Courtesy of Shannon McPherron.

are additional characteristic features associated with soliflucted fine-grained sediments (see also section 2.7 on post-depositional processes).

Slope deposits with typical regular alternating matrix-rich layers and open-work layers known as grèzes litées can form impressive sequences that are several meters thick (Bertran et al., 1995) (Figure 2.31). Formation of these sheets probably reflects the seasonal alternation of slow movement which acts as a conveyer belt. During the dry season, frost creep produces the thick stony front and during the humid season rapid movement (gelifluction) produces sheets of fine material which progressively bury the front (van Steijn et al., 1995). This alternating movement probably explains the complex pattern of grading features that includes normal and reverse, but also alternating grading in the same depositional unit. Nonetheless, such sequences are characteristic and easy to recognize.

Microstructure The most characteristic microscopic feature of soliflucted sediments is the platy to granular microstructure due to ice-lensing (Figure 2.32a) (see section 2.7.2 on diagenesis for details) (van Vliet-Lanöe, 2010; Bertran and Texier, 1999). Thin vesicular crusts are often observed at the surface, which are derived from the collapse of the soil aggregates during periods of oversaturation (melting of ice). Accumulations of fine textured mineral material on the upper surfaces of larger grains (cappings) are very often observed (Harris and Ellis, 1980; Courty et al., 1994b) (Figure 2.32b).

Orientation Analysis

Important and detailed studies of clast orientations of solifluction sediments in archaeological sites have been conducted by Lenoble and Bertran (2004), Lenoble et al. (2008), and Bertran et al. (2010), who made similar measurements under experimental conditions. Solifluction gives rise to artefacts that have preferentially oriented along the slope, even if the slope gradient is very weak. Clasts are frequently imbricated, but they have various dips, which can be observed in both open-work and matrix-rich layers. However, lobe fronts may not exhibit such well-oriented fabrics but rather imbrication, which is often well developed (Bertran et al., 1997).

Effects on Archaeological Materials

Solifluction may have a strong influence of the preservation and position of archaeological materials. Artefact post-depositional movements may increase dramatically on slopes when solifluction adds to sorting processes (Bertran et al., 2010). However, it has to be taken into consideration that the slow documented rates of solifluction often do not cause erosion of previously deposited archaeological deposits. On the contrary, it has been suggested that site burial by sheet-like solifluction prevented archaeological levels, lithic artefacts, and bones from undergoing alteration (Todisco and Bhiry, 2008b). Therefore, timing of solifluction in relation to the human occupation is critical for correctly interpreting artefact patterns. Furthermore, although solifluction can certainly redistribute archaeological materials depending on the intensity of the processes, the material may remain at the same occupational horizon albeit with a disturbed pattern. Whereas these processes clearly occur in many sites in the Dordogne (France), their ability to completely displace artefacts and features is problematic, as shown by numerous intact hearths within them (A. Turq, personal observation).

Figure 2.31 Stratified éboulis (*grèzes litées*) from UP, Laugerie Haute, France, interbedded with dark occupational layers. These deposits are inclined toward the direction of the slope, which explains why they are horizontal in this strike view. Scale: height of profile is about 3.6 m.

2.3.4 Debris Flows and Mudflows

Definition, Processes, Occurrences

Debris flows or mudflows develop when a loose archaeological deposit on the top of a relatively steep slope is mobilized in the form of thick slurries by the addition of water (Selby, 1993: 303). The fundamental differences between debris flows or mudflows, and water and hyperconcentrated flows is that in debris flows solid material (stones, lithics, pottery, fine-grained sediment, organics) and water move together as a single semi-plastic body (Coussot and Meunier, 1996); solids may constitute 45–80% by volume of the total mass (Costa, 1988). When deposited, there is no separation of the different fractions of sediment as in other type of fluids. Consequently, large stones and suspended clay are deposited together, resulting in a poorly sorted, non-bedded deposit known also as diamicton (Figure 2.33) (Mazza and Ventra, 2011; Bertran et al., 2012). Mudflows are fine muddy debris flows with relatively few clasts larger than sand size (Figure 2.33b).

Relatively steep slopes over short distances (range of a few meters) are occasionally produced in the life history of a site. In constructed sites, gradual collapse of buildings always produces cones of material that are prone to remobilization by rain water, including raindrops (Friesem et al., 2011, 2014a,b). Collapsed stony construction debris, with secondary infilling of fine-grained sediment, is often the locus of debris flow development because they always produce sloping accumulations. Loose deposits that have been accumulated on the top of these collapsed accumulations are also prone to mobilization. These loose sediments can be regular occupational deposits produced during the secondary use of the area above the collapsed constructions or they can be weathered, disintegrated, and accumulated by a variety of processes if the area is abandoned.

Sites that are close to hillslopes covered with colluvial soils are also vulnerable to debris flows, but these debris flows normally do not carry archaeological materials unless there is occupation upslope. Any steep change in the relief of a site can also be the locus of debris flow formations (Karkanas, 2015). Such a relief can be inherited from the bedrock relief of the site, removal of bedrock, and disruption of the soil cover by quarrying of the bedrock, or they can be produced by previous erosional episodes. Water flows or slides can produce steep slopes that can later remobilize accumulated sediments. The flanks of tells are also areas where debris flows can be produced during the life of the settlement, but they can also occur after abandonment. Sediments accumulating in cave entrances are also prone to remobilization by rainwater or even sea spray in sea caves, and as a result, debris flows are often encountered inside caves, particularly near the entrance below the cliff face (Macphail and Goldberg, 1999; Mallol et al., 2009; Dirks et al., 2010; Karkanas and Goldberg, 2010a; Karkanas et al., 2008, 2015a).

Intensive rainstorms or rainfall in melting snow can provide the necessary water (Selby, 1993). More rarely, however, river floods or overland flows into sites with high relief can trigger debris flows. Slope failures often start as slides or debris avalanches, but as they become quickly saturated with water, they are transformed into a downslope surge of debris. Debris flows can be quite fluid, but more often they are viscous fluids. They can be described as catastrophic flows and they always form

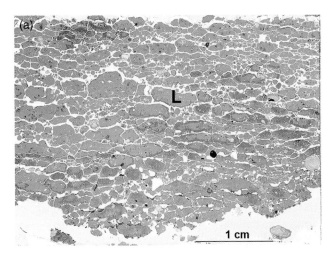
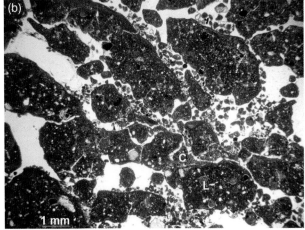

Figure 2.32 (a) Scan of thin section showing ice-lensing (lenticular microstructure) from freeze-thaw activity. Note finer-grained subrounded aggregates are found in between the lenticular sediment aggregates (L). From Karkanas et al. (2008). (b) Detail of ice-lensing with silty coating (C) and fine textured mineral material on the upper surfaces of larger lentoid aggregates (L). PPL, MP, Dadong Cave, China. From Karkanas et al. (2008).

Figure 2.33 Examples of debris and mudflows. (a) The uppermost layer (DF) is composed of boulders floating in an unsorted gravel-rich matrix. The lower bedded sequence (WF) has characteristics more typical of water flow, such as crude imbrication (arrow) and obstacle-induced, fine-gravel sorted tails at the back of a boulder (S). Flow is towards the left. MP, Dadong Cave, China. From Karkanas and Goldberg (2017b). (b) The wavy dark layer (MF) is a typical mudflow having formed at different stages as there are least three layers (1–3) to the left that contributed to its formation. The main source, layer 3, has been cut by a transverse channelized debris flow (C) (flow toward the observer). A lobate head (PL = plug) of coarse sediment is shown at the front of the mudflow (see also Figure 2.13c for a detailed image of the plug). The dark colour of the sediment is due to the high content of burnt remains. MSA, Site PP5-6, South Africa.

suddenly by failure of slope material (Coussot and Meunier, 1996). As they move downslope they may incorporate material along the slope, depending on the viscosity of the fluidized sediment, the slope type and angle, and distance of travel. Clay content plays a significant role in the development of debris flows since clay quickly loses its cohesion when saturated with water.

Recognition

Field All debris flows share some common features that are the product of the distinct mechanisms of flow (Blikra and Nemec, 1998; Nemec and Kazanci, 1999; Blair and McPherson, 2007; Pope et al., 2008). They have a lobate head followed by some wave forms behind. The lobe is characterized by a snout and lateral deposits (levees) at the sides of a median channel (Figure 2.34). When highly concentrated, slurries cannot move any further, mainly because they meet a horizontal plane and they rapidly deposit their material as dense lobate bodies (Figures 2.13c and 2.33). However, these features are rarely observed in archaeological sites because (a) the relief of debris flows is usually not preserved unless the site is abandoned afterwards, (b) the geometry of the whole body is only gradually or partly revealed because archaeological deposits are often excavated in small trenches, and (c) the sloping geometry of a debris flow is difficult to follow along its entire extent. Despite these factors, debris flow deposits can be recognized by their inclined, sharp, and undulating lower contacts (Figure 2.33). They are also elongated and have a generally lenticular form. Indeed, deposits formed on a sloping surface are candidates for debris flow accumulations. Preferential concentration of coarse elements such as large bone, stone, pottery, and

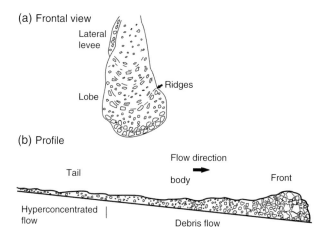

Figure 2.34 Sketch of debris flow in profile and frontal view showing the different parts. Note the concentration of coarser particles in the front and its relationship to hyperconcentrated flow at the tail. Adapted from Ritter et al. (2011).

lithics along elongated diffuse belts represents deposition along the snout and lateral levees.

The most distinguishing feature of debris flows in the field is their texture and fabric (Costa, 1988; Blikra and Nemec, 1998): they consist of an almost uniform distribution of all sizes from clay to boulders (Figures 2.33 and 2.35). The largest clasts are floating in a matrix of sand, silt, and clay. They may also contain pieces of wood and other light materials. It follows then, that debris flow sediments are extremely poorly sorted; they lack stratification, but contacts between successive debris flows may be distinct. Therefore, if bedding is present it should separate different debris flow events.

Debris flow deposits do not have well-developed sedimentary structures, such as those described in water and hyperconcentrated flows (Costa, 1988; Coussot and Meunier, 1996; Blikra and Nemec, 1998). Because of buoyant forces, large stones may be concentrated at the top of the deposit, forming reverse grading (upward coarsening); poorly developed normal grading is also observed. In deposits with small water content, large stones or archaeological objects have random orientations, with some of them being vertically orientated (Figure 2.35). In sections along the axis of development of the debris flow, 'trains' of clasts are also observed (Figure 2.35). In more fluid flows, large objects and stones may show a poorly preferred orientation (Figure 2.36). Lack of internal stratification and sorting differentiates debris flow from water flows. Lack of clast segregations (e.g. gravel lenses) or macroscopic imbrication of clasts and the presence of a clayey matrix distinguish debris flows from hyperconcentrated flows, although they are often spatially interrelated (Figure 2.34). Dry fall deposits have a different internal structure and fabric. Rock and debris falls are often coarse-grained,

lack a clayey matrix and show characteristic upslope grading (see section 2.3.2).

Microstructure The fine-grained matrix of debris flows is relatively homogeneous, with coarser particles floating in a silty clay matrix (porphyric related distribution; Stoops, 2003); numerous vesicles are common due to the escape of air. In both cases, however, these features are not indicative of only debris flows (Bertran and Texier, 1999). Nevertheless, lack of regular microscopic lamination and grading, as well as poor sorting, distinguish debris flow from water-flow processes. Weak microscopic lamination is locally observed in debris flows of high-water content, however. They take the form of laminae of silt and clay almost a few grains thick or the form of stringers of sorted silt and fine sand embedded in a massive mud deposit (Figure 2.37). These features are usually transient, and they quickly lose their organizational structure laterally because of localized liquefaction (Bertran and Texier, 1999; Karkanas and Goldberg, 2013). In addition, since they are locally developed, they usually do not have a macroscopic

Figure 2.35 Debris flows. (a) 'Trains' of large clasts floating in an unsorted finer matrix of a debris flow. Note the wavy appearance of the base (dotted line) of the gravel-rich part. The dark colour is due to the high content of burnt remains. (b) Detail of the debris flow showing different orientations of large clasts (marked with lines) as a result of buoyant forces. Chits are 1 cm. MSA, Site PP5-6, South Africa.

Figure 2.36 Scan of resin-impregnated slab of a debris flow. Note the semi-plastic movement of clasts flowing inside the fine-grained matrix. Arrows mark major flow lines. Palaeolithic, Stelida, Naxos Island, Greece.

expression. Porphyric-related distribution is often encountered in anthropogenic sedimentary structures, such as mud floors, bricks, etc. However, such structures often show a dense matrix almost devoid of pores (except for vesicles), and a porphyric-related distribution with an almost homogeneous aspect (Gé et al., 1993; Karkanas and Efstratiou, 2009; Friesem et al., 2011, 2014b).

Distinct features that may be encountered in some debris flows are rotational features or inclined preferred orientations of elongated clasts and microscopic tiling (Figure 2.38). Rotational features are shear deformation features produced during the semi-plastic movement of the material. They encompass various turbate structures, from arcuate grain arrangements to circular and almost concentric clast alignments (Karkanas et al., 2015a). Note that rotational features can also be observed in man-made mud constructions because shear deformation is also produced in wet application of mud materials (see sections 3.4.1 and 3.4.2). Water escape (hydrofractures) and fluidized features (flame structures), and small-scale folding and faulting have been also described in some debris flows in pure geological contexts (Phillips, 2006; Menzies and Zaniewsky, 2003), but until now they have been rarely recognized in archaeological contexts, perhaps by oversight. Water escape features are shown by completely filled fractures with pure silt, sand, or clay. Flame structures are irregular, elongated injection features of fine-grained material (Figure 2.39).

Orientation Analysis

Fabric analysis, the study of orientations of natural clast and artefacts, is also of help in assessing debris flow processes in archaeological sites (McPherron, 2005). However, most studies are of natural deposits (van Steijn and Coutard, 1989; Bertran et al., 1997). Debris flow processes have been identified through fabric analysis in the Middle Stone Age site of PP13B, in South Africa (Bernatchez, 2010), which were in agreement with the micromorphological results of the site (Karkanas and Goldberg, 2010a). Clasts in debris flow deposits are often oriented parallel to the slope direction, similar to the orientation found in deposits moved by solifluction and in contrast to the orientations found in most water flows (see also section 2.4). These deposits often have lower orientation strengths, however, when compared to solifluction deposits (Bertran et al., 1997). Nevertheless, oblique orientations of clasts are often observed in lateral levees whereas transverse orientations are found mainly in lobe fronts. Weak imbrication may also be observed, particularly in clast-rich debris flows and frontal lobes.

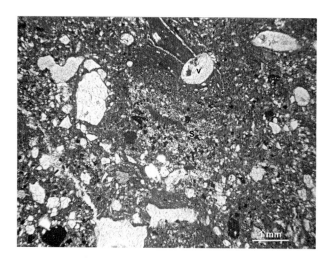

Figure 2.37 Photomicrograph of a debris flow showing transient features of water flow of sorted sand, silt, and clay pockets (S) in an otherwise unsorted matrix. Vesicles (V) due to water escape are also shown. PPL, BA, Agios Charalambos Cave, Crete, Greece.

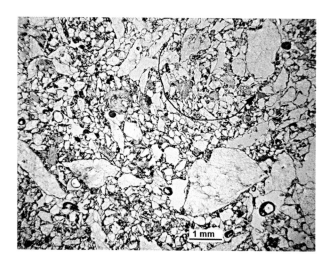

Figure 2.38 Photomicrograph of rotational feature (delineated with arrow) from the sandy mudflow shown in Figure 2.33b. PPL, MSA, Site PP5-6, South Africa.

Figure 2.39 Flame structures produce by liquefaction of thinly-bedded sand and silts. Arrow points to the water escape movement. LP, Marathousa 1, Megalopoli, Greece.

However, most of the time the long axes of clasts are inclined parallel to the local slope surface.

Effects on Archaeological Materials

The catastrophic nature of debris flow affects the archaeological record in several ways. The original position of the archaeological objects will be destroyed, and larger objects will be concentrated at the head and levees of individual flow bodies. Non-cohesive fluidized debris flows are capable of eroding the substrate and thus older archaeological material can be incorporated in their contents. In contrast, cohesive debris flows are often non-erosive and tend to blanket the previous topography. However, their development on inclined surfaces will bring into contact assemblages of different times or periods. Transportation of thick dense debris flows as single semi-plastic masses can often result in destabilization of layers with distinct content. Burnt remains which are rich in light charcoal and ash can be particularly affected and moved almost *en masse* to a new position (Karkanas et al., 2015a). Therefore, archaeological objects appear to belong in one cultural phase but their original pattern has been totally destroyed.

An illustrative example is this of cave PP5-6 (South Africa), where distinct dark elongated lenticular layers are observed in between aeolian sands in the upper part of the sequence (Karkanas et al., 2015a). The inclined lower contacts of these layers are sharp and undulating, and the geometry of the bodies appears somewhat wavy (Figure 2.33b). Some of these layers contain exclusively Howiesons Poort artefacts, and when initially exposed they were seen to directly overlie much older layers (see also Chapter 4). However, the dark matrix of this layer – rich in homogeneous dispersed fine-grained burnt remains – suggested that it was reworked without having affected the underlying layers. Only after exposing larger parts of this sequence was it realized that it originated several meters uphill where *in situ* thick burnt remains had been destabilized by rain water and moved downwards and deposited as debris flow sediment.

On the side of the fortified Bronze Age site of Palamari, Greece, fallen construction stones were observed floating in a homogeneous sandy silt matrix (Karkanas, 2015). The limestone blocks dipped away from a wall remnant below a so-called destruction layer. The overall geometry of the layer containing the blocks was a gently inclined tabular body with a diffuse lower contact. The arrangement of the floating construction blocks, their inclination, and concentration in clusters with tails are typical characteristics of debris flow deposits (Figure 2.24). Obviously, the materials of the destroyed wall were lubricated by rain water and moved downwards away from their original position. The enclosed archaeological objects were derived from a higher stratigraphic position and the formation of the debris flow implies a considerable time gap between the abandonment of the building to which the wall belonged and the rebuilding of the area.

In some Palaeolithic sites, the recognition of debris flow sediments has sometimes led to erroneous interpretations that archaeological materials were totally reworked sequences without any possibility of finding *in situ* archaeological objects and features (e.g. Bertran, 1994). Often, successive debris flows built amalgamated sequences without obvious separation of the distinct events. This is particularly problematic in archaeological sites because the small area of the excavated trenches prevents full exposure of the geometry of the deposits and therefore there is a high chance of missing contacts separating similar deposits. It is very probable that ephemeral but important occupational episodes can be blanketed and preserved in between episodes of cohesive-debris flow that are not erosive. The bottom line here is that a general assessment of the process that deposited a sequence is often misleading if particular attention is not paid to the details of the evolution of this process.

2.4 Water Flows in Sites

Water flows are very different from mass movement processes. The solid fraction (sediment concentration) in water flows is generally less than 50% by volume (Costa, 1988). The velocity of the coarsest clasts that are pushed and rolled on the bed (bedload) differs considerably from that of the water with fine-grained suspended sediment that flows around them (Coussot and Meunier, 1996); therefore, the water–sediment mixture in water flows does not move as a single mass as in debris flows and other mass movements. Water flows are characterized by very low viscosities, a turbulent flow, and separation of the suspension and bedload fraction. Water flows are generally divided into normal water flows, which have less than 20% of sediment concentration, and hyperconcentrated flows, which can reach up to 50% of sediment load; each one can display slightly different sedimentary features.

2.4.1 Shallow Water Flows

Deep water flows in the form of streams and rivers are not foreign to archaeological sites. Sedimentary features related to deep water flows are generally obvious even to a person with no training because they usually comprise thick sequences of well-rounded gravels and well-sorted pure sand and silt bodies. However, shallow water flows may not be readily defined in the field because they form thin, restricted bodies of sediment with ill-defined sedimentary structures.

Two main types of shallow water flow can be recognized: (a) sheetwash or overland flow, which can be characterized as unconfined (non-channelized) water flows on slopes, and (b) channelized flows, which are confined inside rills, gullies and similar small transient channel features. Another type of water-laid deposit described here is small-scale standing water bodies such as ponds and similar features.

1) Sheetwash/Overland Flows

Definition, Processes, Occurrences

When rainfall rate exceeds the capacity of soil to absorb water then the soil surface becomes saturated and the ponded water moves down the slope as overland flow, sheetwash, or sheet flow. Rain will become overland flow if it falls on soils that are already saturated by prolonged rainfall, are impermeable because of high clay content, or are hydrophobic as a result of burning (Burns and Gabet, 2015); poaching of pasture soils by animals (Schofield and Hall, 1985; Beckman and Smith, 1974; Selby, 1993) will also lead to overland flow. Normally, however, overland flow is generated in semiarid environments and is a rare phenomenon in areas with undisturbed vegetation (Hassan, 1985). As a consequence, a sheet of water is developed along the slope but in reality water is concentrated in small pathways. Even a very small slope gradient can produce sheetwash.

Human occupation and activities often disturb the vegetation cover and therefore it is expected that sheetwash will be a dominant process in archaeological sites. Indeed, in urban and generally in architectural sites, overland flow will affect outdoor areas such as streets and yards, which are normally barren. Along the same lines, the floor of caves and rockshelters are prone to sheetwash, particularly in cases where floors slope towards the back of the cave.

Recognition

Field Sheetwash transfers the finer surficial material downslope, but under certain conditions when the slope is steep and the substrate is supersaturated, the flow is capable of transporting pebbles and more rarely larger clasts (Blikra and Nemec, 1998). Sheetwash deposits are usually thin and lenticular with flat tops and weakly erosional bases. Indeed, the lack of clear channelized features is characteristic of this type of water-flow deposits.

On steep slopes, the flow is very shallow and movement of sediment is controlled by saltation, rolling, and creep, so most of the sediment is carried as bedload with little material going into suspension. Such deposits often show a faint to distinct plane-parallel stratification and an ill-defined sorting (Nemec and Kazanci, 1999). On gentle slopes to almost flat areas some of the material is transferred in suspension and therefore sorting is better and stratification more well-developed (Figure 2.40). In the latter cases the sediment will be in the range of sand to clay (Selby, 1993: 232; Kittel, 2014; Majewsky, 2014). However, the majority of the clay fraction is found as clay aggregates and not as single clay particles (Alberts et al., 1980). In addition, individual layers are thinner and follow the pre-existing topography without greatly eroding the substrate. In such gentle slopes, rainfall on fine-grained substrates (soils, loess) produces individual layers (laminae) 1–2 mm thick that look homogeneous in composition (Mücher et al., 2010). Rhythmic alternation of sand and sand/silt layers has been also reported (Figure 2.10c). Such layering indicates either fluctuation of the intensity of sheet erosion (Majewsky, 2014) or, more likely, separation of the bedload (forming the coarser increment) and the suspended load (forming the

Figure 2.40 Laminae of beige sand with a few interbedded brown mud stringers deposited by very shallow water flow. Some crude grading is observed, and overall the sediment is relatively well sorted. Chits are 1 cm. MSA, Site PP5-6 South, South Africa.

finer increment) (Blair and MacPherson, 1994). Massive fine-grained sloping layers with relatively crude sorting could be related to higher content of sediment carried by the flow, probably resulting in hyperconcentrated flows (see below).

Microstructure Lamination is the most characteristic microscopic feature of fine-grained sheetwash deposits (Mücher et al., 2010; Bertran and Texier, 1999) (also see Chapter 5). Laminae can be moderately to very well differentiated in size and content depending on the intensity of the rainfall and the sediment load (Figure 2.41a,b), and thus very good sorting may be occasionally observed. Nevertheless, only horizontal and subhorizontal bedding (Figure 2.41a), and no cross bedding, are observed (Mücher et al., 1981). In cases of very high rain intensities, sediment may be non-laminated and composed of loose aggregates produced during transport (Mücher and De Ploey, 1977). Therefore, it appears that sheetwash deposits are mostly washed out of free clay particles and the material consists of loosely to dense packed rock and mineral grains or clay-rich aggregates (Figure 2.41b) (Karkanas and Goldberg, 2013). Lighter material, like charcoal or organics, can also produce distinct laminae (Kourampas et al., 2009). Occasionally, silty intergrain aggregates and bridges, and irregular coatings around clasts reflect translocation of fine particles by water percolation when the surface runoff loses its competence (Texier and Meireles, 2003).

In addition, the presence of vesicles with escaping air is often observed as a result of turbulent shallow flow.

Effects on Archaeological Materials

Artefacts on slopes can move under the influence of sheetwash. This includes direct movement by overland flow as well as indirect movement through erosion of the supporting finer matrix. Small solid anthropogenic particles are more easily entrained by overland flow and travel farther than larger particles; this effect increases as slope angle increases. Where gradients are generally low (less than 5°), overland flow has little effect on artefacts more than 20 mm in maximum dimension (Fanning and Holdaway, 2001; Macphail et al., 2010). Surface runoff washing over hearths or hearth debris redistributes charred particles and may produce discrete laminae of charcoal (Kourampas et al., 2009; Sherwood, 2001).

Given the restricted transport capacity of sheetwash processes, small rounded to subrounded fragments of artefacts in slope deposits are obviously redeposited and therefore they document the *terminus post quem* of deposition. Large and more angular fragments of archaeological objects are more likely deposited after formation of underlying sediments, and document the *terminus ante quem* of layer deposition. Therefore, often the archaeological objects within slope deposits document the stratigraphy of deposits, and they often indicate a chronological relation between sheetwash deposition and human activity at a nearby site (Kittel, 2014).

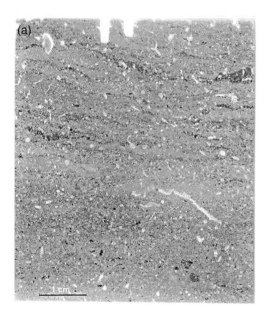
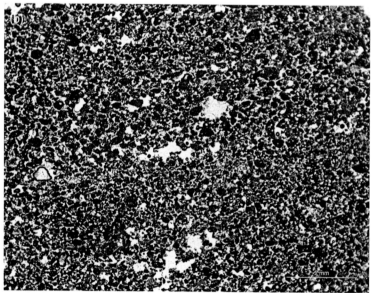

Figure 2.41 Sheetwash. (a) Thin section scan of sheetwash sediment: massive (lower part) to crudely laminated, moderately sorted sands and silts with interbedded laminae of phosphates (from Karkanas and Goldberg, 2010b). MSA, Site PP13b, South Africa. (b) Photomicrograph of crudely bedded, moderately sorted, rounded to subrounded clay aggregates, typical of shallow sheetwash. PPL, BA, Pelekita Cave, Crete, Greece.

2) Shallow Channelized Flows

Definition, Processes, Occurrences

When slopes are effectively steep, sheetwash becomes unstable and starts to incise into the loose sediment (see also section 2.7.1 for a review of erosional features). Concentration of sheetwash in rills and shallow channels enormously increases the power of water to entrain sediment. Rills are areas where flow concentrates in narrow channels of a few centimetres to a few tens of centimetres wide because of natural topographical features and soil roughness (Selby, 1993: 232). They are usually discontinuous temporal features and occur in slopes steeper than 2–3°. Snow melt usually produces a dense net of rills even in very low gradient slopes and is particularly active in periglacial environments (Wilkinson and Bunting, 1975). Rill erosion is the most important agent of sediment transport in barren slopes, and in steep and gentle sloping areas of settlements.

Recognition

Field Shallow channelized flow deposits are found as incised channel fills with trough-shaped sharp erosional bases (see also section 2.7.1). Individual layers will be thin and lenticular, and they are much longer than wide or deep. Their upper contact is planar in contrast to most mass movement deposits, which often attain a more curved upper surface. It is important to note here that shallow channels may also encounter other types of deposits such as debris flows, grain flows, and other mass movement deposits (see sections 2.3.2 and 2.3.4); thus the basal shape of the deposits alone does not define the depositional process.

In very steep slopes, flash floods can selectively leave a coarse lag at the bottom of the channel (Figure 2.42). On steep slopes, shallow channel flow deposits are coarse and gravelly, whereas on gentle slopes up to 5–6°, sediments are much finer. Flow deposits in rills with 0.25 m water depth can normally entrain clasts up to the size of medium pebbles (~2 cm) for 2° slopes, whereas for 5° slopes the size of the largest clast that can be entrained is very coarse pebble size (~6 cm) (Blair and McPherson, 1994; Costa, 1983). Thus, particles eroded from rills are much coarser than those eroded from interrill areas. However, for smaller water depths, the maximum size of clast that can be entrained substantially decreases, and at water depths of ~0.1 m and slopes of only 5–6°, sizes larger than coarse pebble (~3 cm) cannot be moved. Obviously, on steep slopes much larger clasts can be moved: on 20° slopes, for example, small boulders can be moved within water depth of 0.25 m. Since the steeper the gradient, the longer the potential transport distance, some rounding of clasts is observed in channel deposits of long steep slopes.

Shallow channel flow sediments are moderately sorted and therefore show faint to distinct stratification, with individual layers exhibiting different clast sizes (Figure 2.43a,b). Due to the greater water depth of such flow deposits both bedload and suspended load are encountered, and turbulence is much enhanced (Wilkinson and Bunting, 1975). Hence, small sequences comprising alternating layers of gravel, sand, and silt are typical. Mud layers resulting from deposition of the suspended load can cap the sequence. Sorting of individual layers is often crude, and the rounding of gravel is not well developed. Nevertheless, sandy layers can be better sorted and are often laminated. Sets of parallel layers may cross-cut each other at very low angles (low-angle cross bedding: Figure 2.14b). Coarse layers contain interstitial fine-grained sands and silts, thus showing a bimodal grain-size distribution. Such a grain-size distribution is related to large discharge fluctuations in fluvial regimes where the coarse clast framework is filled with fine-grained sediment during the waning stages (Texier and Meireles, 2003). Typical well-rounded gravels of high-energy fluvial deposits are lacking, as are well-developed elliptical and discoidal shapes. Nevertheless, normal grading in gravelly and sandy parts is frequent (Figure 2.43), and occasionally cross-stratified sands may be encountered (Texier and Meireles, 2003). All these features are spatially restricted and thick extensive layers are not observed. Because of the intermittent deposition of flow deposits on slopes, sediments are exposed for most of the time and not quickly buried, so sediment will tend to be oxidized and generally have warmer hues than typical fluvial deposits.

Microstructure Microscopic fabrics of the fine-grained increments are not much different from those observed in unchannelized flows. However, grading features are more developed and microscopic cross-stratifications

Figure 2.42 A lens of subrounded gravel lag formed during a flash flood. Note also the underlying poorly sorted angular coarse sand. EBA, 'Lower Town' of Mycenae, Greece. Camera lens cap for scale.

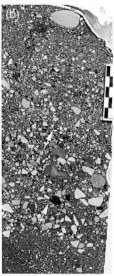

Figure 2.43 (a) Channel cutting through reddish clayey substrate and filled with fine gravel and coarse sand. Lenses of different clast sizes are visible. (b) Crude stratification and faint grading (arrows) are evident in the resin-impregnated slab of the same sediment fill. A red rounded pottery fragment is shown in the middle of the slab. Protogeometric to Hellenistic, 'Lower Town' of Mycenae, Greece.

are occasionally observed (Figure 2.44). Sorting is also better expressed and therefore typical features are clean sand laminae. It is important to stress that flash flood sediment characterizing channelized deposits and their intermittent character will result in uneven development of sedimentary microfabrics; thus, fabrics may vary across sedimentary layers.

Orientation Analysis
Transport of clasts by rill flow is thought to generate fluvial patterns, that is, transverse orientations, perpendicular to the slope directions. However, several studies have shown that random distribution of the orientation can also occur in rill deposits, although there are multiple secondary factors that can influence this pattern (Bertran et al., 1997). Preferred orientation (anisotropic fabrics) of long axes of bone, on the other hand, has been observed in very low energy processes and probably very shallow rill processes. Long bones show the clearest orientation pattern, followed by scapulae and pelves, and axial elements (Cobo-Sánchez et al., 2014).

Effects on Archaeological Materials
Obviously, depending on the flow conditions of channelized flows (slope, water velocity) most archaeological objects will be affected and transported downwards and deposited as concentration lags on the channel substrate. Some rounding of artefacts is expected, particularly if transported for long distances. Rills are readily erodible, and thus unstable and transient features; they frequently change position and thus affect large areas of the landscape. As a consequence, rilling can possibly destroy

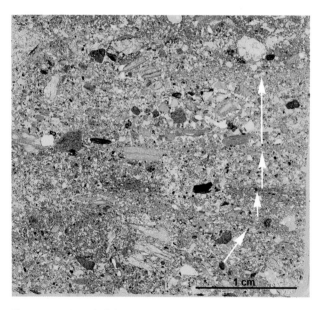

Figure 2.44 Detail of thin section scan showing graded laminae (arrows) and crude cross-stratification in the lower part. Laminae consist of poorly to moderately sorted sand grading to silt. EBA, Palamari, Skyros Island, Greece.

archaeological sites and completely rework and redistribute their archaeological content, particularly during abandonment phases. High-velocity, channelized flows are capable of destroying standing walls and other constructions by undermining their foundations. Normally building stones will not be moved further but mudbricks can be occasionally entrained. Nevertheless, the nature of shallow channelized flows normally is not so destructive: the width of small channels is generally minor and not very aerially expansive except in badlands; archaeological remains cannot move very far from the original source, and there is the possibility of reconstructing aspects of their context at least in rough fashion.

3) Puddles, Shallow Ponds, and other Standing Water Bodies

Definition, Processes, Occurrences

Ponds are standing bodies of water usually much smaller than a lake, that is, smaller than about one hectare. Small depressions can form temporary ponds after spring snow melt, or during rainy seasons (vernal ponds). Ponds also form in flat areas between dunes (interdunal), in abandoned river channels, and can be part of a cave, wetland, floodplain, coastal plain, seasonal lake (playas), or oasis. Pond deposits inside caves are often called backswamp deposits (Bosch and White, 2004; Martini, 2011). In general, pond deposits are shallow, can contain marsh and aquatic plants and animals, and generally lack intensive wave action in the shoreline. Ponds were attractive hunting places and butchering sites throughout the history of human kind (e.g., Kuman et al., 1999; Overstreet and Kolb, 2002; Le Tensorer et al., 2007; Holliday, 1997; Holliday et al., 2007; Zerboni, 2011).

Puddles are very small accumulations of water up to a few square meters in size formed by pooling in a depression by rainfall. Puddles can form in any site, including outdoor areas of urban sites and therefore are indicative of open-air environments. They are also indicative of the intensity of use of this area, since preservation of the related sedimentary features is related to the intensity of human activities and animal traffic (Heimdahl et al., 2006). They can also form in caves from dripping water (Goldberg, 2000a). Hydraulic structures such as wells, artificial ponds, and occasionally cisterns and managed floodplains (fish and mill ponds, irrigation canals and reservoirs, etc.) are standing bodies of water that are often filled with water-laid sediment (Brown, 1997; Karkanas, in press; Dart, 1986; Huckleberry, 1999; Purdue et al., 2010). More rarely, collapsed man-made structures can collect rainwater and sustain temporal water bodies (e.g., collapsed underground tombs, Karkanas et al., 2012).

Sedimentation in a standing water body includes quiet settling of sediment-laden water occasionally interrupted by coarser, storm-derived sediments from channelized flows and sheetwash processes entering the water body. In relatively large ponds abundant organic debris can come from subaerial sources, or local, high organic productivity also associated with rich aquatic fossils content.

Recognition

Field Puddles are ubiquitously categorized as having a very shallow basin-like substrate. Often, they are characterized by undisturbed, planar lamination of the sediments, which indicates that sedimentation was calm and occurred in a stagnant water body (Figure 2.45a). Well-preserved laminae likely indicate deposition in a place that was not walked upon. Water cover would have also protected the sediments in the pond from disturbance (Heimdahl et al., 2006). On the other hand, the presence of undisturbed laminae could also indicate high sedimentation rates that could bury and protect the particular sedimentary features. Homogeneous, massive fine-grained sediment indicates a body of very shallow water that is disturbed by rain-splash, or simply a heavy rainfall with rapid inputs (Figure 2.45b). In addition, intensity of bioturbation is an indication of the rate of deposition.

Nonetheless, sedimentary sequences related to all types of standing water bodies will be associated with basin-like features and more rarely with silting up phases

Figure 2.45 (a) Puddle deposits consisting of a coarse, unsorted sandy sediment capped with fine laminated silts (L) appearing as sheet-like fissility. EBA, Mycenae, Greece. (b) Massive fine-grained sediment (MF) filling a cistern at the 'Lower Town' of Mycenae (Protogeometric to Hellenistic). Relatively coarser sediment at the middle (C) represents a flash flood input into stagnant body of water. Scale is 30 cm long.

in channels representing abandonment. In excavations, identifying basin geometry of small constructed features such as wells, irrigation reservoirs or collapsed structures is straightforward (Figure 2.46). However, in large ponds it is difficult to identify the geometry of the substrate unless the periphery is revealed. Nevertheless, the upper and lower bounding contacts of the water-laid sediments will be sharp and horizontal; layers, if present, will be planar and continuous except for the area of feeding canals, where microdelta features may occur (Figures 2.43 and 2.46b). These features collectively represent undisturbed sedimentation in a low-energy environment.

Sediments are well sorted and fine grained. Anthropogenic material is generally rounded (e.g. fine bone, pottery), but angular stone chips can also occur because transport distances may not be long. Sedimentary structures in all types of standing water bodies are dominated by features of suspended particles and an absence of bedload sedimentary structures. Thus, silty and clayey sediments predominate (Figure 2.45b). Some coarse-clastic deposition is limited to pond margins, but scattered fine gravel within muddy facies is deposited during severe flood events. Coarse-grained layers can form inclined bedsets on the peripheral deltaic features spreading towards the centre of the standing water body (Figures 2.45b and 2.46b). Roundness and sorting of coarser particles is an indication of the size and energy of the feeding channels and, as a consequence, probably the size of the pond itself. In cases of wave action at the margins, sand layers with ripple and very small-scale cross-stratification may be encountered. Wave action washes out clay and silt, and therefore their absence

is a good indication of shoreline processes (see also section 2.4.3 on high-energy water flows).

If sedimentation occurs in a water pond that is turbulent, the sediment will be homogeneous (Heimdahl et al., 2005). In layered sequences, individual layers are well sorted and can be graded or not. Laminae can consist of pure clay, silt, sand, or their mixtures, and laminae can consist of alternating coarse and fine couplets. Repetitive fining upward (graded) sequences of sand, silt, and clay are often encountered. These are interpreted as the result of pulses of flooded conditions and settling in a standing body of water so that differential sediment settling can occur (Knapp et al., 2004; Karkanas et al., 2012). On the other hand, sequences of parallel-laminated silt to clay deposits with no grading suggest deposition in a quiet, very low energy, subaqueous environment without pulses of sediment-laden water entering the pond (Martini, 2011). Some beds can be characterized by soft-deformation due to slumping processes, with convolute laminations and water escape pipes. These features are characteristic of gently sloping marginal pond areas or as the expression of loading due to heavy animal traffic or fallen stone blocks (e.g. inside a cave environment).

Often, fine, plane-parallel organic-rich laminae occur that trace the base of the depression, the low topographic gradient of the original depositional interface. The lateral continuity of thin organic-rich laminae points to quiet bottom waters (Mazza and Ventra, 2011). Thick peat layers form in larger ponds, as well as in lakes and lagoons (Reading, 1996). Remnants of roots on the substrate of the peat deposits imply *in situ* decay of organic matter and a generally shallow environment at

Figure 2.46 (a) Remnants of a well fill showing alternating dark and white laminae. This alternation probably reflects seasonal variation in water content. During wet periods organic-rich debris entered the well, whereas during dry periods evaporation and precipitation of calcareous deposits prevailed. Classical to Roman, Molivoti, Thrace, Greece. (b) Fan-like coarse sediments (CS – lower contact outlined with dotted line) feeding fine-grained pond deposits to the left (FS) in a cistern (see also Figure 2.45b). A flash flood event probably destroyed the stone structure and moved downwards masonry stones. Scale: distance between the two nails is 1 m. Protogeometric to Hellenistic, 'Lower Town' of Mycenae, Greece.

the margin of the water body (Galloway and Hobday, 1996: 339). However, drifted detrital organic matter can form peat layers that may reflect longer distant transport and may not be related to the coast at the area of deposition. Dark, massive fine-grained sediments particularly enriched in organic matter are called gyttja and are deposited in swampy areas (Stahlschmidt et al., 2015). In all cases, sedimentation in standing water bodies is characterized by individual layers exhibiting non-erosional basal contacts, persistent lateral continuity, plane-parallel layering, good sorting, occasionally normal grading, and fine-grained sizes (Figures 2.47a,b).

Another type of sediment in relatively large ponds but also in lakes and lagoons is chemical deposits (Pavlopoulos et al., 2006; Ismail-Meyer et al., 2013; Hampton and Horton, 2007; Mallol, 2006), which include mainly calcareous sediments (marls) and evaporite sediments (e.g. dolomite, gypsum). Shallow calcareous chemical sediments are characterized by massive or layered light grey to beige marl sequences with remarkable lateral continuity (Figure 2.48). Commonly associated with bedded organic-rich layers (peat or gyttja) and loose, highly porous calcareous tufas are aquatic plants, with charophytes especially common (Macphail, 1999). Layering is not observed in shallow environments due to intensive subaqueous bioturbation (Ismail-Meyer et al., 2013). Shells of molluscs and shellfish are frequently encountered. Sand-rich marly layers are indicative of beach sediments forming at the pond margins; they are well sorted and can be massive or show laminations, ripple, and small-scale cross-stratification (Le Tensorer et al., 2007).

Figure 2.47 (a) Laterally extensive, thinly bedded muds. Their reddish colour and local development suggest deposition in an ephemeral lacustrine environment. EP, Kokkinopilos, Epirus, Greece. (b) Well-developed, thick and thin, plane-parallel bedding of clayey and dark organic rich (D) sediment, typical of lacustrine environments. Pliocene, Schimatari area, Greece. Section is ca. 2.5 high.

Figure 2.48 Image of lacustrine marl deposits (M) overlain by fluvial gravels (F). Individual parallel beds of calcareous marl can be followed for several tens of meters. Late Pleistocene, Ancient Lake Lisan, Jordan Valley.

Microstructure Puddles typically contain microscopic sedimentary crusts (Pagliai and Stoops, 2010). These are frequently composed of several microlayers, each one showing coarser particles at the bottom that gradually become finer upward; planar voids and vesicles are observed in the finer increments (Figure 2.49a). At the margins of the puddles, cross-stratified laminae occur as overwash, microdelta features. Crust fragments are often found dispersed in sediments, suggesting the past occurrence of small stagnant water bodies (Figure 2.49b).

Massive mud layers in the field often reveal microscopic stratification and lamination. It is generally believed that massive mud layers are deposited by settling of fine silt and clay in very low energy to standing water bodies. However, it has been shown recently that under certain conditions high-energy flows can deposit massive layers of mud that contain microscopic cross-stratification (Schieber et al., 2007). Thus, the study of fabrics at the microscopic level may reveal the actual depositional processes responsible for the formation of 'massive layers'. Massive muddy layers can be also result of intensive bioturbation, something that can be easily deciphered through microscopic analysis (see section 2.6.2). Homogeneous deposits of moderately sorted single-grained sands and silts to massive clayey silts are associated with water stagnation and have been observed in man-made hydraulic structures (Figure 2.50) (Purdue et al., 2010; Karkanas, in press). These probably resulted from episodic dense turbulent mud flows entering the structures during rain storms.

Typical laminated ponded sediments with normal graded laminae show fines distributed throughout the laminae, but each successive increment has one less coarse grade (Figure 2.51a). This type of lamina is produced from suspensions where all sizes are carried and settle out gradually (Reineck and Singh, 1980: 118), typically during flood events.

Another type of microscopic structure observed in pond deposits is rhythmically laminated facies that consist of alternating coarse and silty clay laminae that do not grade to each other (Figure 2.51b) (Shunk et al., 2006). The lower laminae of the couplet are relatively poorly sorted sand that may also contain rip-up clasts and horizontally bedded organic-matter fragments. Rip-up clasts are eroded particles of pre-existing sediments and are informative about sediment source (Karkanas and Goldberg, 2013). The upper laminae of the couplet are composed of more uniform silt and clay. Such rhythmites have been interpreted to result from alternating high- and low-energy depositional processes. Continuous inflow

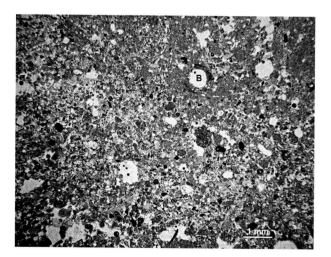

Figure 2.50 Photomicrograph of lower massive to crudely bedded silty sand overlain by mud pond deposits. PPL. Bioturbation pores (B) indicate frequent exposure and reworking. Protogeometric to Hellenistic, 'Lower Town' of Mycenae, Greece.

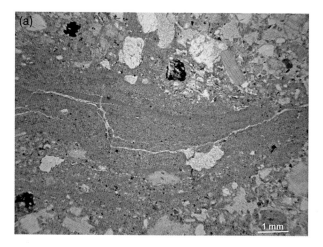

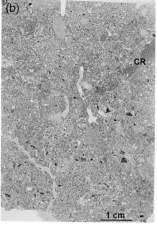

Figure 2.49 (a) Photomicrograph of crudely stratified clayey crusts formed in small puddles. PPL. (b) Thin section scan showing eroded, rip-up crust fragments (CR). Such crusts are normally evidence of subaerial exposure and typically represent outdoor facies. EBA, Palamari, Skyros Island, Greece.

into the standing water body is suggested for the fine-grained laminae that were interrupted during flood periods of deposition of higher discharge of sediment particles, soil clasts, and organic debris (Karkanas et al., 2012).

In chemical calcareous pond deposits, the sediments are usually composed of carbonate micrite (= microcrystalline calcite, <4 microns in size), which precipitated as a result of the metabolism of aquatic micro-organisms and bacteria. Layers may contain a large quantity of algal remains from stonewort (*Characeae*) (including stems and fruits/oogonia), algal filaments, diatoms, insects, and fresh water ostracods. Irregular planar voids (fenestral fabric) and corrugated lamination are characteristic of microbial mats often associated with aggregates of gypsum crystals (Figure 2.52a,b) (Mallol, 2006;

Le Tensorer et al., 2007; Ismail-Meyer et al., 2013). Nevertheless, still water bodies in cave environments display clastic micrite embedded in clay with occasional sand grains, which form massive sequences (Karkanas et al., 2008; Angelucci et al., 2013).

Orientation Analysis

Coarse archaeological objects are not normally deposited during the formation of pond deposits unless they were thrown in. Those objects appear flat, lying on the depositional surface with random orientations of their long axes. Large clasts and archaeological objects entering a standing body of water through flash floods will be oriented parallel to the flow direction, similar to the orientation in deposits moved by solifluction and in contrast to the orientations

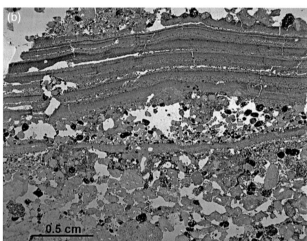

Figure 2.51 Photomicrographs of pond sediment. PPL. (a) Graded laminae with fines distributed throughout the laminae. (b) Rhythmic bedding with alternating coarse and fine laminae that do not grade into each other. See text for interpretation. EBA, Barnavos chamber tomb, Nemea Valley, Greece.

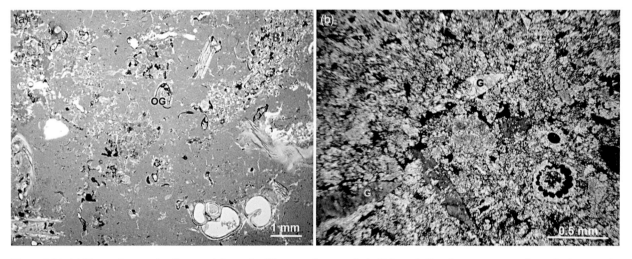

Figure 2.52 (a) Photomicrograph of lagoonal dense fossiliferous calcareous (micritic) mud. Also shown are some charophytic oogonia (OG) and burrows filled with lime fecal pellets (B). PPL, Holocene, Marathon, Greece. (b) Charophytic lime mud with diamond-shaped gypsum crystals (G) suggesting saline conditions and strong evaporation. PPL, Holocene, Marathon, Greece.

found in water flows (see also sections 2.3.3 and 2.4.3). This is because the clasts tend to float and glide along the flow entering the water body and not roll along the substrate.

Biological Evidence

Since all large standing bodies of water are rich in faunal and flora remnants, their identification and analysis is of prime importance in order to reconstruct the conditions of the particular environment and its palaeoecology. Macrobotanical remains, phytoliths, charcoal, and pollen are all used for reconstructing the palaeovegetation of a site when preserved. Pollen is the most widely adopted and versatile method in the reconstruction of Quaternary environments. However, pollen records not only the local vegetation conditions immediately surrounding the water body but often that of the catchment area. Therefore, knowledge of pollen production and dispersal, pollen sources, their deposition and preservation, and the relationship between fossil pollen and former plant communities is required (Lowe and Walker, 2015).

Macrobotanical remains in ponds and lakes normally reflect the immediate surrounding environments but they are mostly found in a decayed state and thus often lose their taxonomic identity. Charcoal in lacustrine deposits occurs in the form of microcharcoal, which is also not of taxonomic value. However, microcharcoal has been used recently used as a proxy for the intensity of natural and anthropogenic fires and other landscape changes in the catchment area, and even as an indication of population expansion or cultural-economic transitions (Li et al., 2010; Turner et al., 2010; Sánchez Goñi et al., 1999; Sorensen, 2017). However, caution must be used in applying such data at the local, site-specific area.

Opal phytoliths are relatively stable materials and therefore can survive in oxidizing environments where all other palaeobotanical materials will normally decay. However, phytoliths do not provide so much detailed taxonomic information as does pollen, for example, except for grasses (Piperno, 1988).

Diatoms are very sensitive to changes in water chemistry and therefore they are used extensively as proxies of salinity and trophic status (Mees et al., 1991). In this light, they are very helpful in defining variations in coastal environments and palaeohydrology of salt ponds and lakes from which past precipitation levels can be inferred. They may also reflect on eutrophication produced by anthropogenic activities (e.g. agriculture, deforestation and clearing, etc.).

Ostracods are also good indicators of salinity. However, they are abundant mostly in marl deposits. Nevertheless, important information on water quality and palaeoclimates can be provided by careful study of ostracod assemblages (Lowe and Walker, 2015; Macphail et al., 2010).

Other biological analyses of importance are foraminifera in coastal areas, molluscs in marine and fresh water environments, and insects, which are particularly helpful on palaeoclimates (see Lowe and Walker (2015) for an extensive review).

Effects on Archaeological Materials

The very low energy of pond deposits does not affect the distribution of pre-existing archaeological remains. In most cases, coarse anthropogenic remains and generally any occupational phase found in pond deposits will be associated with periods of exposure and therefore will be accompanied by features related to soil formation, and physical or anthropogenic post-depositional disturbances (see section 2.7).

The clayey nature of most pond deposits and often their high sedimentation rate can result in quick burial and protection of fragile archaeological materials such as organic remains and bone. However, they may also enhance downward water circulation, thereby leading to enhanced decay rates if the materials are not directly sealed and covered by clayey deposits. Swamp and generally sediments rich in organic remains will buffer oxidation reactions, thus protecting the destruction of organic materials of archaeological interest (Glob, 1965; Renfrew and Bahn, 1991). However, in waterlogged environments the macroscopically preserved organic materials have been altered at the macromolecular level (Weiner, 2010).

2.4.2 Hyperconcentrated Flows

Definition, Processes, Occurrences

Gravelly and sandy deposits exhibiting intermediate features between mass flow and water flow deposits could be considered the product of hyperconcentrated flows (Table 2.2). Hyperconcentrated flows are streamflows containing large amounts of sediment, 20–47% by volume (Costa, 1988). They are also called mud floods, and non-cohesive or turbulent mudflows. However, in hyperconcentrated flows water and sediment are separate components of the flow as in water flows, in contrast to debris and normal mud flows (see section 2.3.4); therefore sediment is transported as bedload and suspended load (Coussot and Meunier, 1996). Nevertheless, hyperconcentrated flows carry in suspension large amounts of clay and silt, and occasionally sand. They are the product of [an] 'extraordinary form of flood in which an extremely large volume of sediment with a wide range in grain-size is moved and deposited in a short period of time' (Smith, 1986: 2). However, very small-scale flash floods can also carry large amounts of sediment relative to the total volume and hence produce hyperconcentrated flows. Centimetre to metre-scale typical mudflows running over small slopes in archaeological sites can produce small fans (Heimdahl et al., 2006). During their waning stage, these sediments are often transformed downslope to small-scale hyperconcentrated flows. Strictly speaking, hyperconcentrated flows could be categorized as a variety of high-energy water flow, but their conflicting field features (see section 2.4.3) make them difficult to be recognized in the field, and therefore they need special attention.

Hyperconcentrated flows can be typical of some rivers in flood, but they can be also part of slope deposits, which themselves may be channelized or not (Benvenuti, 2003). In several cases, there are lateral variations of debris and mudflows when the latter become diluted by water escaping from the main body of the debris flow. Hence, in archaeological sites they can be formed in the same environments where debris flows, sheet flows, and channelized flows occur (see sections 2.3.4 and 2.4.3).

Recognition

Field Hyperconcentrated deposits in the field have the form of tabular-bedded units. Absence of extensive scouring at their base implies unconfined flows (sheet floods), but association with channel forms also occurs quite often. Overall, sediments have a coarse sandy texture with distinctly less silt and clay than in debris flow deposits (Figure 2.53). They also show poor sorting but better than in debris flow deposits. Sandy deposits appear generally massive but they also exhibit faint horizontal stratification. They are characterized by the absence of cross-stratification and the dominance of horizontal bedding (Figure 2.53b). This stratification is defined by relatively poorly sorted, relatively coarse- and fine-grained strata (Figure 2.53a). Contacts between strata are difficult to define, and internal grading is rarely observed; inverse grading can be observed in some cases (Pierson and Scott, 1985; Smith, 1986; Wells and Harvey, 1987; Costa, 1988; Benvenuti, 2003; Ventra et al., 2013).

In hyperconcentrated flood flows, gravel deposits are rarely imbricated and the space between them is occupied by poorly sorted, very coarse-grained sand and pebbles (Figure 2.54). Inverse grading is also observed in

Figure 2.54 Hyperconcentrated flow in an arid stream (wadi) in Israel. Sediment is poorly sorted with subrounded gravel floating in a fine-textured matrix; overall there is internal organization of the different sizes, with faint stratification that differs from the chaotic arrangement of debris flows.

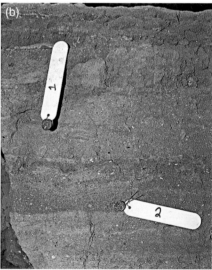

Figure 2.53 Examples of hyperconcentrated flows. (a) Poorly sorted sediment with crude bedding made of alternating coarse and fine layers with angular clasts. Hellenistic, ancient Messini, Greece. (b) Massive sands with poorly defined stratification. Tags are 7 cm long. BA, Pelekita Cave, Crete, Greece.

subunits. However, lack of extensive matrix-support argues against debris flow emplacement. This texture suggests rapid deposition of sediment with a wide range of grain-sizes (Pierson and Scott, 1985; Smith, 1986; Pope et al., 2003, 2008). Gravels are subrounded to rounded but generally less rounded than in fluvial strata of the same magnitude.

In some cases, shallow channel flows may share several sedimentary structures with hyperconcentrated flows. Nevertheless, the latter lack cross-stratification and internal grading, and do not form graded sequences ending with mud caps (Figures 2.15a, 2.42, and 2.43). Deposits of stratified coarse sands are typical hyperconcentrated flows.

Microstructure Studies on the microstructure of hyperconcentrated flows are generally lacking. Deformation structures typical of debris flows are not observed, and the fabric of the sediment is more similar to typical water-flow deposits. Microlaminations will develop where silt layers alternate with sand and fine gravel. Because of the high sediment content of these flow deposits, rounded rip-up clasts (mudclasts) and small-scale laminar undulations will be present, particularly where the flow overrides an irregular surface or encounters a gravel clast (Figure 2.55a) (Lachniet et al., 1998). Friesem et al. (2014b) suggest that some crudely sorted laminae occasionally showing inverse grading may represent hyperconcentrated flows formed in small fans along the base of degrading mudbrick walls. In general, features that will point to rapid deposition of

suspension load (sand to clay) without much internal sorting should be assigned to hyperconcentrated flows (Figure 2.55b).

Orientation

Clasts typically show bimodal orientation: some large clasts (e.g. cobbles and boulders) have their long axes perpendicular to inferred flow direction, whereas smaller clasts (e.g. pebbles and small cobbles) have their long axes parallel to the flow. This fabric suggests that the flow is capable of carrying both small cobbles and pebbles above the bed but that larger clasts are transported by traction as bed flow (Pierson and Scott, 1985; Benvenuti, 2003). Obviously, any large archaeological object with some elongation will follow this pattern; however, specific studies on archaeological materials affected by hyperconcentrated flows are lacking. Light anthropogenic fractions such as bone are expected to show orientations with long axes parallel to flow but heavier stone artefacts will probably show rolling orientations with their long axes normal to flow.

Effects on Archaeological Materials

Hyperconcentrated flows can be catastrophic if occurring in volumes at a large scale. In terms of their final product, they are not different from any other high-energy water flow. Indeed, in alluvial environments close to rivers such deposits are probably evidence of extreme floods with devastating results.

Archaeological objects materials found in hyperconcentrated flows are obviously reworked, the scale of

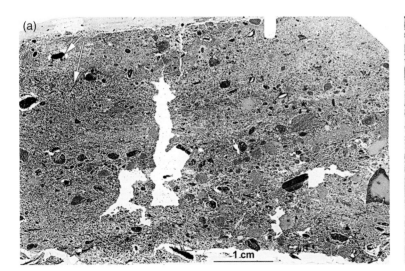
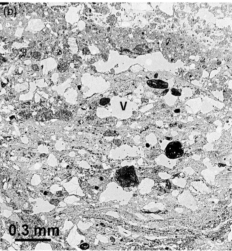

Figure 2.55 (a) Scan of thin section showing rounded mudclasts embedded in a sandy silt matrix also rich in black organic matter (arrows). The sediment shows some stratification but the coexistence of different textures suggests rapid deposition from hyperconcentrated flows. LP, Marathousa 1, Megalopoli, Greece. (b) Detail of thin section scan showing crudely bedded silts with some bands rich in reddish mud clasts derived from decay of mudbricks. The presence of several vughs (V) suggests deposition by slurries with high sediment content. Recently abandoned village of Kranionas, Greece.

which depends on the magnitude of the event. The latter is related to the thickness and the extent of the produced strata. In thick deposits, archaeological materials will be rounded and mixed assemblages from different periods are expected to be the rule.

2.4.3 High-energy Flows

Definition, Processes, Occurrences

Archaeological sites are often encountered in alluvial settings or coastal environments that are subjected to high-energy river flows and wave processes. Fluvial and coastal deposits have been extensively studied, and their sedimentary structures and architecture are very well known (Reading, 1996; Reineck and Singh, 1980). These types of deposits are not normally part of an archaeological site and often constitute only the substrate or post-abandonment phases. In some cases, fluvial deposits also form part of cave sequences that are close to rivers or are part of subterranean fluvial systems (Bosch and White, 2004; Angelucci et al., 2013). Archaeological objects inside high-energy water flows will be extensively reworked and therefore are only rarely studied in detail. Nevertheless, some fluvial deposits that bear materials of human early history are of special interest given the rarity of these remains. In several areas, Lower and Middle Palaeolithic (and Stone Age) archaeology in particular is characterized by large numbers of derived assemblages occurring in secondary contexts within river terrace gravel and sand deposits. A great example is the artefact assemblages of the Thames River terraces (UK) (Hosfield, 2011). Undisturbed occupational phases associated with high-energy water flows should always be related to sedimentary stasis and pedogenic processes, features that are discussed in sections 2.7.1 and 2.7.3.

Fluvial and coastal sediments show a vast array of sedimentary structures and textures, each one pointing to a specific microenvironment of a river plain or a coast (Reading, 1996; Reineck and Singh, 1980; Nemec and Steel, 1984; Scholle and Spearing, 1982; Picard and High, 1973). These types of sediments are readily recognized in the field when encountered in an archaeological site. Given that the detailed analysis and identification of their various depositional microenvironments have limited importance for interpreting archaeological sequences, we present here only some basic characteristic features. These features are more valuable for establishing end members of water flow processes to which all other types can be correlated and compared.

Recognition

Field The most characteristic feature of fluvial deposits is the frequent appearance of complex sequences of alternating gravel-, sand-, and clay-rich layers as expressed by intercalated planar to lenticular beds of various thicknesses (Figure 2.56a) (Nemec and Steel, 1984; Costa, 1988; Cant, 1982; Collinson, 1996; Picard and High, 1973). This sediment architecture is a direct manifestation of the broad fluctuations of energy during transport of particles by water currents inside a river channel and the adjacent floodplain. Sediments are often stratified, and very well-sorted sediment appears in massive layers. Fine lamination and cross-stratification are typical for fine-grained sediments, whereas imbrication and more rarely cross-stratification are observed in gravelly deposits (Figure 2.56b). Individual layers are often very well sorted

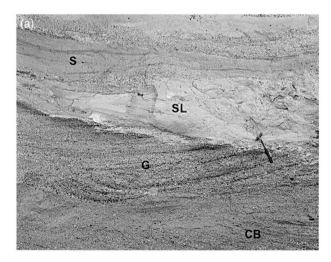
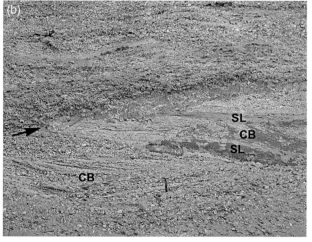

Figure 2.56 Examples of fluvial deposits. (a) Typical complex architecture of fluvial sediment with alternating well sorted, planar (G) and cross bedded rounded gravels (CB), lenticular bodies of bedded silts (SL), and laminated sands (S). (b) Large-scale cross bedding of sandy rounded gravels (CB) with intercalations of silty beds (SL). Arrow points to a large channel erosional surface. Pleistocene, Aliakmon River, Grevena area, Greece.

to the extent that one of the most characteristic features is the identification of layers consisting of pure sand or well-rounded gravel (Figures 2.5). Internal normal grading (i.e. within a layer) and also sequence grading are often observed (Figure 2.13a). The latter is identified as a sequence of beds, with each one showing finer grain-sizes; thus, sequences starting at the base with channel gravels, overlain by horizontal and cross-stratified sand layers and capped by stratified silts and overbank massive muds are typical in fluvial environments (Goldberg and Holliday, 1998). Gravels are very well rounded and often show well-formed roller shapes (elliptical) (Figure 2.5). Cut-and-fill features that form by channel erosion and subsequent filling are often observed. Nevertheless, ephemeral fluvial systems are characterized by sedimentary structures that are less organized. Arroyos, wadis, and similar terms describing channels with temporary or seasonal flows and fills are often comprised of poorly sorted sediment and less-rounded forms; however, they are still generally better sorted and rounded than all other water-lain sediments (Figure 2.57). Sedimentary structures in ephemeral streams are not much different from larger fluvial systems, although they are smaller scale and less well-developed (Picard and High, 1973). Sediments of ephemeral streams have redder hues as the result of prolonged exposure and oxidation.

Beach deposits show somehow different sedimentary structures and textures from those in fluvial deposits because of the rhythmic and repetitive nature of waves (Nemec and Steel, 1984; McCubbin, 1982; Reading and Collinson, 1996). Coastal deposits also include lagoon, aeolian, and deeper marine sediments, some of which are discussed in other sections. However, deep water sediments are not discussed in this book since anthropogenic deposits are not found in such environments. Beach strata are always well stratified and individual sand or gravel layers are laterally persistent and texturally uniform. Indeed, one of the most diagnostic features is thin horizontal sand layers or laminae that can be followed for metres without any serious change in their grain-size (Figure 2.58a) (Massari and Parea, 1998; Zecchin et al., 2004). Gravels are generally very well rounded and show both rolling shapes (elliptical/rods) and sliding ones (discoidal). A characteristic feature of beach deposits is shape shorting, with layers dominated by disc/blade or rod/sphere shapes (Figure 2.58b) (Nemec and Steel, 1984). Beach sediments are devoid of silt and clay, and cut-and-fill features are lacking, hence the architecture and facies associations of beach deposits are totally different from those of fluvial sediments. These aspects of the deposits are the result of the dispersive nature of the dominant wave process of building the beach; it does not form channels but does wash out fines to the deeper parts of the sea or lake body.

Microstructure Most fluvial and beach sedimentary structures are readily seen in the field, although some details of finer-grained sediment, such as massive mud sequences, are better studied at the microscopic level. Sand units are made up of laminated clast-supported sand that is often conspicuously devoid of fines. Microstructure is mostly single grain but some infiltrated fines are also observed forming bridge and other intergrain features. Sand-sized bone, charcoal, and shell are well rounded, and elongated grains will show preferred orientation and imbrication (Figure 2.59a). Overbank floods, however, can transport

Figure 2.57 Fluvial sedimentary sequence from an ephemeral stream (wadi) showing general rounded to subrounded gravels (G) overlain by sands (S) and finally muds (M). Although not as well structured as other perennial fluvial environments (compare with Figure 2.56), they still retain a typical fluvial architecture.

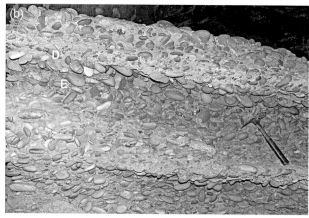

Figure 2.58 (a) Pleistocene beach deposits consisting of thinly bedded sands with considerable lateral persistence. Klein Brak, Mossel Bay, South Africa. (b) Pleistocene beach conglomerate with shape sorting of gravel in layers dominated by discoidal (D) and ellipsoidal (E) shapes. Glentana Bay, Mossel Bay, South Africa.

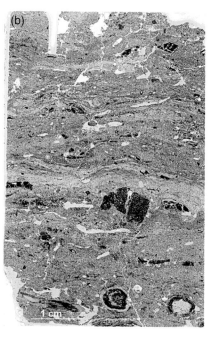

Figure 2.59 (a) Thin section scan of beach sands showing stratified, sorted sands and well-rounded bioclasts (shell fragments). Great Brak, Mossel Bay, South Africa. (b) Thin section scan of overbank laminated sands incorporating black charcoal pieces, dark mud construction fragments and brownish decayed pieces of wood. Neolithic, Szeghalom-Kovácshalom, Körös region, Hungary. From Parkinson et al. (in press).

fine occupational remains within the fine-grained fraction (Figure 2.59b). Normally, graded laminated sequences and microscopic cross-stratification are recurrent features. Alternating coarser and finer beds can form planar or intercalating lenticular bodies; finer beds can be massive even at the microscopic level or show faint stratification (Angelucci et al., 2013; Woodward et al., 2001).

Orientation

High-energy water flows show a preferred pebble orientation of long axes transverse to current flow and upstream imbrication of the intermediate axis. This is a typical fluvial rolling fabric, which characterizes most fluvial deposits that have well-developed fabrics (Stow, 2010). However, with increasing flow intensity – as expected in alluvial-flash flood deposits – bimodal orientations with some parallel to the flow have also been observed (Lindholm, 1987; Boggs, 2006; Hein, 1984). Seaward dipping imbrication is often prominent in beach deposits (Nemec and Steel, 1984).

If the grains have a streamline or teardrop shape, the blunt point of the grain will point upstream (Boggs, 2006), and thus stone artefacts with such shapes will also show similar orientations. Concave forms like shell and probably some bone lie with their concave side down when carried in the bedload; this is considered the most hydrodynamically stable rest position. However, if carried in suspension, they are deposited commonly concave-up (Lindholm, 1987). Only dry bone is observed to float (Coard, 1999), and therefore in such cases a concave-up position in fluvial deposits might be expected.

Similar fabrics of current–transverse orientations have been confirmed in fluvially modified stone artefacts (Schick, 1992). In general, a strong anisotropic fabric is observed in archaeological assemblages modified by water flow (Toots, 1965; Isaac, 1967; Voorhies, 1969; Schick, 1984; Domínguez-Rodrigo et al., 2014). Bones seem to be far more sensitive to water disturbance than lithics, and it seems that bones tend to orient themselves parallel to the current direction. However, preferred orientation should not be directly assumed to be the result of bone transport alone and anthropogenic or carnivore processes might impart an anisotropic fabric (Domínguez-Rodrigo et al., 2014). The experimental work of Kaufman et al. (2011) also showed that in general the majority of bones move by sliding rather than rolling; this explains preferred orientations that are parallel to the current.

Effects on Archaeological Materials
High-energy water flow is probably one of the most destructive processes of the archaeological record in producing mixed and reworked assemblages. Fluvial processes are also responsible for removing considerable parts of sites (e.g. Shiqmim, Israel (Goldberg, 1987)). Lithic artefacts and pottery will behave in the same way as sedimentary lithoclasts. Bone and lighter anthropogenic material (e.g. charcoal, plant and food remains) will show similar deposition patterns as natural shell, fossil, and organic matter.

Archaeological assemblages associated with bedload gravel and sand are more prone to high-energy hydraulic modification, whereas suspended load (silt and clay) contexts indicate low-energy settings and hence less modification. Moderately disturbed and relatively *in situ* (*sensu lato*) archaeological remains can be found in relatively low energy slackwater fluvial sediment (Wu et al., 2012). Ephemeral exposure surfaces with evidence of very brief occupation can be buried and protected when sedimentation rates are high (Angelucci et al., 2013; Macphail et al., 2004; Skaarup and Grøn, 2004; Bell et al., 2000).

Extensive abrasion is a good indicator of long transport distances, but it is certainly not a direct correlate (Hosfield, 2011). Rounding may occur as a result of water flow over stationary items and not only by intensive water transport (Shackley, 1974, 1978; Petraglia and Potts, 1994; Thompson et al., 2011). Mallol (2006) suggested that the abraded portions of the archaeological assemblages of the Lower Palaeolithic site of 'Ubeidiya are a result of prolonged surface exposure rather than high-energy transport from a distant source or to wave reworking at the shoreline of former of the 'Ubeidiya Lake. Nevertheless, the presence of sharp or fresh archaeological material would point to an assemblage for which discard was broadly contemporaneous with the accumulation of the fluvial deposits, with little or no transport. Preservation of such an assemblage would also require rapid burial by lower energy sediments. In this respect, although the archaeological remains might not be strictly *in situ* the age of the enclosing fluvial deposits will also date the archaeology.

Hydraulic processes can distort size distributions of cultural artefact assemblages by removing certain size fractions, depending on flow velocity (Petraglia and Potts, 1994; Schick, 1984, 1987; Boaz and Behrensmeyer, 1976). Under slow to moderate flow conditions an appreciable number of small artefacts will be moved downstream and leave coarser elements as a lag. At higher velocities even the heaviest artefacts will be transported for long distances. Concentration of artefacts by high-energy flows occurs in meanders, along river bars, depressions, obstructions (boulders and vegetation), and at confluences with tributaries and alluvial fans (Petraglia and Potts, 1994; Schick, 1984, 1987; Hosfield, 2011).

According to Voorhies (1969) the lightest bone elements (vertebrae, ribs, and pelves) are the first to be removed by water flow, followed by scapulae and phalanges, limb bones, calcanei, and astragali; in higher water velocities, mandibles, tibiae, crania, and maxilla are the last to move and may remain as lag deposits. Apart from density and shape of bones, however, there are several patterns that affect hydraulic transport and sorting, such as the age of the animal, dry or water-saturated bones, and also the anthropogenic modifications of the original assemblage (Coard, 1999; Kaufmann et al., 2011, Domínguez-Rodrigo et al., 2014). Therefore, differential anatomical representation varies according to bone type: if the most cancellous (spongy) grease-bearing bones are underrepresented it implies their removal by high-energy hydraulic modification (if human interference, e.g. grease and marrow extraction, can be excluded).

2.5 Aeolian Processes

Definition, Processes, Occurrences
Wind, water flows, and mass flows are the three groups of agents that transport and deposit clastic sediments in archaeological sites. Aeolian processes are important in a set of different environments, such as beaches, glacial outwash plains, and cold and hot arid deserts. All these settings are characterized by the lack of vegetation, and

as a consequence aeolian accumulations have long been used to extract palaeoclimatic information. Sandy aeolian dune fields are characteristic features of deserts, and fine dust-sized sediment deflated from aeolian provinces accumulates as loess with the most spectacular examples coming from China, Central Asia, and the central plains of Europe.

Wind energy and sediment supply are important factors in the formation of aeolian deposits. However, aridity, wind intensity and direction, vegetation cover, and overall stabilization of the substrate also regulate aeolian processes. The range of grain-sizes transported by wind is generally limited to sand size and smaller, although very rarely wind is capable of transporting pebbles up to 2 cm (Reading, 1996: 128).

Air has a low density and viscosity, and therefore it is a very efficient transporter of the finer-grained sediment particles. Only particles coarser than about 0.1 mm (very find sand) can normally be picked by suspension in wind, while particles coarser than 1.00 mm can be moved by exceptionally strong winds only as bedload (creep and saltation) (Reading, 1996: 15). Suspended loads carried by winds are called dustloads or dust storms. Sand transport takes place at relatively high wind velocities.

Coastal archaeological sites and sites in hot or cold arid regions – and especially deserts – will normally contain aeolian sand deposits. Increased rates of sedimentation will be closely related to climatic changes, typically following aridification phases. When not related to large flat desert areas, aeolian deposits are generally found next to major alluvial valleys, where they originate from aeolian deflation of the adjacent valley floors (Frederick et al., 2002). In coastal sites, enhanced aeolian sedimentation is observed during transgression or regression phases when extensive parts of the shelf are revealed and the exposed sea bottom sediment becomes the major source of aeolian sand (Bateman et al., 2011; Karkanas et al., 2015a). However, in all cases occupation is often associated with phases of dune stabilization by vegetation cover and therefore increased humidity and waning aeolian activity.

Small amounts of aeolian dust are expected to be a component of most archaeological sites, although they are not readily recognized with the naked eye. On the other hand, extensive thick sequences of aeolian silt, known as loess, cover almost 10% of the earth's land surface (Lowe and Walker, 2015). In temperate regions loess was formed during glacial periods, when the climate was cold and dry; the Gravettian lithic culture, for example, is closely related to loess deposits in Central and Eastern Europe. Nevertheless, loess is formed only when the rate of dust accumulation exceeds the effects of sheetwash or weathering (soil formation) rates (Pécsi, 1990). Loess is also found in low latitude desert areas, such as North

Africa and the Levant, where silt and clay are deflated from large open expanses (Pye, 1995; Coudé-Gaussen and Rognon, 1988).

Recognition

Field Similar to water transport, aeolian transport of sand results in deposition of bedforms ranging in sizes from centimetre-sized ripples to dunes hundreds of meters high called draas or megadunes. Ripples are good indicators of local wind direction as their crests are orientated perpendicular to the wind, and the steeper lee slopes on the downwind side (Pye and Tsoar, 2009). Areas of predominantly aeolian sand without dune forms are called sand sheets; these often occur in inter-dune areas or peripheral areas of the dunes.

Although not exclusively diagnostic of aeolian transport, certain granulometric attributes of sands are often characteristic of this process. Aeolian dune sands consist of moderately to well-sorted grains in the size range 70–250 μm. Indeed, this type of aeolian sediment often produces the best sorting in comparison to all other types of clastic sediments. Relatively poorly sorted, coarse sands are observed in aeolian lag deposits, sand sheets, and inter-dune deposits in which the medium and fine sand fractions have been partly removed by winnowing. We would like to stress once again, however, that sorting is a comparative description variable. Poorly sorted sand is better sorted than a poorly sorted gravelly mud and far better sorted than the overwhelming majority of archaeological deposits, which consist of mixtures of different sized clasts. Therefore, aeolian deposits when encountered in an archaeological site will be comparatively one of the best-sorted deposits (Figure 2.60).

Aeolian sands tend to have lighter colours relative to fluvial deposits of the same region. This is because aeolian sands normally have undergone more cycles of sedimentation and reworking, and thus become more mature sediments dominated by resistant minerals. They predominately consist of quartz that in coastal settings can be enriched in white calcareous bioclasts. Quartz grains of aeolian origin show well-developed frosting that appears as a dull surface, which is marked by very small, irregular pits produced by the bombardment of sand grains during transport (Reineck and Singh, 1980).

Aeolian silt and clay (loess) is composed mainly of 10–70 μm particles, and fine-grained aeolian dust consists mostly of material finer than 10 μm (Pye and Tsoar, 2009: 76). Aggregates consisting of fine silt and clay-size grains are also present in loess, which can in some cases can be attributed to very localized penecontemporaneous surface reworking of freshly deposited loess (Rognon et al., 1987). The diameters of almost all aggregates consisting of fine silt and clay-size particles fall in the range 10–50 µm, and most of these have diameters of between 10 and 20 µm. Dune sands normally do not contain large amounts of silt and clay because fine particles are carried away in suspension. However, the silt and clay content of dune sands may increase by post-depositional accumulation of allochthonous dust. In cases where dunes are mixed with periodically filled playas (e.g. Mojave Desert, CA; Lake Mungo, Australia), sand-sized clay pellets can be blown out of the lake basins and accumulate with the quartz sand fraction; such clay grains can be dispersed and thus yield a silty sand with interstitial fines between the sand grains. Clay will be also produced by soil-forming processes including *in situ* weathering of the sand grains (see section 2.7.3).

In addition to sorting, a widely expressed feature of all aeolian dune sediments is large-scale, often steeply inclined cross bedding (Figure 2.61a). The slip face of a dune displays well-developed foreset laminae, which dip downwind at a maximum angle of about 35°. Such steeply dipping foresets are produced mainly by grain-flow deposition on the slip face. Cross bed sets form when individual foreset laminae (usually 1–5 cm thick) are grouped together. Sets are defined by bounding surfaces that may be planar or curved. The cross bed sets thus formed may be tabular, wedge-shaped, or trough-shaped in cross-section. Cross bedding is not unique to aeolian deposits, as fluvial and shallow marine deposits also show extensive cross-stratification. However, the characteristic picture of numerous sets of cross beds of

Figure 2.60 Finely laminated to bedded, well-sorted clean yellow aeolian sand deposits (S) interbedded with numerous darker thick and thin occupational layers (arrows). MSA, Die Kelders, South Africa. The lower part of the profile is 2 m across.

Figure 2.61 (a) Cross bedded aeolian sands in an archaeological site. The colourful appearance is due to sand input from the surrounding occupational layers and soils. Aeolian reworking has produced scours and vertical cuts (arrow). A borrow, filled with homogeneous sand, is shown in the upper right. Site PP5-6 South, South Africa. (b) Aeolianites with characteristic large-scale trough and planar cross bedding. Danabay, Mossel Bay, South Africa.

various thicknesses, multiple orientations, and dipping (up to the angle of repose of 35°) is unique to aeolian sand dunes (Figure 2.61).

In archaeological sites, dune sequences with widespread and well-developed cross-stratification are likely to form the substrate of the site or the sediment that caps the phase of abandonment. More often, aeolian sediment intercalated with archaeological deposits will be represented by relatively thin layers of massively sorted sand without visible laminated structures. However, in contrast to similar sized fluvial sands, aeolian sands will follow the previous topography, thus blanketing substrate irregularities and forming inclined beds. Moreover, aeolian sands may be thinly interbedded with archaeological deposits without disturbing them (Figure 2.60).

The primary textural and mineralogical characteristics of aeolian dune sands are commonly modified significantly by post-depositional changes. These include physical reworking by surface processes, bioturbation, compaction, weathering, pedogenesis, and cementation. Dune systems can exhibit evidence of modification by surface wash, erosion, sliding or slumping, although the basic forms of the dunes may still be readily recognizable. Carbonate-cemented aeolian sands in which primary aeolian depositional features (e.g. cross bedding and grain-size lamination) are readily visible are called aeolianites (Figure 2.61b). In such cases, carbonate cementation of an aeolian dune formed during a single phase of aeolian sand advance has occurred within the vadose zone (above the water table). Where grain-size lamination is pronounced, the finer sand layers often become preferentially cemented. However, in cases where bedding of cemented aeolian sands is obliterated by soil-formation processes, they are more appropriately described as calcretes (see soil-forming processes, section 2.7.3).

In cold arid deserts violent winds often result in intensive niveo-aeolian sedimentation. Although typical cross-bedded medium sands are observed, beds of fine gravel chips can be transported on the surface of the ice mainly by saltation and within the snow mantle during violent blizzards (van Steijn et al., 1995).

Typical pristine windblown loess has a pale yellow colour and is massive and not stratified; it forms characteristic vertical standing profiles (Figure 2.62) (Pécsi, 1990). Loess also tends to cap irregularities of the substrate. Loess sediments are very often redeposited by water and gravity flow processes and altered by soil-forming processes; thick exposures of loess and intercalated palaeosols are widespread in Central and Eastern Europe, Central Asia, and China. In addition, loesses are typically modified by cryogenic features during cold glacial periods (Cremaschi et al., 1990; Pécsi, 1990; Händel et al., 2009;

Figure 2.62 Impressive thick sequence of beige loess intercalated with different types of brownish palaeosols (white labels) that actually define stratigraphy in an otherwise homogeneous silty deposit. Mende locality, Danube, Hungary.

Lisá et al., 2013). Stratification and increased sand and clay content are normally related to redeposited and altered loesses, although sandier sections have been attributed to increased aeolian activity (Lisá et al., 2013). Archaeological deposits tend to be related to upper horizons in palaeosols although occupation horizons in pristine loess deposits have been also reported (Händel et al., 2009).

Microstructure Aeolian sands are made up of well-sorted sand-sized grains with well-developed horizontally or inclined laminated sand (Figure 2.63). Thin laminae rich in heavy minerals (those with densities >2.9 g/cm3) may alternate with thicker ones poor in heavy minerals. Often sediment particles are inversely graded (coarsening upward), which is a product of the grain-size sorting of sand by wind; it is a feature that is very rarely observed in fluvial sands (Draut et al., 2008).

Aeolianites are cemented with calcite crystals that are precipitated in the vadose zone. Pendulous cement crystals are formed on the underside of grains where

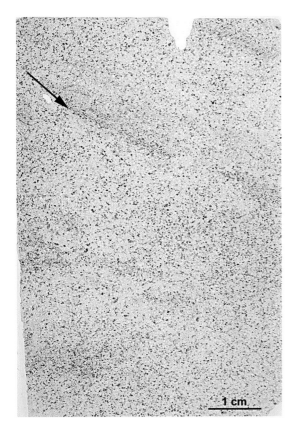

Figure 2.63 Thin section scan of aeolianite with very well sorted sand and laminae rich in heavy minerals (arrow). Pleistocene, Klein Brak, Mossel Bay, South Africa.

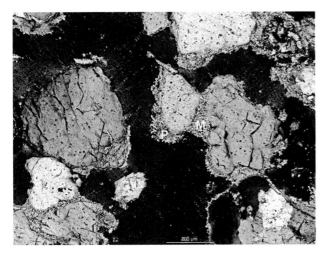

Figure 2.64 Photomicrograph of aeolianite cementation with meniscus (M) and pendant calcite (P). Pleistocene, Klein Brak, South Africa.

droplets of water are retained by gravity and surface tension. Meniscus cements may also form at points of grain contact where moisture is retained by surface tension (Figure 2.64). In some aeolianites, all of the porosity may be filled with calcite (Gardner, 1981).

Scanning electron microscope (SEM) studies of aeolian transported quartz yield specific features of grain surface micromorphology. Grain collisions due to wind impart upturned plates and disk-shaped concavities across the whole grain surface. These features in association with very well-rounded and low relief grains are very indicative of modification during transportation in mature aeolian sands (Bull and Goldberg, 1985; Krinsley and Doornkamp, 2011).

In loess, quartz grains are predominantly subangular; feldspars and carbonates tend to be subrounded to round. Heavy minerals are very often observed and have been used as sourcing and stratigraphic attributes (Pécsi, 1990). Typical loess, which is homogeneous and has massive microstructures punctuated with some voids, has an isotropic non-oriented fabric. Clay minerals are aggregated. Typical loess is well sorted, in contrast to the irregular grain-size distributions in reworked 'loess-like' deposits.

Orientation

Only very light and small archaeological objects will become oriented by wind action. Orientation fabrics thus can be better studied under the microscope. The preferred orientation direction of wind-deposited sands is approximately parallel with the wind direction, but the same preferred orientation is observed in water currents.

Effects on Archaeological Materials

Aeolian activity is not capable of moving archaeological objects larger than sand sizes. Therefore, only fine chips of lithic assemblages and light organic material such as charcoal and seeds and ashes will be readily moved – and commonly removed – by wind action. March et al. (2014) in experimental fires noted extensive dispersion of ashes and charcoal due to the action of the wind. In middens, fine fragments of white calcined shells are also winnowed out by the action of the wind (Villagran, 2014).

On the one hand, aeolian activity tends to quickly bury and protect previously deposited archaeological sediments. On the other, loose aeolian sand and loess is prone to sliding and slumping, and can be easily eroded and reworked, and as a consequence any included archaeological structures may be destroyed. Indeed, primary loess tends to collapse when the material is wetted and subjected to an applied load. Glacially deposited loess is often affected by solifluction and cryoturbation (Cremaschi et al., 1990; Händel et al., 2009; Lisá et al., 2013). Moreover, occupation deposits occurring within fills can be heavily diluted by blown sand from local deflating soils that may record environmental or land use changes (Mikkelsen et al., 2007).

2.6 Biological Sediments and Processes

Biologic sediments result from the accumulation of organic material or biologic activity. Geological examples are siliceous oozes from skeletal debris (radiolarians or diatoms) and shell accumulations (coquina). Natural biological sediment such as turf is used as constructing material or can be the substrate of an occupation. In this section, we analyse typical biological sediments and processes encountered in archaeological sites that are of importance for understanding site-formation processes. Among the most important organic components of archaeological sites are animal and human refuse. In post hunter-gatherer societies, animal dung is one of the most important components of the sediments, yet survives with difficulty and is challenging to identify. Carnivore coprolites are frequently found in Palaeolithic caves and the remains of guano can be observed directly microscopically or indirectly by phosphate diagenesis (see section 2.7.2).

A by-product of the accumulation of organic matter in archaeological sediments is bioturbation, the reworking of sediment and soils by animals and plants. Fauna and flora produce mechanical disturbances in the sediment but also induce biochemical alterations. Darwin (1881) was one of the first who recognized the effect of burrowing animals on soils.

2.6.1 Dung, Coprolites, and Guano

Animal faeces, also collectively referred as dung, are frequently encountered in archaeological sites. Typically, coprolites are fossilized faeces but the term often refers to relatively intact desiccated human and animal faeces in archaeological excavations. Large amounts of dung appear in the archaeological record only after the domestication of animals. They often form thick accumulations in all kind of livestock enclosures (Shahack-Gross, 2011). In addition, dung is used as building material, for fuel, and for other purposes (Reddy, 1998). Carnivore coprolites are more characteristic of Palaeolithic sites (Horwitz and Goldberg, 1989; Rodriguez et al., 1995; Macphail and Goldberg, 2012) where often humans and carnivores were competing for shelter.

Guano is a by-product of bat, bird, and pinnipeds (seals in general) excretion. As a primary component of archaeological sites, guano is predominately found in caves (Courty et al., 1989; Shahack-Gross et al., 2004a; Rellini et al., 2013; Bergadà et al., 2013). Guano decay releases large amounts of phosphate in the surrounding sediment (Hutchinson, 1950) that severely affects archaeological materials, such as ash, shell, and other calcareous materials (Karkanas et al., 1999, 2000; Karkanas, 2010a).

Recognition

Field Herbivorous dung is composed mostly of organic matter and includes a small amount of inorganic materials such as opal phytoliths and fine detrital mineral grains. In contrast, carnivore coprolites consist of inorganic material, mainly phosphates in the form of apatite, the mineral that also forms bone. Herbivorous faeces can be recognized by their composition and shape, and can be related to the type of animal that produced them (Courty et al., 1989, 1991; Macphail et al., 1997; Taglioretti et al., 2014; Macphail and Goldberg, 2010; Courty et al., 1994a). In general, horse and cattle dung has more large voids and consists of relatively long undigested plant fragments (Figure 2.65a), whereas sheep and goat dung has smaller voids and contains large

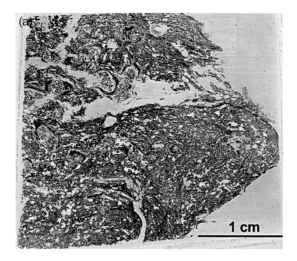
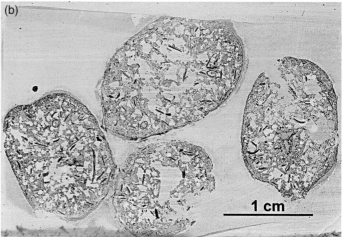

Figure 2.65 Thin section scans of modern (a) cattle dung and (b) sheep dung. Plant content is generally coarser and more 'fresh' looking in cattle dung.

amounts of amorphous organic matter and finely chopped plant fragments (Figure 2.65b). In general, browsers tend to produce a higher proportion of large particles in their faeces than grazers and intermediate feeders. This is related to the fibre digestibility and selective particle retention of the different categories of animals (Clauss et al., 2002). When preserved, pellets of ruminants such as sheep and goat dung are readily recognized in the field from their oval or spherical shapes and their dark and harder outer smooth coating. Herbivore dung is readily degraded unless it is either quickly buried or preserved in special environments which favour organic preservation, such as very arid, desiccated, or waterlogged conditions (Shahack-Gross, 2011). Fragments of dung can be identified on the basis of similarities with whole coprolite pieces. Generally, they will appear as masses of fibrous-rich organic matter with smooth outer surfaces. When trampled by animal traffic they become very dense masses of parallel-aligned plant fragments embedded in amorphous organic matter with an overall brownish colour (Macphail et al., 2004; Macphail and Cruise, 2001). When dry, they break with a fibrous fracture (additional information about dung related to human activities is presented in section 3.3.2).

Coprolites of omnivores, such as pig, contain more amorphous organic matter and phosphates, but bear coprolites, for example, can be rich in seeds and identifiable organic remains. Carnivore coprolites appear in the field as generally hard masses of whitish grey, reddish and light yellow colours with a harder outer smooth coating and spiral or annular markings. Some of the shapes of human coprolites include cylindrical masses, segmented pellets, and amorphous pats; human coprolites in archaeological sites are most concentrated in cess pits and have yellowish to orange colours and an amorphous fine fabric.

Microstructure Herbivore dung is characterized by large amounts of phytoliths in addition to relict plant fragments and amorphous organic matter (Figure 2.66a) (see also section 3.3.2). Calcium oxalate druses and the characteristic dung spherulites are also regular components of relatively fresh dung (Brochier, 2002). Spherulites (Figure 2.67a) are inorganic microscopic spheres of calcium carbonate with a characteristic extinction cross under crossed polarized light (Canti, 1999; Shahack-Gross, 2011). They differ from starch grains in having higher interference colours. In addition, starch grains have a bubble-like appearance in plane-polarized light (Figure 2.67b). However, spherulites are not always present (Lancelotti and Madella, 2012), and will readily dissolve under mild acidic conditions (Canti, 1999). Sheep and goat dung is also characterized by a convolute fabric (Figure 2.66b), high values of amorphous organic matter, and a vughy porosity. Amounts of detrital mineral grains and occasionally diatoms or sponge spicules are also present, having been incorporated with the grass and water the animal was eating and drinking.

Carnivore coprolites are characterized by a pale yellowish amorphous phosphatic matrix that can contain dissolving bone fragments as well as hair, phytoliths, pollen, and mineral matter. In addition, vesicles produced by escaping gas (Figure 2.68a) are a frequent feature. They also have an outer denser cortex and show iron-manganese staining and dendrites (Rodriguez et al., 1995; Horwitz and Goldberg, 1989). Human coprolites have been identified on the basis of certain viable bacterial spores and parasite eggs (Figure 2.68b). They also tend to include a wider range of diverse components such as charcoal, cracked and ground seeds, various types of shell fragments, bird eggshells and feathers,

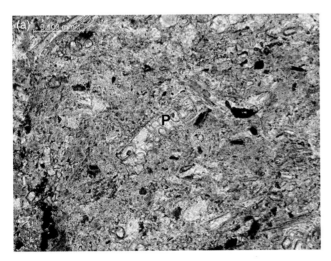
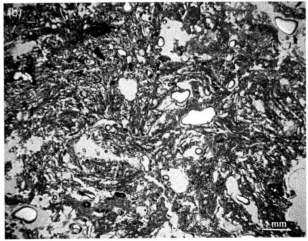

Figure 2.66 (a) Photomicrograph of phytoliths (P) inside amorphous dung organic matter. PPL, Neolithic, Koutroulou Magoula, Greece. (b) Photomicrograph of convolute fabric of plant remains in sheep dung. PPL, Neolithic, Leontari Cave, Attica, Greece.

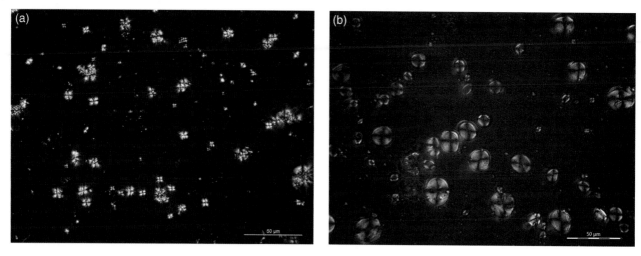

Figure 2.67 Photomicrographs of dung spherulites (a) and starches (b). Although both have a characteristic cross in crossed-polarized light, spherulites are generally smaller and have higher interference colours.

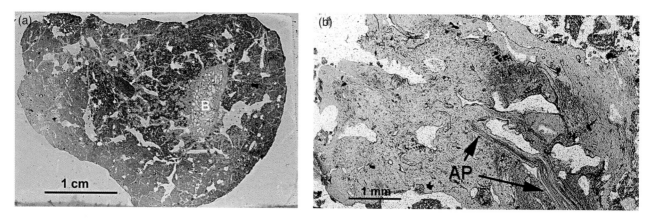

Figure 2.68 (a) Thin section scan of a fragment of Palaeolithic hyena coprolite. A bone (B) is embedded in yellowish phosphate matrix. Klissoura Cave I, Greece. (b) Detail of thin section scan of human faeces. Articulated phytoliths (AP) are embedded in an amorphous phosphate matrix with elongated subrounded to rounded voids (vughs). Bronze-Iron Age, East Chisenbury, England. Courtesy of Richard I. Macphail.

insect chitin, bone fragments, mammal hair, and plant fibres (Reinhard and Bryant, 2002).

The characteristic microscopic microstructural appearance of fresh guano is bedded spongy or aggregated, with a vughy porosity (Cullen, 1988; Shahack-Gross et al., 2004a; Karkanas and Goldberg, 2010b; Bergadà et al., 2013). Fruit bat guano consists of dark to black fibrous fragments of amorphous organic material (Figure 2.69). In contrast to dung, guano does not contain recognizable plant fragments, phytoliths, calcium oxalate druses, or dung spherulites. Most other types of guano contain various coarse inclusions, and depending on the diet of the animal include insect scales, chitin, and amorphous plant remains. Alteration products of guano are gypsum and phosphate minerals, such as apatite and other complex aluminium-rich phosphate minerals (Karkanas et al., 1999, 2000; Shahack-Gross et al., 2004a) (see also section 2.7.2), which often occur as nodules and

void infillings. Apatite and other phosphate minerals can be recognized in thin section by their microcrystalline, amorphous appearance (Figure 2.70a,b).

Chemical and Mineralogical Analysis

Biomolecular analysis using sterol content, bile acids, and DNA has been used to identify animal dung (Bull et al., 1999; 2005; Poinar et al., 1998; Shillito et al., 2009). Elevated $\delta^{15}N$ and phosphate values are also characteristic of dung. In general, $\delta^{15}N$ values of sediment in livestock enclosures are 9–10‰ heavier than those of control soils (Shahack-Gross et al., 2008a). Phosphate values in pristine dung – although higher than background levels – vary significantly depending on the animal type. Sheep dung has higher values than cattle dung (McDowell and Stewart, 2005), with carnivores having the highest values. However, phosphate alone cannot be used as a proxy for dung since it can have entered the soil

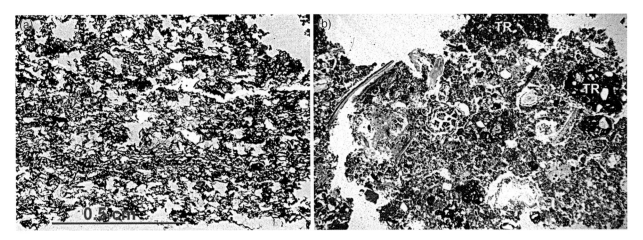

Figure 2.69 (a) Detail of thin section scan of fresh bat guano consisting of organic matter with a spongy fabric. Adritsaina Cave, Greece. (b) Photomicrograph of moderately decayed modern fruit bat guano showing terra rossa aggregates (TR) mixed with undifferentiated organic matter. Kebara Cave, Israel.

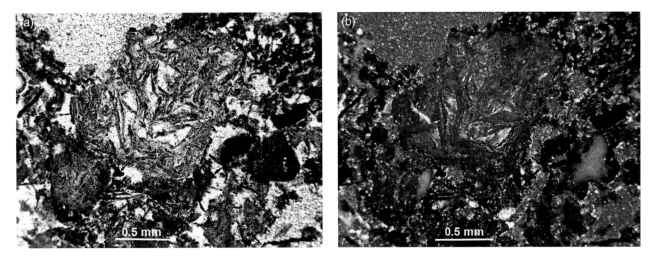

Figure 2.70 Photomicrographs of guano altering to (a) yellowish (PPL) and (b) amorphous and isotropic phosphates (XPL). Mesolithic, Schisto Cave, Piraeus, Greece.

via several sources, such as food processing, fine bone inclusions, bedding, and ashes.

Mineralogical analyses have been conducted for identifying phosphate alteration minerals in dung and decayed guano. Gypsum and apatite are the first minerals to be formed in relatively fresh guano. Note that the same minerals have been observed in animal dung deposits (Brochier et al., 1992; Shahack-Gross, 2011). Certain aluminium-rich phosphate minerals such as taranakite, montgomeryite, and leucophosphite are typical minerals found in altered guano-rich archaeological deposits of caves (Goldberg and Nathan, 1975; Karkanas et al., 1999, 2000; Karkanas, 2010a; Shahack-Gross et al., 2004a).

Effects on Archaeological Materials

Animal dung in deposits that post-date hunter-gatherer societies should be considered archaeological material in itself because it is the product of a human-induced process. Not only does dung reveal details of animal husbandry practices but it also provides excellent evidence of diet and datasets for palaeobotanical reconstructions.

Nevertheless, the decay of dung and guano induces certain chemical reactions in archaeological deposits that produce alterations that may distort the original stratigraphic context. Such changes may bring about volume changes, destroy interfaces, and alter or destroy materials of archaeological importance such as bone, shell, and wood ash (Karkanas et al., 2000; Karkanas, 2010a; Weiner, 2010) (see section 2.7.2 on diagenesis).

2.6.2 Bioturbation

Bioturbation is a biological process that affects archaeological layers after their deposition and may continue after burial, if they are shallow enough. In this sense, bioturbation is also part of the post-depositional processes. Bioturbation is one of the most important soil-forming processes and it can affect all types of archaeological deposits and sites almost immediately after their deposition; similarly, it may also affect raised structures such as mudbricks. The large amounts of organic remains accumulated by occupational activities are particularly attractive to larger animals, as well as microfauna. Whereas in architectural sites both indoor and outdoor sediments are affected, exterior deposits are also affected by root growth and decay, and tree falls (Wood and Johnson, 1978; Johnson, 1990). Roots also colonize rockshelters and the light zone of deep caves.

Tunnelling activity by large and small animals produces mixing of sediments. Among the best known soil dwellers are carnivores and rodents (e.g. fox, rabbits, porcupines, badgers, gophers, moles, rats and mice); for microfauna the effects of earthworms, termites, ants, and cicadas are particularly noteworthy. Ingestion of soils and excretion of casts by earthworms and other microfauna can also lead to severe mixing of the sediments and displacement of artefacts (Courty et al., 1989; Stein, 1983).

Root propagation brings about mechanical disturbance, although generally mixing and compaction are limited to the immediate vicinity of the root. Root respiration – with or without the help of associated microorganisms – results in chemical modifications of the surrounding sediment by alteration and production of new mineral phases.

Recognition
Field Tunnelling by large animals occurs in the form of burrows also known as *terriers* (in French) and krotovinas (Figure 2.71a). They are typically circular to elliptical channels about 6–10 cm across depending on the animal size (Bocek, 1986; Lim et al., 2007; Pietsch, 2013; Pietsch et al., 2014). However, foxes and rabbits can make larger holes, 15–30 cm across (Figures 2.8 and 2.71b). Tunnel depths also vary and can be several metres deep, although typically they are limited to the first half metre of the soil or sediment (Bocek, 1986; Balek, 2002).

Burrows are generally passively filled with soft material coming from the surrounding soils, but it can also come from stratigraphically higher layers or soil horizons, and thus the infills can be quite different from the surrounding sediments. As a consequence, the textural contrasts between the fill and the surrounding deposits makes it easy to recognize burrows in the field. Illuviation during flooding events can also produce laminated fillings in burrows (Lim et al., 2007).

Excremental casts of invertebrates can also add extraneous material, although they are normally difficult to discern with the naked eye (some are visible with a hand lens). Nevertheless, welded earthworm casts produce a relatively loose granular structure that is typical of some topsoils (Figure 2.72a). Earthworm burrows occur as near-vertical stripes about 1–3 cm wide (Canti, 2003a). As in the case of large animal burrows, those from earthworms may be filled with material of the overlying sediment and soils, and thus they can stand out against the surrounding material (Figure 2.72b).

Careful visual inspection of a soil or sediment lump can customarily reveal numerous small channels produced by soil microfauna. At the field scale, a characteristic feature of intensive earthworm and other microfauna activity is the formation of stone lines that mark the base of old topsoils (Johnson and Balek, 1991; Canti, 2003a);

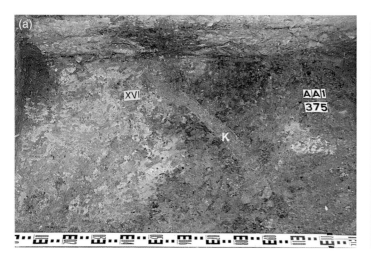

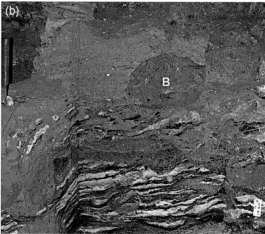

Figure 2.71 (a) Plan view of an animal burrow (K = krotovina). Note the spotted appearance of the surrounding sediment attributed to bioturbation. MPL, Klissoura Cave I, Greece. (b) Large animal burrow (B) above and within sequence of hearths in Kebara Cave, Israel (MP). The excavation square in the centre is 1 m across.

Figure 2.72 (a) A sequence of buried soils (S1 and S2) and fluvial deposits (F1 and F2). The soils are characterized by well-developed A horizons with granular structure because of microfaunal activity. See euro coin for scale. Pleistocene, Grevena area, Aliakmon River, Greece. (b) Faunal activity (burrowing) is evident by the dark spots (arrow) that start at the contact with the overlying layer. MSA, Site PP5-6, South Africa.

they can be faint or readily visible. However, not all stone lines are the result of earthwork activity (see Figure I.2 and section 2.7.1) because buried deflation and erosional surfaces as well as levelled construction fills can produce similar features (Birkeland, 1999). Only stone lines associated with soil horizons can be safely interpreted as bioturbation features.

In cases of intensive bioturbation, complete mixing of the sediment or soil takes place, resulting in a homogenized deposit; microscopic analysis is decisive for clarifying such cases. Bioturbation may be suspected in cases where homogenized deposits are not to be expected, such as the case of pit fills that normally are the loci of repeated accumulations of organic trash.

In contrast to animal burrows, most root traces taper and branch with depth (Retallack, 2001) and they often follow pre-existing burrows, which makes it difficult to make the distinction between the two. Roots are very irregular in width and range from large features several centimetres across to hair-like ones. Under favourable conditions decayed organic matter might be preserved inside the root trace (Figure 2.73a).

In arid and semi-arid conditions repeated wetting and drying cycles produce alternating calcite dissolution and precipitation around roots. Eventually roots become heavily encrusted with calcite that results in the formation of calcareous rhizoconcretions (root pseudomorphs) (Figure 2.73b). In soils and palaeosols, a bluish grey or

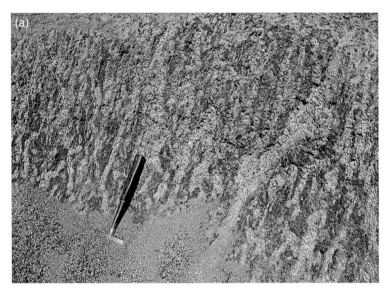

Figure 2.73 (a) Complex system of drab halos formed around decayed roots and burrows. LP, Kokkinopilos, Epirus, Greece. See geological hammer for scale. (b) Fossilized, complete root system (calcareous rhizoconcretions – example with arrows) developed in a fossil dune (aeolianite). Pleistocene, Great Brak, Mossel Bay, South Africa.

greenish grey halo often surrounds a root trace; drab haloes form around root traces in clayey, periodically waterlogged soils (Figure 2.73a). These haloes are often the result of iron reduction by anaerobic bacterial activity in stagnant water around the roots (Retallack, 2001).

Microstructure Burrows are clearly characterized micromorphologically by a loose, aggregated fabric (Figure 2.74a) (Goldberg and Bar-Yosef, 1998; Goldberg et al., 2007). Aggregates can be from different sediment and soil sources. Dumping can also produce similar fabrics so documenting the macroscopic geometry of the feature is critically important.

Passages of microfauna occur as channels and chambers and often have compacted walls. Excrements produced by small invertebrates differ in shape and size depending on the species and age of the animal, and thus they can be used for identifying the type of animal which may have produced them (Courty et al., 1989). Earthworms produce large (tens to hundreds of microns in diameter) mammillated excrements with low organic content (Figure 2.74b). Enchytraeids produce small, dark loose excrements rich in organic matter (Bullock et al., 1985).

The action of fine roots also produces channels and chambers, which can be closely associated with microfaunal activity. However, roots and their associated bacteria induce certain chemical reactions that are readily discerned at the microscopic scale, including calcite or iron accumulation and depletion along root passages (Figure 2.75a). Rhizoconcretions formed by biologically mediated calcification of cells within roots form alveolar septal structures (Figure 2.75b).

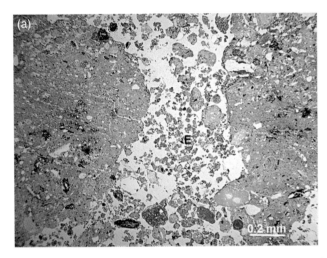
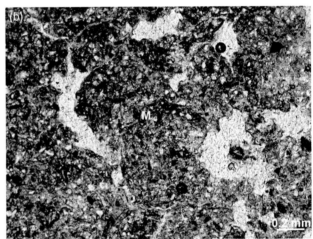

Figure 2.74 (a) Photomicrography of microfaunal excrements (E) in loose infilling of a void. Neolithic, Dispilio, Kastoria, Greece. Their existence in this lakeside site implies a period of exposure and drying of the soil. PPL. (b) Photomicrograph of mammillated earthworm excrement (M). PPL, Classical-Hellenistic, Molivoti, Thrace, Greece.

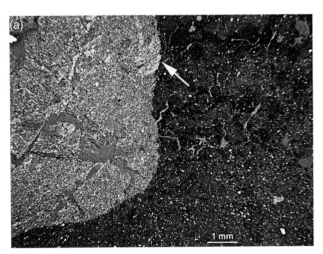
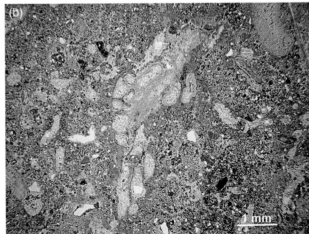

Figure 2.75 (a) Iron depletion along a root passage. Note the discoloured clay feature (arrow) that is continued uninterrupted in the in the drab halo. PPL, LP, Kokkinopilos, Epirus, Greece (from Tourloukis et al., 2015). (b) Example of calcareous alveolar root structure. PPL, LP, Qesem Cave, Israel. From Karkanas et al. (2007).

Effects on Archaeological Materials

Bioturbation is generally considered one of the most destructive agents of the integrity of archaeological sediments. It may lead to extensive reworking of the deposits, and vertical and lateral redistribution of cultural materials. It also tends to obliterate contacts and other aspects of the stratigraphy, and homogenize the deposits (Wood and Johnson, 1978; Stein, 1983; Bocek, 1986; Courty et al., 1989; Canti, 2003a; Morin, 2006).

In particular, tunnelling by small animals results in displacement of small archaeological objects and materials. However, systematic displacement is generally limited to the top 30–50 cm of the surface, and objects larger than ~5 cm are rarely affected (Bocek, 1986); on the other hand, large animals such as foxes and badgers can displace much larger objects (Balek, 2002). According to Canti (2003a: 141) the '…ultimate depth of artefact burial due to worm action in normal soils can only be defined to a fairly broad range, probably 10–25 cm'. The maximum size of displaced objects by earthworms is about 2 mm (Stein, 1983), which was important in accepting radiocarbon dates of centimetre-sized pieces of bone associated with the dating of early pottery in China: they were unlikely to have moved in worm bioturbated deposits (Wu et al., 2012; Cohen et al., 2017). On the other hand, sandy deposits are particularly prone to bioturbation, and cases have been documented of vertical movement of artefacts at the scale of metres (Cahen and Moeyersons, 1977).

As already stated, bioturbation may also lead to false artefact accumulation at depth (stone lines) below the topsoil. Nevertheless, there is also a wide belief that bioturbation not only destroys the archaeological record but also helps to quickly bury artefacts lying on the surface (Morin, 2006 and references therein). In addition,

displacement processes slow down at the bottom of the topsoil where the sediment is more compact. Therefore, quantifying the whole process may reveal some stratigraphic integrity even in sites where mixing is documented (Morin, 2006). The end result is that although surface layers are often extensively reworked careful examination can reveal a 'memory' of occupation (Morin, 2006) or intact 'islands' of occupation through micromorphological analysis.

Tree uprooting in forests and woodlands can produce severe mixing of archaeological layers (Johnson, 1990; Macphail and Goldberg, 1990; Crombé et al., 2015). On the other hand, relatively thick roots can penetrate cultural layers without destroying the integrity of the deposits. Indeed, often contacts continue uninterrupted on the other side of the root cast or trace contrary to the wide belief that roots are major destructive agents of the stratigraphy (Figure 2.76). It is the associated soil-forming processes such as fissuring and cracking due to alternating wetting and drying that produce severe mixing of the soil column.

2.7 Post-depositional Features and Processes

As we stated in Chapter 1, post-depositional processes are those that act on a layer after its deposition and even after its burial by later deposits. All the processes that alter the archaeological sediments after their deposition are collectively referred to as diagenesis. Post-depositional processes include both physical and chemical alteration of the sediment. Loss of water and compaction are also important post-depositional processes under certain circumstances. Nevertheless, chemical or biochemical transformations,

Figure 2.76 A fossilized root cutting (arrow) through the deposits without disturbing either side the horizontal arrangement of layers. MSA, Site PP5-6, South Africa.

such as mineral alterations, oxidation of organic matter, and precipitation of new material in the pores (cementation) are the most destructive ones. Some of these processes are also included in soil-forming processes (pedogenesis) but these diagenetic processes can act in non-soil-forming environments as well (e.g. caves, inside buildings). In soil formation, a set of processes are acting together and result in material that is characteristic of the prevailing environmental conditions acting on a particular substrate.

2.7.1 Erosional Features, Deflation, Lags, Stone Lines, and Pavements

Erosional features are produced by all main agents of transport, namely, water, wind, and ice. Erosional features by water flow develop when surface runoff becomes channelized as the sediment substrate becomes saturated. Rills, gullies, and more rarely tunnels (pipes), streams, and rivers are encountered in archaeological sites. Even the slightest sloping area of an outdoor occupation site is prone to erosion, the intensity of which is governed by the specific atmospheric conditions of the area and the nature of the substrate. Barren, non-vegetated, upland landscapes with sedimentary and soil cover are always crossed by rills that may end up as steep gullies or small streams. In arid environments, storms generate deep erosional gullies along which the flow is channelized (Bocco, 1991). As already described above in this chapter (channelized flows, section 2.4.1), rills are small erosion features, up to a few tens of centimetres across, whereas gullies are ravine-like erosional features up to several meters across and deep (Figure I.1a). Gullies are erosional features that do not permanently contain water. They are often filled by gravity flow deposits and therefore not easily recognized in the field (Figure 2.15a). It goes without saying that permanent water flows, that is, streams and rivers, are the

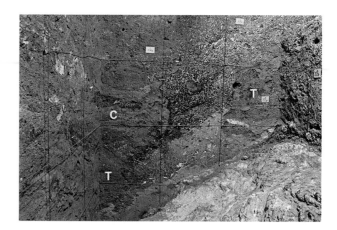

Figure 2.77 A series of underground tunnels, most likely after burrows (T) and channels (C) in the lower part of the stratigraphy at Klissoura Cave I, Greece (MP). Note the irregular and wavy walls of the right tunnel and the incorporation of fragments of the wall into the fill, which indicate that the channelled sediment had been consolidated prior to erosion. Grid is 50 cm.

largest erosional features, but at the same time they are the easiest to identify since they carry typical fluvial deposits.

Erosion by underground water flow usually at the water table produces tunnels and pipes. Natural cracks and animal burrows may also initiate tunnels by infiltrating into and moving loose subsoil layers. Impressive complex tunnels probably formed along burrows have been observed in archaeological caves (Figure 2.77) (Karkanas, 2001); at the site of Dmanisi (Georgia Republic) the human fossils were found within ancient gully and pipe deposits (Ferring et al., 2011).

Wind is not capable of moving coarse particles and objects, but deflation lifts and removes loose sandy and finer particles from the surface. In desert areas, but also often in coastal areas, deflation leaves a cobble stone-like surface behind called desert lag or desert pavement (Figure 2.78). Sand particles moved by wind 'sand blast'

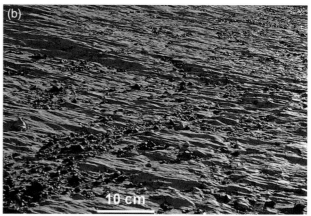

Figure 2.78 (a) Deflation lag of coarse stone in an aeolian coastal environment. Lemnos Island, Greece. (b) Wind-sculptured and polished bedrock with some stone lag concentrations. Western Desert, Egypt.

rock surfaces in a process called abrasion, which shapes and polishes exposed bedrock. Abrasion, however, is active close to the surface because sand grains are lifted only for a short distance. In deserts, abrasion can form impressive and extended erosional surfaces, and features such as yardangs. These are normally not related to the depositional processes that formed a site but rather have participated in forming the overall landscape and geomorphology of the substrate. The same holds true with moving ice that sculpts the underlying bedrock. Although there are characteristic features related to glacier movement, they are not directly pertinent to the interpretation of any overlying archaeological deposits.

Recognition

Field   Erosional features such as rills, gullies, and stream channels are recognized as elongated, straight to meandering features that cut across pre-existing layers of deposits and that form characteristic U and V cross-sectional shapes (Figure I.1a). The banks of an erosional cut tend to develop symmetrically, although local deviations do exist. Sheetwash erosion may produce only a shallow undulating surface that will appear in a profile as a sharp wavy cut (Figure 2.79). Such an erosional feature may truncate more than one contact of the

Figure 2.79 Shallow and wavy erosional cut (arrows) with relatively smooth bottom suggesting that it was formed by low energy processes. Grid is 50 cm. MP, Klissoura Cave I, Greece.

underlying sedimentary sequence and thus provides a definite indication of erosion when clear vertical channels are missing.

Gullies and rills often have highly sinuous, undulating, and rough walls with variable morphology that depends on of the nature of the eroded substrate (Figures I.1.a, 2.9a, 2.15a, and 2.79). Sliding and slumping tend to locally obscure the contacts of the erosional feature, and commonly sharp erosional contacts are not always observed along the whole stretch of a gully. Obviously, fluvial sediments characteristic of channelized flows are associated with rill, stream and river channels (see sections 2.4.1 and 2.4.3).

Shallow rounded small gullies and rills are generally formed in cohesive relatively impermeable clay-rich materials. U-shaped gullies with steep sides and generally flat bottoms are characteristic of silt-rich sediments and V-shaped ones of gravelly and sandy sediments. In moderately cohesive sandy clay sediments, U-shaped forms predominate but they have rounded bottoms (Imeson and Kwaad, 1980). Nevertheless, U-shaped gullies are also formed in the area where subsurface soils are either equal or more erodible than the surface soils while surface deposits are more erodible in V-shaped gullies. At the initial stages of gully formation water flows undercut the banks forming steep walls that tend to cave in (Figure 2.9a, 2.15a, 2.77, and 2.79). Collapse, slumping, and water erosion eventually results in the sloping back of the walls to approximately the angle of repose of the particular type of sediment and its cohesion. Therefore, highly ragged and slumped profiles indicate immature stages of gully formation. Underground tunnels have rounded shapes and elliptical to circular cross-sectional shapes, which upon collapse evolve into gullies (Figure 2.77).

A desert pavement can be recognized as flat stable surface made by interlocking pebble and cobble-sized stones only one or two rocks deep (Bloom, 1998). The stones are polished and often faceted and are called ventifacts. In non-arid but barren landscapes, transient lags of stones are formed by the combined action of sheetwash and wind. These produce rather discontinuous carpets (Figure 2.78) that in profile are recognized as stone lines (Figure 2.80). As described above, stone lines are also produced by bioturbation at the base of soils.

Effects on Archaeological Materials

By definition, erosional features imply that any archaeological material that was part of the affected sediment has been washed away. The amount of eroded material in most instances is unknown, and in the majority of cases it is impossible to estimate except where part of it is deposited as a fill within the erosional feature. Erosion indicates a depositional hiatus and a probable loss of archaeological deposits. A time gap may be

Figure 2.80 Examples of stone lines (arrows). Note that they do not continue along the whole extension of the contact, but can be used as a safe guide for defining it. Scale is 30 cm. EBA, Mitrou, Greece.

associated with an erosional phase but erosion can also happen in a very short period of time.

Deflation differs from erosion in that just the finer fraction of the sediments and the archaeological materials are removed. Fine, sand-sized archaeological objects are winnowed out, such as fine chipped stones in Palaeolithic occupational surfaces where a lag of coarse lithics is left behind. Occasionally, deflation hollows are formed, thus trapping and protecting archaeological accumulations (Goldberg and Macphail, 2006). Deflation also leads to artificial 'concentrations', as the finer sedimentary fraction is removed and hence the density of the coarser particles increases. In the same way, mixing of successive occupational phases can occur (Goring-Morris and Goldberg, 1990) (see also Chapter 5).

2.7.2 Diagenesis

Various physical and chemical processes act on sediment after its deposition. The intensity of the chemical effects is directly related to the amount of water passing through the sediment and the presence of organic materials. Therefore, diagenesis is particularly intensive where sediments are frequently water saturated and contain large amounts of organic matter. Many situations are particularly prone to chemical diagenesis in humid climates: caves with bat guano, pens, open areas where animals pass frequently, dumping areas, and pits filled with organic refuse. All types of wetlands are also affected by alteration processes (Miedema et al., 1974, 1976). Understanding diagenetic processes and reconstructing the palaeochemical environment of a site is a prerequisite for assessing its integrity and the completeness of the archaeological record (Weiner, 2010).

1) Compaction

One of the most important physical processes that may act in archaeological sites and which deserves special discussion is compaction, whereby fine-grained sediments can undergo volume reduction (compaction) after deposition. This situation is particularly important when the sediment contains a large proportion of pore water. During burial, this water is expelled over a period of time and the porosity of the sediment is reduced. Paul and Barras (1998) found that for the simple case of self-weight compression of clay sediments deposited underwater, the mid-depth correction is about 5–10% of the layer thickness for compressible sediments such as clays; it is about 1–2% for incompressible ones such as sand. Organic-rich sediments deposited underwater can undergo much larger compaction, with coastal peat compaction rising up to 50% (Long et al., 2006). On the other hand, compaction in terrestrial environments such as of floodplain muds occurs within a few tens of centimetres of the surface and thus sediments in fluvial planes exhibit minimal compaction with burial. Indeed, several lines of evidence support this conclusion: the preservation of intact, undeformed vertebrate tracks (Nadon and Issler, 1997), and the geometry of depositional features and structures that appear unaffected even in burial depths of kilometres (Plint, 2014; O'Brien, 1987).

An obvious reason for the above is that exposure of clastic sediments on land results in complete dewatering and therefore they compact on the surface immediately after deposition (Nadon and Issler, 1997). According to the equation of Sheldon and Retallack (2001) and the material dependence values of Retallack (2001), the compaction of sandy sediments and soils under burial compaction of 100 m is only about 2% of the original thickness. Furthermore, it is important to stress that even in cases of compaction at depths of several kilometres, sedimentary structures and microfabrics of underwater deposits are not affected (Plint, 2014; O'Brien, 1987), thus probably implying an isotropic, isovolumetric compression. This is very important for interpreting depositional features in archaeological sites which are only buried a few metres below the surface.

Nevertheless, compaction does occur in archaeological sediments (Crowther et al., 1996). It is often observed in filled pits and ditches which show sagging in the middle and form festoon bedding (Figures 2.16 and 2.81). This is apparently the result of artificial filling, which involves rapid deposition of loose sediments that gradually consolidate and sag. Indeed, this was reported and explained already by Cornwall (1958: 60), who in addition rightly stated that 'natural fillings slowly consolidate as they are formed'.

The important consequence of this observation lies in fact that the rate of deposition and the nature of the deposits play a significant role in compaction. Very high rates of deposition prevent exposure (cf. Shillito and Matthews, 2013) and self-compaction at the surface. Dumped deposits consist of loose aggregates that have not been homogenized by depositional agents such as water and wind. Therefore, individual particles have not found their best settling geometry. As a consequence, organic-rich sediments dumped inside pits in a single operation leads to compaction with time. Their gradual accumulation and homogenization with the other types of sediment due to trampling and reworking by natural and anthropogenic processes result in surface compaction and partial consolidation that preclude burial compaction.

The most important cases of volume changes relate to those chemical diagenetic changes that also have important archaeological consequences. Dissolution of minerals and decay of organic matter are the principal processes involved in differential compaction. The most usual process is dissolution of the calcareous component of the sediment. Wood ashes, which consist of poorly crystallized calcite (see review in Weiner, 2010: 170–174), are the first to be dissolved. Rain water will partially dissolve a portion of the ashes, depending on the hydrologic conditions on the site. Indeed, decalcification is a regular pedogenic process, particularly pronounced in humid climates. However, in archaeological sites surface decalcification will not lead to burial compaction. It is expected, though, that in sloping areas underground water flow could lead to differential dissolution and removal of calcite and soluble materials such as calcitic ashes.

Other types of more severe chemical diagenesis resulting in severe compaction have been recorded in archaeological caves where large amounts of organic matter (bat guano) alternated with the accumulation of anthropogenic burnt remains in some Palaeolithic caves. Guano decay releases organic acids and phosphates that in hydrological active environments lead to dissolution of calcite, including ashes. At the same time, these waters react with the mineral phases and precipitate newly formed phosphate minerals. Eventually as the pH of the sediments become more acidic, bone will also dissolve and more complex aluminium phosphate minerals will be formed (Weiner et al., 1993; Schiegl et al., 1996; Karkanas et al., 1999, 2000).

Recognition

Field The festoon-like geometry of pit and ditch fillings is a good sign of compression with burial (Figure 2.81). Sagging of the different filling layers decreases with height until they attain horizontality again. This implies

Figure 2.81 (a) Impressive compaction of ashy sediment (arrow) producing a festoon-like geometry in this gigantic pit. Scale is 4 m long. The wall of the bedrock appears just below the arrow. Neolithic, Promachonas/Topolnitsa, Greece. (b) Compaction and folding of a burnt layer (arrow) that has been detached from above and sagged into the pit below. Scale is 50 cm. BA, Kolona, Aegina Island, Greece.

that burial compaction also affects the overlying deposits up to a particular height.

An unmistakable indication of lack of burial compression in most archaeological sites is the absence of differential compaction of the archaeological deposits, other than pits and similar fast dumping areas (e.g. ditches, trenches). If compaction has modified the sediments, it will affect the various types of deposits differently, with organic-rich ones being influenced the most. Therefore, sagging of contacts in places or inclined beds that otherwise should be horizontal (e.g. floor sequences) may be observed. Indeed, there are cases where such features are observed particularly at the local level where large amounts of organic matter are involved. Hence, burnt layers consisting of charred organic matter will show undulating upper contacts due to compaction by the overlying sediments (Figure 2.81b).

In caves affected by severe chemical diagenesis differential compaction is very pronounced (see also Chapter 5). Very often whole sequences of deposits dip towards the hydrological centre of the cave, which may be the centre in relative deep caves or the entrance in rockshelters. If such dipping is associated with presence of phosphate minerals then chemical compaction is the obvious explanation (Figure 2.82a). Phosphate minerals, when in abundance, can be easily identified in the field as nodules, aggregates, and filaments of whitish to yellowish firm material; they are not as hard as calcareous features and they do not readily effervesce in HCl (Figure 2.82b).

Microstructure So far, microscopic evidence of compaction due to burial compression and chemical diagenesis has not been reported in archaeological sites. Indeed, preservation of microfabrics such as fine layering associated with distinctive activities, straight contacts,

details of coprolite fabrics, and particularly intact bioturbation features in sediments that admittedly have been affected by burial compaction suggest that the process is rather negligible or isovolumetric. Therefore, a relative isotropic accommodation of compression stresses within the pore system is a logical explanation for compacted deposits in pits (Shillito et al., 2011a; Shillito and Matthews, 2013).

Indirect evidence of compaction due to chemical diagenesis can be found in the mineral assemblages, their reaction microstructures, and the chemistry of the sediments. These are discussed in the following sections in which the various chemical diagenetic processes are described.

Effects on Archaeological Materials

The obvious effect of compaction of archaeological deposits is changes in the density estimations of archaeological objects. Correction or decompaction, as it is called, is not an easy task and is beyond of the scope of this book. Nevertheless, if the porosity of the original material is known or hypothesized from modern analogues (e.g. dumped ashes) then comparison with the affected sediment will provide a rough estimate. In layers that laterally sag and thin out due to chemical or physical compaction, the unaffected horizontal part can be used as the original thickness.

An important chronostratigraphic inference can be drawn if changes are recorded in the inclination (dip) of layers due to chemical diagenesis. The interface where the change is observed is an important stratigraphic marker, which also denotes a significant temporal hiatus.

2) Cryogenic Processes

Another important mechanical disturbance, particularly in Pleistocene archaeological deposits, is cryoturbation.

Figure 2.82 (a) Dipping of the lower layers in this image is due to chemical alteration. The upper layers are not affected and therefore a temporal hiatus between the upper and the lower sequence is implied. Grid is 50 cm. MP, Klissoura Cave 1, Greece. (b) Whitish phosphate filaments filling freeze-thaw fractures at Theopetra Cave, Greece (MP).

When the temperature drops several degrees below zero pore water transforms into ice. This transformation produces large stresses in the sediment that lead to a variety of macroscopic and microscopic transformations, particularly when repeated freezing and thawing occurs (Courty et al., 1989; van Vliet Lanoë, 1998, 2010). Enhanced humidity facilitates the formation of cryogenic alterations.

Most archaeological deposits of temperate environments and mountainous areas experienced cryoturbation during the glacial periods of the Pleistocene (Goldberg, 1979a; Karkanas, 2001; Bertran et al., 2008; Mallol et al., 2010). Furthermore, younger archaeological sites in areas which are presently in permafrost areas and generally periglacial environments would also have been affected (Todisco and Bhiry, 2008b).

Prehistoric caves provide particularly good records of cryogenic alterations, with cryoclastism dominating the deposits of several famous Palaeolithic sites of Europe. Freezing of the pore water of the limestone bedrock of the caves leads to fracturing and spalling of the cave walls. This leads to deposition of thick deposits of frost-shattered clasts known as *éboulis* (Figure 2.29 and 2.83a) (Farrand, 1975; Laville et al., 1980; Turk and Turk, 2010).

Recognition

Field Frost-shattered clasts have freshly broken surfaces with sharp edges and are often platy (Figure 2.83a). However, post-depositional weathering and corrosion can blur the broken surfaces and round the edges, producing clasts that are not much different from roof collapse deposits due to weathering. In addition, other actions such as earthquakes can create angular limestone pieces in caves (Woodward and Goldberg, 2001). Typically, frost shattering produces clast-supported fabrics but matrix-supported sequences are also frequent. The overall geometry of the deposit and its relation to other more secure cryogenic features can be used to correctly interpret its formation. If the clasts are found close to faults and shattered limestone bedrock zones, their cryogenic origin is arguable. Sometimes *éboulis* deposits are themselves cryoturbated and therefore can be safely interpreted as the product of frost action (Figure 2.83b).

Large-scale cryogenic features include involutions, injections and folds, and frost cracks and wedges (Figures 2.84). These are easily to identify in the field and are always associated with their microscopic counterparts, such as ice lenses.

Microstructure Freeze-thaw action produces typical platy microstructures consisting of platy aggregates of often-compacted sediment separated by a fissure

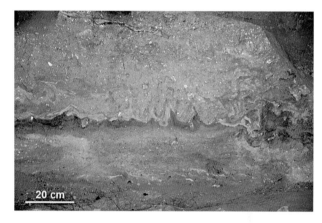

Figure 2.84 Image showing complex involutions, injections, and convolutions in sediments affected by cryoturbation in Theopetra Cave, Greece (MP).

Figure 2.83 (a) Angular platy eboulis of various sizes from boulder to fine gravel. MP, Pech de l'Azé I, France. (b) Image of cryoturbated eboulis with folded appearance. MP, Pech de l'Azé II, France.

network (Figures 2.32 and 2.85) (Courty et al., 1989; van Vliet Lanoë, 2010). Continuous freeze-thaw activity leads to lenticular shapes for the aggregates with coarser fragments also having a lenticular shape (ice-lensing) (Figures 2.32 and 2.85). During thawing periods, cappings of fine particles accumulate on the top of these lenses. Interstitial infilling of loose single-grained sand and silt-sized particles are formed in the same way. In sloping surfaces frost creep (gelifluction) results in attrition of the aggregates and formation of rounded aggregates also capped with fines (Figure 2.32b). Note that often these features are large enough to be seen by careful observation with the naked eye. It is also important to stress that any pre-existing soil fabric will be disrupted due to desiccation caused by ice formation in the areas between the aggregates. Therefore, typical rounded soil aggregates formed by pedogenic processes have a totally different appearance.

Effects on Archaeological Materials

In general, sediments affected by freeze-thaw activity will be mechanically disrupted and their original depositional fabrics will be destroyed. However, formation of platy and lenticular microstructures alone does not destroy the overall stratigraphic integrity of the deposit. Nevertheless, large-scale cryogenic alterations such as involutions, injection, and folding produces deformation of the deposits and the stratigraphic contacts. In some cases, the overall geometry of the deposits – albeit highly deformed – can be followed particularly when stratigraphic markers exist. This has been observed in cases where sequences of burnt layers are folded and convoluted, yet the original layered

structure of the deformed hearths are still visible (Figure 2.86) (Karkanas et al., 2002; Karkanas, 2001). Finally, it should be stressed that cryogenic processes affect pre-existing sediment and therefore the content of these deposits is not necessarily related to the climatic event that formed the cryogenic features, and careful observations of the deposits in the field and thin section are needed to link the two.

3) Replacement, Dissolution, and Precipitation

The main chemical processes that act on archaeological deposits after their deposition are dissolution of

Figure 2.86 Impressive cryogenic deformation of MP burnt layers at Grotte XVI, France. Note red coloration of sediment above the burnt sequence as a result of post-burning chemical alteration and *not* the burning itself. Hanging grid is 1 m.

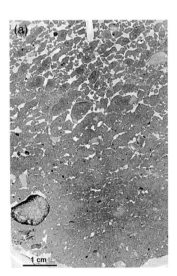 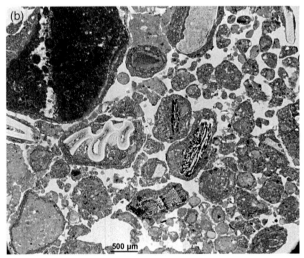

Figure 2.85 (a) Thin section scan of microstructure produced by freeze-thaw activity. At the lower part planar fissures give way upwards to lenticular and sometimes rounded aggregates. MP, Dadong Cave, China. This change marks the intensity of freeze-thaw activity, which decreases with depth. From Karkanas et al. (2008). (b) Cryoturbated and coated grains of bone, limestone, rodent teeth, and limestone. UP, Hohle Fels, Germany.

unstable minerals and precipitation of new ones. Each mineral is stable under a certain set of environmental conditions, which are mostly governed by water chemistry and flow. Oxidation is also a very important process, but water is the critical catalyst because it provides a favourable environment in which the oxidation can occur. When water comes into contact with sediments enriched with various amounts of dissolved material, its chemistry quickly changes and it attains a new equilibrium. If pore water is not replaced by new water it is not capable of reacting further with the sediment. Therefore, the amount and rate of water passing through the sediment dictates the rate of chemical reactions. Porosity and particularly interconnected pores play a significant role with porous gravels and sands by allowing large amounts of water to pass; in contrast, pure clay acts as an almost impermeable material.

Temperature is a key factor that regulates the rate of reaction by speeding up the processes. Obviously, climate and geomorphic processes are hierarchical higher-levelled agents controlling the whole system. As a consequence, wet and warm climates and gently sloping porous substrates will show the highest dissolution rates. Note also that caves are often hydrologically very active environments and therefore prone to high rates of dissolution.

Most of the archaeological materials also consist of mineral phases and, consequently, their stability follows the same rules. When a mineral is unstable it can be replaced by another new stable phase or it will pass into solution. Precipitation of a new mineral may include all (in a new combination) or part of the dissolved constituents; the remaining part can be carried away with the water and be precipitated as a new phase much later in a new environment. Mapping of the spatial distribution of mineral phases in an archaeological site can reveal the cascade of mineral reactions and in particular those minerals which disappear and those which are formed along the same depositional sequence. In this way we can predict which archaeological materials are stable and which could have been dissolved and are therefore absent because they were never deposited or because they were not preserved (Karkanas et al., 2000; Weiner et al., 2007).

Recognition

Most of chemical alterations are not visible to the naked eye. Often, precipitation may result in recognizable features such as the formation of staining, nodules, aggregates, and filaments of various colours. Carbonates and gypsum have whitish colours, phosphates whitish to light yellowish or orange tints, and iron-manganese concentrations dark colours (Figures 1.5b, 2.82b, and 2.87). Carbonate neoformations are generally hard whereas phosphates are relatively softer. Gypsum is

Figure 2.87 Black iron-manganese concentrations (redoximorphic features) along drab halos in a temporarily waterlogged soil. EP, Kokkinopilos, Epirus, Greece.

often powdery and generally soft, but iron-manganese formations are hard and often have a metallic appearance. As a general rule, these chemical features appear with a patterned distribution that reflects precipitation in favourable locations (e.g. voids, fissures, burrows, around bones). In cases where chemical alteration is well developed, the matrix will appear quite homogenized with the new mineral formations clearly discernible. Tilting of deposits that cannot be explained by primary deposition on slopes could be the result of chemical alteration and possible subsidence to produce the observed relief.

Microscopic features that are characteristic of chemical alteration are well-formed crystals with sharp angular outlines that may truncate pre-existing grain boundaries of corroded minerals (Figure 2.88). New alteration

Figure 2.88 Photomicrograph of neoformed gypsum crystals with sharp outlines and fresh appearance; here gypsum formation is tied to the inputs of marine salts and guano. XPL, MSA, Site PP13b, South Africa. From Karkanas and Goldberg (2010a).

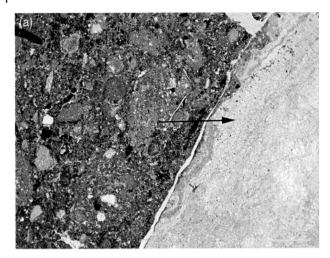 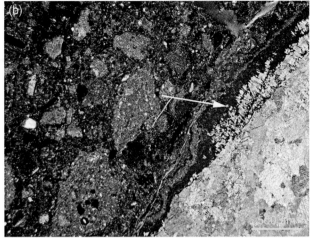

Figure 2.89 Photomicrographs of a phosphate reaction rim around a limestone fragment (arrows). The rim in PPL (a) appears isotropic in XPL (b). Note also ragged remnants of calcite inside the isotropic rim. MP, Denisova Cave, Siberia, Russia.

phases also form rims or reaction coronas on rounded and worn detrital grains (Figure 2.89). Formation of stained areas and nodules are characteristic, but the latter, because of their hardness, can be redeposited, and thus their relation to the matrix has to be demonstrated (e.g. sharp or gradual boundaries; see section 2.7.3 on soil formation). The nature of the neoformed minerals can be identified using microchemical or bulk mineralogical analysis.

Effects on Archaeological Materials

Carbonate Shells, Ash, Dung Spherulites Rainwater dissolves atmospheric carbon dioxide and produces weak carbonic acid. When reacting with organic matter on the sediment surface more carbonic acid is produced. Organic matter is oxidized by microorganisms in the presence of oxygen and during this process carbon dioxide is produced. Thus, biological activity also contributes to the formation of carbonic acid. pH is the most important factor of dissolution of carbonate minerals. Calcite ($CaCO_3$) is stable at pH values above about 8 but undergoes solution under acidic conditions and in the presence of carbonic acid produced by the aforementioned processes. Rainwater is slightly acidic and it will dissolve the calcareous component of the surface sediments. However, in the presence of large amounts of calcite in the sediment and soil, the water eventually equilibrates and becomes alkaline so further dissolution is prevented. This balance is dictated by the water flow through the sediment and in high water flows the buffering capacity of calcite is destroyed. Aragonite, another mineral form of calcium carbonate ($CaCO_3$), is more soluble than calcite in soil and sedimentary environments so its presence indicates that the distribution of calcite

has not been affected by diagenetic processes (Wiener, 2010: 78).

Carbonate materials of archaeological importance (e.g. calcareous shells) are not stable in acidic conditions. White powdery wood ash consists predominately of fine-grained calcite and therefore ash will be one of the first components to dissolve; this is one of the reasons why ash is not often present in open-air site environments. Indeed, when sediments are decalcified, that is, they do not contain any calcite, ash would also have been dissolved away even if originally present.

Another form of calcium carbonate found in several archaeological sites is dung spherulites (see coprolites, section 2.6.2). These are microscopic spherical aggregates often found in stabling remains (Canti, 1997). As they survive burning they are also found in ashed dung. Dung spherulites are more soluble than geogenic calcite (Canti, 1999), but generally the conditions of their preservation are broadly similar to those of wood ash.

Recrystallization implies a change from a more soluble to a less soluble substance. Usually small crystals recrystallize to big crystals. Recrystallization is very important in carbonate sediments. In particular, calcitic ash crystals, due to their very small size, are readily recrystallized to form indurated sediments that resemble natural carbonate rocks (Figure 2.90a) (Karkanas et al., 2007). Ash is normally found in microscopic aggregates, pseudomorphic after plant structures such as calcium oxalate, and they have rectangular, rhombic, or rod shapes (Figure 2.90b) (Wattez, 1988; Wattez and Courty, 1987); unaltered rhombic-shaped calcium oxalates have a red interference colour). The presence of pseudomorphs of the calcium oxalate crystals associated with charcoal or fine charred material can differentiate recrystallized ash from natural geogenic calcite. Furthermore, charcoal, oxidized (burnt) reddish soil aggregates, as well as a close association with burnt bone and other burnt archaeological materials, are further indications.

Bone and Phosphates (also see Chapter 5) Archaeological bone is another important material that can dissolve. The inorganic mineral component of bone is the calcium phosphate mineral, apatite. Bone is stable in sediments with pH above ~8 (Berna et al., 2004), but it will rapidly dissolve in more acidic environments, particularly at pH values below 7; therefore exposure of bone to rainwater will result in slow dissolution. In buried bone, the dissolution rate depends primarily on the water flow (Hedges and Millard, 1995). Experimental observations (Berna et al., 2004) and case studies (Weiner et al., 1993) have shown that bone is stable in the presence of calcite, and thus it is expected that it will be preserved in calcareous environments.

During the decay and oxidation of organic matter phosphates are released in the pore water. In the presence of calcium anions apatite will precipitate. However, with prolonged decay, the pH drops due to the formation of carbonic acid and apatite is no longer stable. Both naturally formed and bone apatite will dissolve and new phosphate mineral phases will precipitate. These are mainly aluminium-rich phosphates (e.g. taranakite, crandallite, montgomeryite, tinsleyite, leucophosphite). This sequence of minerals has been well documented in cave environments where large amounts of guano have decayed and altered anthropogenic sediments (Schiegl et al., 1996; Karkanas et al., 1999, 2000, 2002; Weiner et al., 2002, 2007; Shahack-Gross et al., 2004a)

Silicates (Phytoliths, Diatoms, Lithics) Silicate mineral phases of primary archaeological interest include biogenic forms of opal (hydrous amorphous silica) and flint, chert, or quartz; phytoliths, diatoms, and sponge spicules are all made of opal. Phytoliths are mineralized parts of plants and are found in large quantities in archaeological sites because they survive burning and oxidation. Opal dissolution increases rapidly in alkaline environments and particularly above pH 9 (Krauskopf, 1979). Therefore, in the presence of enough water phytoliths will dissolve in sediments where calcite is present, an indicator of a generally alkaline environment (Piperno, 1988). However, the solubility of opal is high enough in the normal pH range of the soil environment (4–8), and thus biogenic forms of opal will start dissolving if a high rate of water flow is maintained through the sediment (Fraysse et al., 2009).

Flint and chert consist predominately of quartz, which is the most stable silicate mineral. Nevertheless, opal, chalcedony, and calcite occur as minor components in chert and flint, which can dissolve in various chemical environments as described above (Sheppard and Pavlish, 1992; Burroni et al., 2002).

Organic Matter Pollen, charcoal, and other botanical remains are primary constituents of archaeological sediments. In the presence of oxygen they are oxidized and destroyed. During oxidation, reducing conditions are produced because oxygen is consumed by the reaction. Reduction favours dissolution of iron minerals, which are generally very insoluble under oxidizing conditions. Therefore, dissolved, reduced ferrous iron moves away from the decaying organic matter and

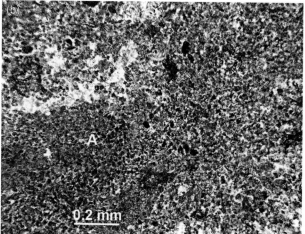

Figure 2.90 (a) Dark grey micritic calcite ash crystals (M) are recrystallized to white large sparitic calcite crystals (S) (compare with b). A partially-ashed grain still exhibits charred remains (C). PPL, LP, Qesem Cave, Israel. (b) Typical pristine micritic rhomboidal crystals of ash (A) with a dark grey appearance. PPL, UP, Klissoura Cave I, Greece.

precipitates as ferric iron when it encounters oxygen. It follows then, that if organic matter was originally present and has since decayed, iron oxide minerals should have been redistributed. Iron containing authigenic (neoformed) minerals can, therefore, in principle, be used to reconstruct past oxidation conditions. Most anthropogenic activities result in accumulation of organic matter and as a consequence oxidation and reduction reactions in archaeological sites are directly related to the degradation of organic matter, which may no longer be present or visible. Oxidation over long periods of time has been found to affect charcoal, a situation that has serious implications for radiocarbon dating (Cohen-Ofri et al., 2006).

Manganese oxides are precipitated under similar to iron conditions in archaeological sites (Arroyo et al., 2008). Decomposition of organic matter releases manganese that in alkaline conditions precipitates in oxidized forms. The very common black encrustations and staining on bones are often the result of this process (Shahack-Gross et al., 1997).

Oxidation of organic matter in waterlogged sediment under reducing conditions favours the formation of vivianite, an iron-rich phosphate mineral (Figure 2.91). Vivianite has been related to the degradation of human and animal waste (Bertran and Raynal, 1991; Gebhardt and Langohr, 1999; Karkanas and Goldberg, 2010b; McGowan and Prangnell, 2006).

2.7.3 Soil-forming Processes

A concise introduction and definition of soil and pedogenic processes has been given in Chapter 1. Here, suffice it to say that soils represent periods of hiatus in the deposition and stability of the landscape. Fossil soils, or palaeosols, in a majority of cases, have been buried by later deposits, but they may also be found at the surface and no longer actively forming in quite the same way as under their original environmental conditions (Retallack, 2001). Buried palaeosols always represent unconformities (Figure 1.6).

Soils have characteristic horizons and pedogenic structures (Figures 1.5, 2.72a, and 2.92). Horizons develop over time and each has its own chemical, physical, and biological properties; in most cases horizons parallel the extant land surface at their time of formation. Each soil type consists of a set of genetically related horizons that constitute a soil profile (Holliday, 2004). It should be stressed once again that horizons do not represent depositional layers but are post-depositional features.

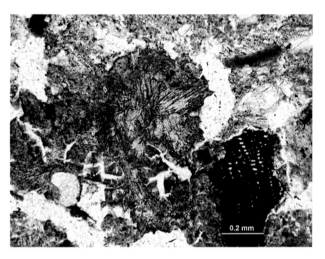

Figure 2.91 Crystals of vivianite with characteristic bluish and greenish pleochroic colours in PPL. In this case, it is related to decay of animal organic matter. Neolithic, Promachonas/Topolnica, Greece.

Figure 2.92 (a) Image of chernozem (Mollisol or Phaeozem) soil with a well-developed ~30 cm thick A-horizon (mollic epipedon) beneath the compact root mat of grasses. Underlying is a bluish-greenish grey gleyed, Bg horizon, characteristic of waterlogging conditions. Grevena highlands, Greece. Note the gradual contact between the soil horizons. (b) Buried soil (palaeosol) with development of angular blocky peds (an example with dotted line). Pleistocene, Ylliki Lake area, Greece.

The most characteristic soil horizons from the surface to the unaltered substrate are the following:

O = Accumulation of organic matter on the surface, such as leaf litter in various stages of decomposition.

A = Part of the mineral soil profile but substantially enriched in humified organic matter coming from the decay of plants and animals living on the surface. Typical processes related to the development of the A-horizon are decay of organic matter, removal of clay and generally fine particles (eluviation), and decalcification due to the presence of organic acids. Bioturbation is often very intensive in this horizon and soil aggregates might represent excrements of microfauna.

E = Eluviation horizon characterized by leaching of clay, humus, iron, and aluminium. It has a light colour and is enriched in resistant minerals such as quartz of sand and silt sizes. The E-horizon forms under acidic conditions often associated with trees.

B = Leaching, illuviation (accumulation of dissolved or suspended soil materials), and weathering of primary minerals form this horizon, processes that result in strong development of a pedogenic structure (see below). Reddening, gleyzation, and formation of calcic horizons are often characteristic of B horizons (Figure 2.93). It is the most conspicuous horizon of well-developed soils and any sedimentary structure or feature of the parent sediment or rock is totally destroyed.

C = Incipient leaching and pedogenic processes affecting the substrate that can still preserve the original depositional structures and fabrics.

R = Unaltered bedrock.

The fundamental unit of soil structure is the ped. Peds are soil aggregates, the clods of earth separated by cracks, roots, voids, burrows, and other soil openings (Retallack, 2001). The soil aggregates are stable enough to ensure free water percolation, thus reducing the possibility of runoff and surface erosion (Birkeland, 1999). Peds are classified according to their shape: crumb, granular, blocky, columnar, prismatic, and platy (Figures 2.92b and 2.94). Peds are well expressed in mature soils, which are represented by thick soil profiles and taken to signify a long time of development.

Soils are categorized into groups based on their properties observed in the field and measured in the laboratory (Soil Survey Staff, 1999). The US soil classification system is based on the existence of diagnostic horizons and some other measurable properties such as texture, mineralogy, moisture, and temperature. Other classification systems include that of the FAO (Food and Agriculture Organization of the United Nations), which is useful for recording global distribution and geography of soils, the French classification system (Krasilnikov and Arnold, 2009), and the World Reference Base (WRB), which is the international standard for soil classification that was established to correlate different soil classification systems.

Soils often constitute the stable surface upon which occupation deposits accumulated. Correctly identifying the type of soil encountered will provide important information for reconstructing the environmental conditions prevailing during the initial occupation and understanding the geomorphology and overall ecological and archaeological landscape (Holliday, 2004). This is especially the case if the soil did not form under the present environmental conditions (e.g. a fossil steppe soil in an area now characterized by forest soils).

Figure 2.93 (a) A truncated well-developed brown B-horizon in a palaeosol displaying white calcareous nodules and rhizoconcretions (arrow). Pleistocene, Aliakmon River, Grevena area, Greece. (b) Reddish, clay-rich B-horizon developed on fluvial gravel-rich sediment. Transition to C horizon is gradual. The overlying layers (their sharp contacts marked with arrows) are not soils, except for the present-day formation of a top soil (A) on the upper layer. Holocene, Aliakmon River, Grevena area, Greece.

Soils can modify archaeological sites after their abandonment: they transform the original depositional stratigraphy by forming soil horizons (horizonation) and homogenize the deposits by mixing them (haploidization). As a consequence, the original archaeological context is modified and or destroyed (Holliday, 2004). Buried soils within occupational sequences are evidence of temporal abandonment of the particular area. It is also possible that this abandonment signals a change in the use of the area, as, for example, from agricultural fields to middens, resulting in the formation of an anthrosol.

Anthrosols are soils that are a result of anthropogenic processes and they exhibit both physical and chemical pedogenetic alterations (Figure 2.95) (Holliday, 2004). Examples of such man-made soils are the Terra Preta soils of the Brazilian Amazon, and the Plaggen and Dark Earths of Post-Roman and Medieval Northern Europe. Terra Preta is a very dark, almost black soil, very rich in midden debris resulting from household trashing (Lima et al., 2002; Holliday, 2004; Lehman et al., 2007). It has elevated levels of calcium and phosphorus. Plaggen soils are characterized as dark-grey humose (organic-rich) sandy soils that result from intensive manuring. They contain artefacts, are relatively high in phosphorus, and have a low pH (Courty et al., 1989; Dekker and De Weerd, 1973; Holliday, 2004).

Dark Earth refers to soils developed inside urban settlements (Macphail et al., 2003; Nicosia et al., 2012; Nicosia and Devos, 2014); they have a very high anthropogenic content, including in some cases human refuse, and are slightly alkaline (Courty et al., 1989; Macphail, 1981). Some of these anthropic soils, especially Dark Earth, often are not particularly affected by pedogenetic processes, although they are intensively weathered and altered (Holliday, 2004: 317).

Almost all of the previously described diagenetic processes are normally part of the soil-forming processes, and thus the fate of archaeological materials in soils follows the same rules. Reddening, silt and clay translocation, clay neoformation, gleyzation, calcification, and salinization are additional soil-forming processes that can affect sediments on the earth's surface. Below, we briefly discuss some fundamentals of the above processes that can be also useful in identifying similar features in archaeological deposits not necessarily subjected to or affected by pedogenesis.

1) Reddening

Iron oxides are the most conspicuous products of pedogenesis because of their bright colours. Hematite is an oxidized iron oxide formed in contact with air and has a red colour, whereas goethite, a hydrated form of iron oxide, is yellow to brown.

Figure 2.95 An example of man-made soil (anthrosol) consisting of one massive and thick dark, organic rich horizon on sandy marly substrate. Pre-Classic Colha, Belize. Excavation grid is 2 m.

Figure 2.94 (a) Well-developed granular soil structure (GR) in a Mollisol. Grevena highlands, Greece. Part of a geological hammer for scale. (b) Prismatic structure developed in clayey loess in the upper part of the profile; at the base is a band of diffuse whitish carbonate. Scale is 1 m. Late Pleistocene, Netivot, Israel.

Iron-rich minerals in reduced states such as siderite have a grey colour, and a variety of authigenic ferrous iron-rich clays have green and grey colours. Note that the rank of hues in the Munsell Soil Color Chart follows this trend: colours in reducing, waterlogged environments are bluish, greenish to olive yellow (PB, BG, G, GY and 10Y to 5Y), aerated and slightly oxidized soils and sediments are yellowish to reddish brown (2.5Y to 2.5YR), and fully oxidized ones are red (R). Oxidation reactions are mainly observed in well-drained soils and in warm climates with contrasting wet and dry seasons (Courty et al., 1989: 165–167). It is important to note here that reddening can be produced by high temperatures associated with human activities such as fireplaces and other pyrotechnological processes. However, it has been observed that fire reddening always affects only a thin layer of the substrate of the burning feature (see also Courty et al., 1989: 169; Aldeias et al., 2016; sections 3.2 and 5.2).

2) Translocation of clay and silt

Downward or lateral percolation of water can move material from the upper horizons (usually the A and E horizons), in a process called eluviation, and deposit it in the underlying B-horizon, the illuviated or argillic horizon. Usually only the very fine fraction is illuviated: mainly clay, silt, fine organic particles or complex organometallic colloidal compounds. Nevertheless, illuviation operates in archaeological sediments that are not necessarily pedogenically altered, and vegetation cover regulates translocation of particles down the soil profile. The translocation of coarse silt-rich particles is favoured by turbulent flow in arid barren environments (Figure 2.96a); similarly, in periodically flooded floodplain sediments, poorly sorted silty clay coatings are produced in voids (Brammer, 1971). In contrast, well-developed, bedded fine clay coatings develop under forest covers (Figure 2.96b) (Courty et al., 1989). Moreover, in caves, dripping water with suspended silt and clay can result in the illuviation of these grains beneath the substrate on which the drops fall.

3) Clay Neoformation

Intensive weathering of silica minerals can produce new clay minerals. As a general rule, montmorillonite clay (smectite group) forms in areas with low levels of precipitation, kaolinite at intermediate levels, and oxides and hydroxides of iron and aluminium at higher levels. In addition, kaolinite is the predominant clay in well-drained soils in warmer and wetter climates whereas smectitic clays are found under poorly drained conditions (Birkeland, 1999). Clay neoformation is associated with silica dissolution and therefore lithic artefacts may be affected during this process (Courty et al., 1989). Authigenic opal formation has been observed in such cases (Figure 2.97) (Karkanas et al., 1999; Weiner et al., 2002).

4) Gleyzation

Under waterlogging conditions, reduction of iron occurs, and a grey to bluish grey coloration of the soil or sediment is observed (gleyzation). Microorganisms exhaust oxygen under these anaerobic conditions and produce the reducing environment and discoloration of the original oxidized soil. However, often reduction and discoloration are incomplete, resulting in a reddish-grey mottling. Around roots in periodically

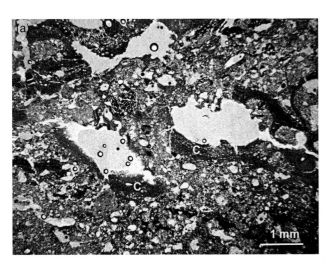 

Figure 2.96 (a) Photomicrograph of compound silty and dusty clay coatings (C) around voids in periodically flooded sediment. PPL, Neolithic, Vashtëmi, Korçë, Albania. (b) Red laminated limpid clay coatings around aeolian sand grains in a well-developed Bt-horizon and probably under the influence of a plant cover. XPL, MP, Site PP5-6, South Africa.

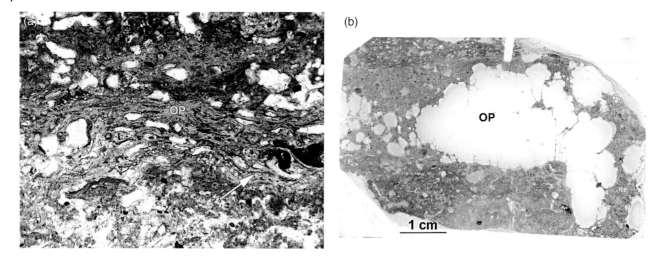

Figure 2.97 (a) Neoformed (authigenic) opal precipitation (OP) (confirmed by SEM/EDS analysis) as a result of phytolith dissolution (arrow). PPL, MP, Theopetra Cave, Greece. (b) Displacive, crenulated, nodules of beige opal (OP) from Hayonim Cave, Israel (MP).

waterlogged soils drab halos form because of reduction of the soil, a process described as surface gleying (see section 2.6.2 on bioturbation, and Figures 2.73a and 2.87). Several other processes may produce surface gley (or 'pseudogley'), including stagnating water on poorly drained surface materials (Retallack, 2001).

5) Calcification

As discussed above, calcification is observed in other, non-soil environments and it can affect archaeological calcareous sediments such as ash sequences. In arid and semiarid climates a carbonate-rich horizon is observed at some depth below the surface when rainfall is insufficient to leach calcite from the soil. These horizons are commonly called calcretes – or formerly, caliche – and in the soil nomenclature are called Bk and K horizons depending on the intensity of the calcification (Figure 2.98). The position of this horizon is related to the depth of calcium leaching and thus is related to precipitation amounts and regimes (summer vs. winter rainfall) and hence climate (Birkeland, 1999). Several stages in calcrete formation are observed. Initially formed are thin calcitic filaments in the matrix and coatings on coarse particles (Figure 2.93a). With time, carbonate deposition increases the formation of nodules, thick coatings, and crusts (Figures 1.5b and 2.98). When voids become cemented and filled with calcite, water percolation is substantially reduced and water tends to circulate at the surface of the calcareous horizon and deposit the characteristic laminated calcareous horizon of calcretes (Figure 2.98a) (Gile et al., 1966).

Figure 2.98 (a) Thick laminated calcareous crust in a Pleistocene calcrete (C). The entire horizon below is cemented with pure calcite filling cracks (arrow). Pinnacle Point, South Africa. (b) Truncated soil horizon with calcareous nodules formed in Pleistocene alluvium in Negev, Israel.

As discussed above, calcareous rhizoconcretions are characteristic features of some desert regions but also of humid coastal dunes rich in calcareous bioclasts (Figure 2.73b). Repeated cycles of dissolution and precipitation of calcite along roots results in their gradual encrustation and the formation of concentric layers of calcite; eventually the root dies. Biologically induced calcification of cells within roots also leads to formation of rhizoconcretions described above (Wright and Tucker, 1991).

Cave environments are particularly noted for calcification processes (see also section 5.2.1). Cave speleothem formations are an expression of this process. Waters passing through the organic-rich soil horizons become enriched in carbon dioxide and then enter the cave through fissures. The new environment is relatively depleted in carbon dioxide and therefore waters will equilibrate by losing carbon dioxide, which results in the formation of calcite (Gillieson, 2009: 116–120). Stalactites, stalagmites, and flowstones are different forms of speleothems deposited by dripping, seeping, or flowing water on the surface of cave walls, floors, and ceilings (Hill and Forti, 1997). Speleothems are slowly formed in increments resembling in many ways tree rings (Figure 2.99). This incremental development has proved to supply the best records of the history of climate on land. The isotopic composition of oxygen and carbon of the calcite of each lamina records the chemistry of waters that precipitate it. This is directly related to climate and the local environmental conditions that produce these waters (Ford, 1997).

The same waters percolating through the pores of the cave sediment will deposit calcite and produce a hard calcareous rock known as cave breccia (Figure 2.99b). Famous cave breccias, such as those within the Cradle of Humankind in South Africa, seal and preserve important early hominid fossils that are millions of years old (e.g. Pickering and Kramer, 2010).

6) Salinization

Salts, such as gypsum and halite, accumulate deep in the soil profile in arid areas with precipitation <300 mm/year (Birkeland, 1999). These salts are more soluble than calcite and are formed in well-drained areas. Salts can come from several sources, including the sea, dust from salt lakes (playas), or pre-existing salts in the rock. Nevertheless, gypsum can form in decayed organic matter, such as guano in caves or as dung in stabling deposits; both are well above the influence of the water table (Figure 2.88) (Shahack-Gross et al., 2004a). An unexpected source of gypsum is the hydration of ash of Tamarisk wood (*Tamariz aphylla*), which contains large amounts of anhydrite, the dehydrated mineral form of gypsum (Shahack-Gross and Finkelstein, 2008); upon hydration, anhydrate is transformed to gypsum.

Another area of salt accumulation occurs in environments with very high evapotranspiration and a fluctuating, high, local, or regional water table (Holliday et al., 2009). In these environments, that is, playa lakes, palustrine environments, wetlands, and marshes, large amounts of various salt minerals can accumulate. A rise in saline groundwater in coastal areas results in loss of original soil structure, gleying, and formation of iron and manganese mottles and organic matter decay. In addition, marine inundation produces decalcification, organic matter oxidation, and ferruginisation (Macphail et al., 2010). In gently

Figure 2.99 (a) Example of thinly laminated speleothem from Koutouki Cave, Paiania, Greece. (b) A hard cemented sediment (cave breccia) containing bone (an example with arrow). An angle grinder was used to remove a sample from the right upper corner. Scale is 20 cm. LP, Qesem Cave, Israel (from Karkanas et al., 2007).

inundated sites archaeological charcoal is spread over larger areas than the original site due to the presence of sodium (Na$^+$) ions which promotes dispersion in saline alluvial gley soils (Goldberg and Macphail, 2006: 162–166).

In desert and coastal areas, salinization can also affect the bedrock or man-made structures, including mudbricks (Courty et al., 1989). Hydration of various salts inside the pores of the bedrock results in volume increase and concurrently stresses that eventually disintegrated the rock.

Recognition of Soils

Field Several field indications differentiate soils from sedimentary deposits (Mandel and Bettis, 2001). Probably the most important is the existence of horizons and particularly their boundaries and geometry. Soil horizons, in general, have a sharp upper contact and a clear to gradual lower contact (Figures 1.5 and 2.93b). The upper A and O horizons of buried soils are often not preserved, usually being eroded or decayed; hence, it is generally the B-horizon that is observed in a palaeosol (Figure 2.93). The palaeosol profile will have a sharp upper contact representing the erosional surface and a gradational lower contact towards the less altered C-horizon, which may preserve original sedimentary bedding and structure. Soils and their horizons will follow the topography and thus might truncate pre-existing horizontal sedimentary bedding. Nevertheless, in sloping areas, sedimentary deposits also follow the topography. Soils and

palaeosols should show lateral persistence and a tabular geometry. Sedimentary deposits, although often tabular, may also show lenticular and trough geometries, and complex lateral variations in fabric (e.g. from gravel to clay).

Nonetheless, the unmistakable identity of a soil is its pedogenic structure. Although experience is certainly needed to recognize them, peds are characteristic structures and when well developed are quite obvious to the naked eye (Figures 2.92b and 2.94). Peds are persistent aggregates that require some effort to be crushed by fingers, but what makes them characteristic is their appearance in a consistent pattern along the entire particular horizon. Granular peds are often formed in the presence of organic matter and consequently will have a dark colour (Figure 2.94a). They are mostly restricted to the surface horizons. Blocky and columnar peds are often found in clay-rich B horizons (Figures 2.92b and 2.94b). Note also that often aggregates of smaller peds coalesce to form larger peds and therefore columnar peds are composed of smaller units.

Other characteristic features of soils are the existence of root traces and in some cases their associated drab halos and fossil roots (rhizoconcretions), discussed above (Figure 2.73 and 2.75a). Reddening is also characteristic of B horizons, but red sediments or reworked and redeposited red soils are not unknown. Stone lines can also help locate a buried soil.

Microstructure Clay and silt coatings, various pedofeatures, and birefringence fabrics (b-fabric) are

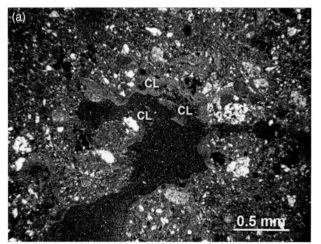 

Figure 2.100 (a) Silty clay void coatings (CL) produced by dripping water in a vadose environment. XPL, BA, Agios Charalambos Cave, Crete, Greece. (b) Clay coating (arrow) formed in burnt remains. B = bone, C = charcoal. XPL, MP, Site PP5-6, South Africa.

 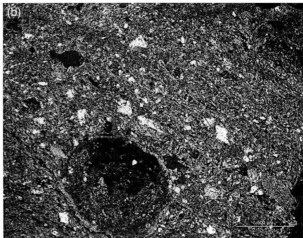

Figure 2.101 (a) Example of cross-striated fabric. Clay particles are arranged into sets of parallel domains normal to each other. XPL, Classical, Molivoti, Thrace, Greece. (b) Parallel striated fabric. Clay particles are arranged in domains parallel to each other in one direction. Note that clay particles are distributed around the large clast at the left bottom, reflecting granostriated fabric. XPL, Neolithic, Vashtëmi, Korçë, Albania.

characteristic of pedogenesis, but not all of them are exclusive to it. Clay and silt coatings are observed around particles and pores (Figure 2.96), and they can have the form of a rim, a pendant, or a meniscus; the latter two are gravitational forms found in a vadose environment (above the water table) (Figure 2.100a). Pure clay coatings are more often unique to pedogenic processes but they can be also found in altered wood ashes (Figure 2.100b) (Courty et al., 1989; Huisman et al., 2012), deep inside the C-horizon, or even in cracks and pores of the unaltered bedrock or the archaeological substrate (Birkeland, 1999). Silt and dusty clay coatings are also observed in pedogenic environments but they can be found in sedimentary and archaeological deposits as well.

Various forms of microscopic nodules and concentrations are observed in soil environments, comprising those of calcite, salt (halite), iron, and manganese. They are all indicative of certain hydrological conditions, but they are not unique to soil environments as they have also been reported inside caves, for example (Karkanas and Goldberg, 2010a; Karkanas et al., 2008).

Striated b-fabrics result from the preferred orientation of the clay particles present in the soil micromass. Several patterns are observed, such as grano- and poro-striated fabrics around particles and pores, respectively, and mono, parallel, and cross-striated in the matrix (Stoops, 2003) (Figure 2.101). However, grano-, poro- and mono-striated fabrics are often observed in all type of environments where shrink-swell stresses produced by wetting and drying are observed (e.g. deep cave environments).

Parallel and particularly cross-striated fabrics seem to be unique to soil environments.

Certain features can be better interpreted under the microscope, as in the case of *in situ* gleying where discoloration by iron depletion can serve to differentiate material that is otherwise similar across the feature (Figure 2.75a). Microscopic observations of nodules can reveal if they have been reworked and redeposited as sediment. Rounded and angular forms with sharp contacts to the surrounding matrix and those with exotic mineralogical content in their mass indicate that the nodule has been transported and redeposited. Alveolar septal calcareous structures are typical of rhizoconcretions, which might have formed in place (Figure 2.75b).

2.8 Concluding Remarks

Our principal aim in this chapter was to provide the reader with the basic characteristics of soils and naturally deposited sediments, and to paint a picture of how they form and how we can recognize them in the field and under the microscope. In order to assist with these goals, we have summarized the above information in Table 2.6. As can be seen, certain observed responses (e.g. sorting, lamination, cross bedding, and clay coatings) can be produced by more than one process, so one has to use an assemblage of processes to hone in on what appear to be the most reasonable interpretation(s).

Table 2.6 Summary of depositional processes observed in a site and their characteristics

Depositional process	Occurrence	Geometry	Bedding	Fabric and structure	Microscopic observations	Orientation analysis	Archaeology
Mass movement							
Slides and slumps	Slope of tells and mounds; caves; large topographic differences; moderately steep slopes especially on softer, unstable substrates	Curved failure planes; rotational movement of soil mass	Massive, disordered; blocks with remnant bedding (upslope tilting)	Chaotic; variable fabric	'Brecciated' microstructure; striated b-fabrics	Random to weakly planar along slope	Original pattern totally destroyed; occasionally *en masse* movement
Rock and debris falls, avalanches and grain flows	Caves, cliffs; steep slopes; eroding sections of site; decay of structures	Cones, sheets (talus)	Stratified; tabular to lenticular	Angular, fresh, coarse grained; open-work to clast-supported; lateral crude sorting, normal or reverse crude grading	Open-work fabric with random orientations	Random along sloping plane; heavier and denser objects moved further downslope	Objects are totally reworked
Solifluction	Periglacial environments; glacial periods	Sheets, lobes, and plugs	Poorly to well stratified; tabular to lenticular	Fine-grained to matrix- and clast-supported, and open-work coarse grained; normal or reverse grading; cryoturbation features and frost cracks	Ice-lensing and capping	Preferentially orientation along slope; clasts are frequently imbricated, but with various dips	Redistribution of objects depending on the intensity of the processes
Debris flow and mudflow	relatively steep slopes (tells, mounds); decayed cones of construction materials; caves; topographic differences	Elongated to broad lobes	Elongated lenticular, tongue-like	Uniform distribution of all sizes from clay to boulders; extremely poorly sorted; lack of internal stratification; normal and inverse crude grading, 'trains' of clasts, vertical orientations	Porphyric related distribution; rotational features; inclined preferred orientations of elongated clasts and microscopic tiling	Preferentially orientation along slope; oblique and transverse orientations in parts of the lobe	Original position of the archaeological objects destroyed
Water flow							
Shallow water flow	Almost flat areas to any slope (see above)	Sheet wash (unconfined flow) – rills to small channels (channelized flow)	Thin, lenticular bedding to plane-parallel stratification and fine lamination; lack of cross-bedding in sheet wash – low-angle cross bedding in shallow channels;	Clay and sand in gentle slopes to pebbles in steeper slopes; ill-defined sorting in sheetwash to good sorting in shallow channels	Lamination and rhythmic alternation of sand and silt; dense packed rock and mineral grains or clay-rich aggregates without clay fines; crude grading in shallow channels	Transverse to random orientations; occasionally preferred orientation along slope	Restricted transport capacity only of small archaeological objects in sheet wash; rilling can partially destroy archaeological sites and rework archaeological content

Standing water bodies	Any flat area	Planar and continuous bedding; laminations	Well sorted silty and clayey sediment; coarse grained restricted to margins; sand layers with ripple and very small scale cross-stratification; massive to graded and rhythmic bedding, chemical and organic deposition	Microscopic sedimentary crusts in puddles; microscopic stratification and lamination in massive muds; graded laminae and rhythmically laminated facies in ponds and lakes; carbonate micrite and microbial mats in chemical deposits	Random orientation on a flat plane; occasionally orientation parallel to the flow direction in land-originated flood episodes	Occupational phases associated with periods of exposure; high sedimentation rate results in quick burial and protection of archaeological materials
Hyper-concentrated flow	Same environments as debris flows, sheet flows, and channelized flows	Tabular	Coarse sandy texture with distinctly less silt and clay than in debris flow; poor sorting; absence of cross-stratification; lack of extensive matrix support; subangular clasts	Mostly massive silts and sands; occasionally microlaminations of silt layers alternating with sand and fine gravel	Clasts typically show bimodal orientation: large clasts (larger than cobbles) perpendicular to flow and small clasts parallel to flow	Can be catastrophic; mixed assemblages from different periods
High-energy flow	Fluvial and coastal environments; phreatic caves	Intercalated planar to lenticular beds of various thicknesses; also thick massive beds	All sizes but well sorted and stratified; lack of clay in beach deposits; fine lamination, cross-stratification, imbrication, and normal grading; well-rounded gravel	Graded laminated sequences and microscopic cross-stratification; single-grain sand laminae; well-rounded sand-sized bone, charcoal, and shell; elongated grains with preferred orientation and imbrication; wavy, lenticular, interbedded, and ripple stratification	Long axes transverse to current flow and upstream imbrication of the intermediate axis; bones tend to orient parallel to the current direction	Very destructive processes; mixed and reworked assemblages
Aeolian	Arid, desert, and coastal environments; loess in periglacial environments	Packages of high-angle cross-stratification of sand; thinly bedded sands; planar bedding in interdune areas; massive loess beds	Well sorted very fine sand and silt; only rarely coarse sand; planar and trough cross-stratification; finely laminated and graded sand; structureless silt in loess	Finely laminated, normal and reversed grading, and microscopic ripple and cross-stratification; laminae rich in heavy minerals; quartz grains are predominantly subangular in loess	Very light and small archaeological objects oriented parallel to wind current	Fine chips of lithic assemblages and light organic material will be moved; quick burial and protection of previously deposited archaeological sediments, but easily eroded and reworked – any included archaeological structures may be destroyed

(Continued)

Table 2.6 (Continued)

Depositional process	Occurrence	Geometry	Bedding	Fabric and structure	Microscopic observations	Orientation analysis	Archaeology
Biological sediments							
Dung, guano	Areas with animals, caves	Piles and lenses of organic matter	Isolated masses of coprolites, planar, lenticular layers and accumulations	Finely or coarsely chopped organic matter; dense fibrous masses; pale or brownish yellow phosphatic masses	Relict plant fragments, seeds, amorphous organic matter, phytoliths, and dung spherulites; amorphous phosphatic matrix with dissolving bone fragments; guano has bedded spongy or aggregated microstructure, with vughy porosity; alteration products of gypsum and phosphate minerals		Decay of dung and guano alters or destroys materials of archaeological importance (e.g., bone, ash shell)
Bioturbation (Note: Can also be considered as a post-depositional effect)	Top soil and generally at and below any surface	Very localized features (millimetres to decimetres) to homogenized layers fading out downwards	Elliptical, rounded, and tunnel features; often aggregated appearance with repeated burrowing	Burrows passively filled with massive soft material; laminated fillings during flooding; granular structure in topsoil; root traces tapering and branching with depth; calcareous rhizoconcretions and drab halos	Loose, aggregated fabric; calcite or iron accumulation and depletion along root passages		Destructive agent of the archaeological integrity; extensive reworking, vertical and lateral redistribution of cultural materials; elimination of contacts; fragmentation of bone and charcoal
Post-depositional features and processes							
Erosional features, deflation, lags, stone lines, and pavements	Particularly effective in arid and non-vegetated environments	Surface planar features; elongated, straight to meandering cuts (rills, gullies, streams, rivers)	Channel with U and V cross-sectional shapes; lenticular or thin sheet-like lags	Sinuous, undulating, rough or smooth erosional walls with variable morphology; coarse lags devoid of fines, stone lines			Archaeological materials are washed away; only finer fraction is removed during deflation

Diagenesis (mechanical, chemical)	Areas rich in organic matter and humid environments	Sagging and large-scale deformation; localized to pervasive alteration features	Festoon features; calcareous, gypsum, phosphatic, and iron-manganese accumulations	Involutions, injections and folds and frost cracks and wedges; shattering; filaments, nodules, and masses of various colours	Ice-lensing and capping; staining; nodules, aggregates, and filaments; amorphous phosphatic features; new well-formed crystals, alteration rims	Blurring of contacts and depositional characteristics (e.g., composition, texture, and bedding); cementation; destruction of bone, ash, chert, and organic matter; volume changes and deformation
Soil-forming processes	Earth surface	Follows the topography (up to ~2 m deep)	No true bedding; formation of horizons grading downwards	Enrichment in humified material, leaching and translocation and illuviation of clay; alteration of minerals and neoformation of clay; oxidation (reddening); formation of soil aggregates (peds); gleyzation, calcification, and salinization	Pedofeatures (silt and clay coatings); birefringence fabrics; microscopic nodules and concentrations of calcite, iron and manganese, and gypsum; iron depletion features	Transform of original stratigraphy and homogenization of deposits by mixing; original archaeological context is modified and/or destroyed

3

Anthropogenic Sediments

3.1 Introduction

The overwhelming majority of human-made deposits in a site can be thought of as a type of clastic sediment since they are mainly the result of the rearrangement of naturally produced sedimentary particles by human activities, such as construction. They also include anthropogenic particles such as fragments and pieces of artefacts. A special type of sediment found in several archaeological sites is produced by animal husbandry, and normally includes large amounts of organic matter mainly in the form of dung remains. However, it also contains syn-depositionally trapped clastic sediment that often represents the only vestiges remaining after decay of the organics. Remains of matting and roofing are additional organic-rich anthropogenic sediments that occasionally contribute to the formation of archaeological deposits. Burning is by far the most prominent activity that adds to sediments in pre-industrial sites, albeit that it is often not recognized when their remains are dispersed within the rest of the clastic sedimentary matrix. From a point of view of depositional dynamics, all archaeological objects and human-produced or induced materials – irrespective of their size and nature – can be considered to be constituents that contribute to the formation of the fabric of the deposit.

The formation of a clay floor produces a deposit consisting of a mixture of naturally derived particles (clay, sand, and gravel) and, in some instances, anthropogenic particles. The plastic nature of the floor, and its emplacement and setting lead to the formation of fabrics that resemble in many ways those of coherent debris and mudflows. The deposition of burnt remains – although themselves the product of chemical reactions – are actually accumulations of individual particles (ash crystals and charcoal) that build up at the base of a fireplace by gravity; natural fires can sometimes produce similar deposits. Several other activities, such as dumping and sweeping, can be considered as analogues because the final fabric follows the same principles as those of natural clastic deposits formed by gravity processes. Human actions add kinetic energy to the gravitational forces acting on the particles.

Nevertheless, several differences exist between natural and anthropogenic sediments. The most important one is the geometry of the deposited body and its spatial relationship to standing man-made structures. Thus, mudflows do not occur as planar bodies with parallel boundaries (i.e. are not mistaken for floors), and accordingly when they are found inside a construction they must be related to an opening that points to the source of the material. Natural sediment associated with gravity processes is always produced in sloping areas and forms fan-like structures; in contrast, dumped fills are not necessarily related to a pre-existing slope. Furthermore, fireplaces are constrained lenticular bodies of burnt material that often display repeated burning events, either within the deposits at a microscopic level or by stacking, thus they cannot be mistaken for natural fires. In sum, human activities produce unique combinations of fabrics, depositional body geometries, stratigraphic contacts, and associations.

There are cases where a naturally formed deposit has been later modified by human processes and hence it would be difficult to disentangle the complex history of the processes. An example would be the levelling of decayed mudbrick and other construction debris for creating a construction fill previously produced by geogenic processes during a period of abandonment. Obviously, identifying evidence of the geogenic process would indicate such abandonment before the new phase of levelling and construction, and therefore it is important to know the whole history of the depositional processes.

In Chapter 2, we discussed the principles of formation of natural deposits in archaeological sites and noted that the dynamics of geogenic deposition are well constrained by natural laws of gravity and flow. As a consequence, knowing the formation processes of natural deposits provides us with a more robust approach to recognize unknown sedimentary fabrics and structures, which most likely are created by the vast array of human activities.

Reconstructing Archaeological Sites: Understanding the Geoarchaeological Matrix, First Edition. Panagiotis (Takis) Karkanas and Paul Goldberg.
© 2019 John Wiley & Sons Ltd. Published 2019 by John Wiley & Sons Ltd.

As we state throughout this book, anthropogenic sediments are the product of human activities on a site. They may be the product of intentional actions associated with the manufacture of constructions or earthwork (e.g. clay floors, daub constructions, matting, bedding), the by-product of craft and other well-defined activities (e.g. fireplaces, middens, pits, debris associated with knapping or metal casting), or the product of actions produced during general activities such as trampling, scuffing, and discarding.

The first category – constructions – includes actions where the final product is predetermined and the fabric and microstructure are intentionally produced. In this way, the sedimentary characteristics are relatively well known and can be more easily analysed based on ethnographic observations, and ethnoarchaeological and experimental studies.

The actions of the second category – by-products and leftovers – are relatively well defined, but the final product relating to the sedimentary characteristics of the deposits are not well constrained and thus difficult to categorize and study. Nevertheless, there is a large body of studies relating to the formation of some of these deposits, such as hearth and midden facies (Simpson and Barett, 1996; Goldberg and Bar-Yosef, 1998; Courty et al.,1989: 107–111; Canti and Linford, 2000; Canti, 2003a; Meignen et al., 2007; Mallol et al., 2007; Balbo et al., 2010; Villagran et al., 2011; Shillito et al., 2011a; Shillito and Matthews, 2013; Shahack-Gross et al., 2014; Karkanas et al., 2015a).

The third category – general activities – is actually related to actions that include unintentional ones or activities that result in products that are not predictable or are highly variable. As a result, little is known about their sedimentary characteristics, although lately experimental work has created some good reference material (Rentzel and Narten, 2000; Miller et al., 2010; Benito-Calvo et al., 2011; Banerjea et al., 2015a).

This chapter presents the current state of knowledge on the characteristics of the sediment produced by human activities. However, this body of knowledge should be viewed as a work in progress that is continually being enriched with new experimental, field, and laboratory studies; it is expected that our understanding of these sediments will be constantly improving. Some of the described features are often successfully recognized in the field based on empirical familiarity, experience, and logical reasoning. Often, however, small-scale natural processes can be readily mistaken for various anthropogenic fills. Details of the stratigraphy of anthropogenic sediments are not recognized and are regularly lumped together under a generic term, such as a 'floor' whereas in reality such features are composed of a complex of several microscopic layers that include proper floors, resurfacings, foundation (i.e. construction fill) layers, and occupational debris

trapped in between the floors. Such a latent microstratigraphy can indicate not only details of the technology and manufacture of constructions but can reveal household processes and the differential use of space that reflect cultural behaviours (e.g. Karkanas and Van de Moortel, 2014).

In this chapter we summarize and review the characteristics of these deposits and pay special attention to their geometry, fabric, and overall structure. We also briefly present some key features of their microscopic fabrics, as these are often mirrored in the macrostructure and fabric of the sediment, and therefore help in providing a better understanding the appearance of the sediment in the field. They also demonstrate the potentials of a more thorough and detailed sedimentary analyses that will help us decipher the microstratigraphy of anthropogenic features.

3.2 Burnt Remains

Fire plays an important part in the archaeological record, and remains of fire go back over a million years (Berna et al., 2012). They are found in open-air sites, within domestic spaces (fireplaces), within and between buildings, as well as among industrial areas associated with constructed features such as kilns and burned rock ovens.

Recognition of burnt features in the field and whether they are intact/in place can range from straightforward to deceitful. Even careful field observations, such as colour, can lead to mistaken interpretations. Red colour can be produced by diagenesis and developed long after sedimentation (see also section 2.7.2). At a number of prominent prehistoric sites in particular, red colour constituted a major line of evidence to infer the past existence of hearths and fire, which was revealed by subsequent studies to be unfounded (Goldberg et al., 2001; Weiner et al., 1998; Stahlschmidt et al., 2015).

Small-scale (~1–2 m^2) combustion features are typically – and almost colloquially – referred to as *hearths*. This word can denote different things to different people (Mentzer, 2014). A common notion of 'hearth' relates to a stone, brick, or concrete fireplace used for cooking food (https://en.wikipedia.org/wiki/Hearth#Archaeological_features) or activity associated with metal (http://www.merriam-webster.com/dictionary/hearth). In an archaeological context, a hearth is considered a place for making purposeful fire, whether built on an unstructured/untreated surface or within a constructed space such as adobe or brick ovens (http://archaeology.about.com/od/methods/fl/Hearths-Archaeological-Evidence-of-Fire-Control.htm). Similarly it is defined as 'the site of an open fire…represented by ash, charcoal and discolouration… [and possibly] slight structural additions such as clay flooring or a setting of stones around it.'(Bray and Trump, 1982: 109). Finally, from prehistoric perspectives the term 'hearth' can refer to a domestic, intact fire feature undisturbed by processes such as water flow and containing '...some or most of its original structural or compositional elements (e.g. organic matter and overlying ash) (Dibble et al., 2009: 185) or '...relatively small surface features used for short-term dry-heat cooking, warmth, and light....' (Black and Thoms, 2014: 203). In light of this broad usage, here we use hearth to mean *a purposefully prepared, constructed (e.g. lined with clay or rocks), or exploited feature (e.g. a fire built on an unprepared substrate) used in a domestic (i.e. non-industrial) context that should have many of the original elements present that characterize a fire, including ashes, charcoal, and perhaps reddened substrate* (Figure 3.1).

Hearths serve a variety of purposes. For the modern Hadza, for example, combustion features serve as (1) household hearths in camp, (2) sleeping fires inside their huts, (3) temporary communal cooking hearths in camp, (4) tuber roasting fires away from the camp, (5) meat cooking fires away from the camp, (6) special monkey-roasting fires, (7) torch-making fires to get honey away from the camp, (8) fires to light cigarettes, (9) night and day ambush hunting hearths, (10) fires for straightening arrows and curving bows, and (11) burning grass fires (Mallol et al., 2007: 2036).

Additional uses of hearths are for warmth, light, smoking meat, keeping away of insects, grease rendering, heat-treating flint, hide processing, modifying wood for tools, protection from predators, and clearing sites of refuse (Rolland, 2000; Sandgathe et al., 2011a; Wadley et al., 2011).

Archaeologically speaking, a hearth represents a relatively short-term event that can indicate a single short-lived fire or one that existed over a longer period of months where it was maintained; it can also be the locus of repeated fires made over returned visits to the site. For hunters and gatherers, it may be the focus of other

activities (Binford, 1996), whereas for later periods it may be associated less with domestic and more with industrial activities that last for longer periods.

Function aside, a fire from a simple hearth is made by placing the material to be combusted (e.g. wood) on a suitable substrate, which can include a natural material (soil, sediment, bedrock (Karkanas et al., 2004; Sherwood and Chapman, 2005; Sherwood et al., 2004) or remains from a previously used hearth; in the latter case, the fire can be started on unmodified hearth remnants of a previous fire or on those which have been spread out in preparation for the subsequent fire (hearth rake-out). The surface can be flat or concave to bowl shaped, if the substrate has been excavated or if a basin-like surface was exploited (Figure 3.2; Goldberg, 1977). In addition to wood, peat, grasses, and

Figure 3.2 Bowl-shaped hearth in a sand substrate (S), also covered with sand (S). Quick burial resulted in preservation of its original structure with a reddened substrate (R) overlain by ashes with charcoal (black) and reddish brown burnt bone (with arrow). Note that the lower part of the ashes is darker than the upper because of incomplete combustion at the lower part. Epipalaeolithic, Moshabi XIV, Gebel Maghara, Northern Sinai, Egypt.

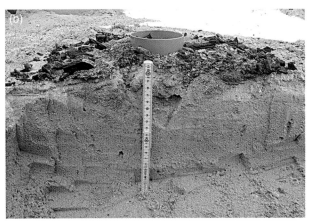

Figure 3.1 (a) Open hearth in an olive field for burning pruned branches. Ash is concentrated in the middle, whereas charcoal is mainly in the periphery and under the ash. The cut in the middle reveals a darkened soil substrate. (b) Experimental hearth on sand showing a well-developed reddened substrate (see Aldeias et al., 2016).

herbaceous vegetation (e.g. sedges) are commonly burned either as a means to start a fire or they may be present in animal dung, for example, which is a widely known fuel worldwide and still used today (Miller, 1984; Shahack-Gross, 2011).

If the fire continues to complete combustion, and all the wood is consumed, the remainder is white/grey ash, which is composed of calcite (see also section 5.2). On the other hand, if combustion is incomplete charcoal remains in addition to ash.

Burnt remains are also associated with constructed fireplaces and other thermal installations such as ovens and kilns. They are usually made of clay-rich sediment with or without masonry stone and have various shapes and forms according to their use, for example baking (usually dome-shaped), roasting (flat, platform-like), pottery (updraft and downdraft kiln types), and lime (large stone kiln constructions). Stone-rich combustion features are also common, including hearths and earth ovens. Stone hearths date back to the late Pleistocene in Europe and continued in use until the Holocene (Thoms, 2009). A classic example of early stone hearths is from Abri Pataud (Dordogne, France) where in the Late Aurignacian and Perigordian layers, for example, depressions contain river cobbles that are fractured, and reddened or blackened (Movius, 1966) presumably by heat; as noted by Thoms (2009) other types of hearths and hearth-like features of similar Pleistocene age and younger occur as concentrations of fire-cracked rock in the Middle East, southern Africa, Australia, and the Far East. Constructed clay hearths are also known from the Aurignacian in Klissoura Cave I (Greece) (Figure 3.3) (Karkanas et al., 2004).

In Western North America, earth ovens are widespread and their aspect of slow cooking among heated rocks serves to extract more calories from a given food and makes it more digestible when contrasted to short-term cooking above flames or coals (Black and Thoms, 2014). Moreover, many of these earth ovens have produced charred remains of underground storage organs and bulbs among others (Black and Thoms, 2014). Stone heating elements – fired *in situ* or elsewhere – were mainly used in closed pits and mounds (Thoms, 2008, 2009). Circular pebble hearths made of river pebbles or field stones are also known in the Levant (Gur-Arieh et al., 2013).

Recognition

Field A simple hearth in profile will commonly look like this from the base upward (Figures 3.2, 3.4) (Courty et al., 1989; Mentzer, 2014):

- A basal substrate that is commonly reddened by the heat, although the reddening is not always present and its thickness can vary because of the character of the substrate (e.g. composition), the environment (water content), and the duration of the burning event.
- A thin basal layer that may include reddened aggregates of the underlying substrate.
- A black, charcoal-rich layer mixed with overlying ash and material from beneath it. The black, charred material results from burning of the organic matter in the soil or charcoal produced by the fuel (Mallol et al., 2013; March et al., 2014).
- At the top is grey-to-white calcitic ash, with darker colours representing a greater presence of charcoal incorporated into the ash. Ashes are often dotted with reddened, burnt soil lumps mostly derived from soil attached to the fuel (Figure 3.5a).

Fresh ash is white, loose, and friable, and consists of microscopic, silt-sized calcite ash crystals; in the field it

Figure 3.3 Bowl-like, constructed clay hearth inside ashes. Aurignacian, Klissoura Cave I, Greece.

Figure 3.4 Sequence of single hearths each one having a reddened substrate (R), a dark layer, rich in charcoal (D), and a white ash cap (A). MSA, Site PP5-6, South Africa. From Karkanas et al. (2015a).

Figure 3.5 (a) Resin impregnated slab of hearth remains dotted with reddened soil lumps (arrows). MP, Hayonim Cave, Israel. (b) Resin impregnated slab of an intact hearth structure with bedded ashes (A), burnt shell (S), and bone (B) lying on a reddened and oxidized substrate (S) consisting of remains of a previous hearth. MSA, PP5-6, South Africa.

has a dusty appearance when dry, and is not altered or recrystallized. Charcoal has a conchoidal fracture, characteristic of the way brittle materials break. This feature differentiates it from decayed plant material relatively readily.

Commonly, fires are built one on top of another, either directly or on a substrate of redistributed combusted products (hearth rake-out), and repeated burning produces ash layers that show crude internal bedding (Figure 3.5b). Sequences of relatively intact hearths appear stratified with alternating black, grey, whitish, and sometimes orange or reddish layers. Individual hearths appear lentoid in profile but overall sequences of hearths can form continuous layers covering large areas of some prehistoric sites (Figure 2.8) (e.g. Kebara, Theopetra, Grotte XVI; Meignen et al., 2007; Karkanas et al., 2002, 2015a).

In addition to charcoal and ash in the case of these relatively simple hearths, the reddened components described above can appear as inclusions, depending on the setting. Additional inclusions can comprise bones that were thrown into the fire as refuse before, during, or after the fire. Bones that are completely burned may indicate simple tossing into the fire or they may have been used for fuel; in the latter case, highly fragmented pieces of bone would occur, particularly spongy bone, since cortical bone is not an effective fuel source (Théry-Parisot and Costamagno, 2005). In fact, the presence of abundant spongy bone in prehistoric fireplaces argues for its use as a fuel. Calcined bone, with its characteristic whitish appearance, may indicate high temperatures and/or prolonged time of burning (above 600–700°C; see Nicholson, 1993; Snoeck et al., 2014) or that, in the case of humans, purposeful cremation as is commonly observed in later sites (an exception is the cremated

bones excavated in Lake Mungo (Bowler et al., 2003)). On the other hand, the presence of unburnt bones shows that they accumulated after the fire was extinguished, whereas partially burnt ones (e.g. the ends of long bones) may indicate different former positions within the fire, for example at the edge of the hearth or possibly part of the fuel.

Assessing whether bones have been burned *can* be straightforward, but often it is difficult to evaluate and, if bones are burned, to what degree. Change of bone colour is often used to portray the heating of bone. In the field, for example, Stiner et al. (1995) used field experiments to relate bone colour to the relative degree of burning (Table 3.1; Stiner et al., 1995: 226). Additionally, they showed that bones that are heated to beyond 'color code 4' tend to become smaller and even powdered when affected by processes such as soil compaction and trampling (Stiner et al., 1995: 229).

Table 3.1 Visual appearance of different degrees of burnt bone (modified from Stiner et al., 1995: 226).

Color code	Description
0	Not burned (cream/tan)
1	Slightly burned; localized and less than half carbonized
2	Lightly burned; less than half carbonized
3	Fully carbonized (completely black)
4	Localized, less than half calcined (more black than white)
5	Less than half calcined (more white than black)
6	Fully calcined (completely white)

Other diagenetic factors can also mimic heating by darkening the surface of the bone. One common process is the precipitation of manganese (Mn) oxides (also iron oxides), which frequently leads to the conclusion that bones have been burnt. A quick field protocol that can help identify the presence of Mn is to place a few drops of 30% hydrogen peroxide (H_2O_2) on the bone, which will effervesce vigorously if MnO_2 is present.

The recognition/identification of hearths involves a number of issues, including those that are taphonomic in nature, such as visibility, obtrusiveness, and degree of preservation. The first involves whether they are originally present or not as a result of human activities (e.g. clearing: Mallol et al., 2007) or whether or not they have been modified by any of a number of post-depositional processes. These include *bioturbation, trampling,* and *diagenesis* (see also section 2.7).

Bioturbation mixes material and results in a chaotic/heterogeneous mixture of the original hearth components (e.g. ash, charcoal, burned bone). Turnover and mixing can be quite deep, on the order of centimetres to decimetres, as in the case of burned rock middens (Thoms, 2015; Black and Thoms, 2014). Bioturbation results in a loose, crumbly, or granular soil structure with aggregates of mostly hearth residues (Figure 3.6).

Trampling (see section 3.5.3) will also result in the redistribution of hearth materials, but in contrast to bioturbation the structure is generally more compact (Figure 3.7). In addition, trampling of a hearth will produce a distinct layer and surface whose upper boundary with the overlying sediment will be distinct. Bioturbation, on the other hand, can produce more blurred zones, with irregular upper and lower boundaries.

Diagenesis and soil formation (pedogenesis) can modify or dramatically alter the original mineralogical and organic composition of the hearths (see section 2.7.2).

Ash – when loose – is fine-grained (silt size) and in the presence of water readily recrystallizes and becomes lithified, eventually producing a very hard rock that preserves the general structure of the hearth remains (Figure 3.8) (Shahack-Gross et al., 2008b). The colour generally remains the same but chroma is less vibrant. If not recrystallized, ash can be partially dissolved or compacted under the load of the overlying sediment, which becomes deformed, slumps, or produces mud flows if saturated with water (Figures 2.33b and 3.9) (Goldberg, 2000a; Karkanas and Goldberg, 2010a; Karkanas et al., 2015a).

Ashes are not stable in acidic conditions. Thus, in areas with acidic soils (e.g. spodosols), locally where the bedrock is acidic (e.g. granite), or when dripping water in caves is undersaturated with respect to calcite, the ashes will be dissolved, leaving no traces of the original hearth, except for the presence of other burned materials and relics (e.g. charcoal, heated clay). Moreover, both calcitic ash and bedrock can be mineralogically transformed,

Figure 3.7 Parts of intact hearth remains separated by trampled homogenized ashes (arrows). MP, Kebara Cave, Israel.

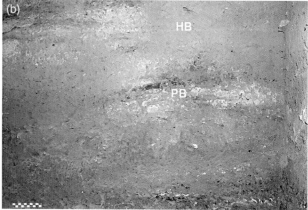

Figure 3.6 (a) Plan view of bioturbated ashes with filled burrows clearly visible (arrow). (b) In the profile, parts of the ashes are heavily bioturbated, producing a homogeneous massive aspect (HB), but in others the general structure of the burnt remains is still apparent (PB). MP, Klissoura Cave I, Greece (see also Figure 2.8).

Figure 3.8 (a) Strongly cemented hearth remains (H), bioturbated (B) and trampled (T) in places. Chisel marks are apparent. MP, Lakonis Cave, Greece. (b) Strongly cemented bedded ashes (A) from LP Qesem Cave, Israel.

Figure 3.9 Deformed and slumped burnt remains. Calcitic ash has been dissolved, which led to the slumping. MSA, Die Kelders, South Africa.

thus changing the appearance of the original material. This is not a rare phenomenon in prehistoric cave sites, where various types of phosphate minerals form by the reactions of phosphate-rich guano with ashes and clays (Karkanas et al., 2002; Schiegl et al., 1996; Weiner et al., 1995) (see section 5.2). These reactions can result in the simple replacement of ashes by phosphates (the minerals apatite/dahllite or leucophosphite) without destroying the original stratigraphic aspect and colour of hearths (Figure 3.10); similarly, the growth of nodular phosphates (e.g. taranakite) can displace and essentially 'explode' the original bedding and structure of the deposits, including hearths (Schiegl et al., 1996; Karkanas et al., 2000).

Likewise, organic matter and charcoal break down and can be totally or partially removed, thus eliminating or subduing the darker colour of the original hearth; such destruction is particularly significant for older prehistoric

Figure 3.10 Mostly intact ash remains that are completely phosphatized in which the original calcite ash has been transformed to leucophosphite (Fe-Al-phosphate) as confirmed by FTIR analysis. MP, Theopetra Cave, Greece.

sites. Agents of such breakdown include (a) biological activity in the substrate with the actions of small insects (e.g. mites) that ingest and excrete organic matter, and (b) repeated burrowing by cicadas and worms that breaks up charcoal and organic matter, which results in micron-sized specks of organic matter that ultimately produce a very blurred aspect of the original charcoal-burning event (Figure 3.11).

In addition, oxidation of organic matter can be manifested not only in a net loss, but also by the production of iron oxides and sediment reddening, most likely through the action of iron-reducing bacteria. Iron in reduced form is soluble and can be moved elsewhere, where it is oxidized and ends up as insoluble reddish hematite (Karkanas et al., 2000, 2002). Such reddish colouring has been observed in several Palaeolithic sites (e.g. Pech de l'Azé IV, Roc de Marsal, Grotte XVI, and Theopetra) and can be mistaken for reddening by fire (Figures 2.86 and 3.12). Nevertheless, in most cases this reddening is indirectly related to the presence of burnt remains, or organic matter in general.

Combustion features can be reworked at the place of original burning (raking out) or the combustion products can be remobilized to a different location. The former case is illustrated at the Palaeolithic site of La Ferrassie (Dordogne, France), where the original combusted components (charcoal and ash) of what appears to be a fire can be found constrained within small (~10 × 10 cm depressions): there is no stratified structure as in undisturbed hearths, and charcoal and burned bones are intimately mixed together. A similar type of 'modified' hearth also occurs at Pech de l'Azé I (Soressi et al., 2013). In addition, some small-scale reorganization can take place by cryoturbation and ice-lensing, which are best documented and revealed in thin section (see sections 2.7 and 5.2).

On the other hand, hearths can be substantially modified and become unrecognizable as distinct features.

Combustion products can be displaced to varying degrees and distances (centimetres to metres) from their original location by natural and human processes. If the ashes of bones, for example, are sufficiently concentrated, they may give the impression in the field that a hearth is still intact (Figure 3.13).

Burnt remains when reworked and mixed with sediment impose an overall grey tint to the deposit. Indeed, in arid or semiarid environments – including Mediterranean ones – a grey colour in the sediment can be immediately translated to the presence of finely comminuted and intermixed burnt remains. In more humid environments, decayed plant material may also be responsible for the grey colours. It is not far from the truth to say that in

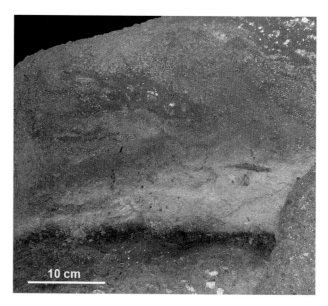

Figure 3.12 Reddening above ashes in intact hearth remains due to diagenesis (post-depositional chemical alteration) (see also Figures I.1b and 2.86, and section 2.7.2). MP, Theopetra Cave, Greece. From Karkanas (2017).

 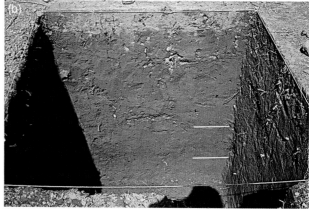

Figure 3.11 (a) Burrows (with arrow) homogenizing burnt remains from pithouses. Late Holocene, Keatley Creek, Canada. (b) Burnt layer (between lines) blurred by cicada burrowing. Trench is 1 m. Archaic, Woodland, Virginia, USA.

Figure 3.13 Accumulation of bedded burnt remains that are out of place. No original structure of intact hearths is recognized except partly at the far left. Compare with Figures 3.2, 3.4, 3.8, and 3.10. MSA, Site PP5-6, South Africa.

many archaeological sites of sedentary, preindustrial societies a considerable portion of the sediment consists of finely dispersed burnt remains (assuming the sediment is not decalcified and ash has not been dissolved away).

Dumping of hearth materials is one such activity (see sections 3.5 and 5.2). Dumped materials can be recognized in the field by their massive or weakly bedded nature, the uniform but heterogeneous distribution of ashy components, and by their very loose nature if not altered (e.g. cemented). Dumps can include all or some of the original hearth elements, but typically ash, charcoal (if combustion is incomplete), and possibly burned bones, seeds/plant remains, and ceramics with a rather chaotic appearance. At Kebara Cave, for example, the remains of an ~80-cm thick accumulation of bedded ashes occur along part of the back wall of the cave; at the time of excavation, they were tentatively assigned to

repeated and long-term burning, but they are now better interpreted as Neanderthal ash dumps (Figure 3.14a) (Meignen et al., 2007). Similar dumped accumulation of massive, centimetre-thick beds of ash and bone can be found in Upper Palaeolithic layers at Üçağızlı Cave, which appear to be ash middens (Figure 3.14b) (Goldberg, 2003; Kuhn et al., 2009; Baykara et al., 2015).

Other disruptive hearth activities can include maintenance activities such as sweeping, removal of fuel for recycling, raking out the hearth in preparation of the next fire, or destruction. Some of the remains of these processes (e.g. sweeping) may end up in the dumped material described above (see section 3.5). Depending on the depth, lateral extent, and energy of the activity, more than one hearth can be truncated and displaced. Deposits that have been modified by rake-out are horizontal, generally loose, weakly bedded mixtures of hearth products

Figure 3.14 (a) Thick, monotonous bedded, essentially pure ashes dumped at the back of Kebara Cave, Israel (MP). (b) Thin beds and thick accumulations of dumped ashes at UP Üçağızlı Cave I, Turkey. Ashes are homogenized and do not show any indication of *in situ* features.

(e.g. bone, ashes, lithics) or any other debris that might have been in previous fires (Figure 3.15). In the field, they can be mistaken for intact, possibly trampled, hearths because of their richness in bone, ash, and aggregate components. In the altar of the sanctuary of Zeus at Mt. Lykaion, Peloponnese, Greece, 'something like 700 m^2 is covered in ash and pulverized bone from burnt animal sacrifice' (Romano and Voyatzis, 2014: 579). The area of intense burning, ~1.5 × 0.9 m, contains about 0.30 cm of ash mixed with fire-cracked limestone, charcoal, seeds, and large amounts of burnt animal bone fragments. Calcined bone is found in large quantities in the sand- and silt-sized fraction of the burnt remains together with ash aggregates and fat-derived char (Mentzer, 2014; Mentzer et al., 2017). The altar represents an area of repeated burning events beginning in the sixteenth century BCE and continuing to the late fourth century BCE (Mentzer et al., 2017).

Hearths can also be reworked – or totally removed – by natural processes. Wind (deflation) can carry off the lighter, ashy components, for example, leaving a lag of the heavier fractions (bones, lithics, soil aggregates). Water can erode hearths completely and redeposit even heavier fractions elsewhere. In Kebara Cave aggregates of Middle Palaeolithic hearths (phosphatized ashes, aggregates of clay and charcoal were redeposited further back into the cave during the Upper Palaeolithic as finely laminated sheetflow deposits (Goldberg et al., 2007) (see section 5.2).

In addition to the fired clay constructed elements of thermal installations, the arrangement of stone and burnt remains can often be used to identify them. At the Neolithic (PPNB) site of Kfar Ha'Horesh (Galilee, Israel),

what appears to be a lime kiln was suggested by the presence of weakly demarcated irregular circular to oval features composed of stones of burnt and fire-cracked limestone, many of which were severely heated. These were distributed as piles within and surrounding poorly defined shallow depressions (Goren and Goring-Morris, 2008).

Hot rock features are morphologically diverse and occur as pits, platforms or heaps of stones; they range in diameter from ~2 m up to 10+ m across (Figure 3.16). A typical earth oven is vertically organized, and its structure is shown in Figure 3.17. Once open, and if not reused and left exposed, which can be months to years, organic decomposition takes place and root growth begins to take place that mixes materials, including those which may have been discarded there (e.g. bones, plants, lithic debris). With time, the material sags, leaving a depression. In more dynamic environments, however, such as floodplains and along the base of slopes, ovens can be readily buried (Black and Thoms, 2014).

Due to their size and coarse nature of the stones within them, burnt rock features usually are not transported laterally over large distances (> metres). Stones can be moved locally, however, by small-scale slumping or by gravitational settling, or by humans during removal of the food or during reuse. Such activities can result in a superposition of finer, organic-rich sediment on to the basal stony layer and may mimic the lime plaster kilns superficially. However, the layering of stones in earth ovens tends to conform to the shape of the underlying excavated depression (see figure of Thoms, 2015), whereas in lime kilns the stones do not. In addition, in sandy terrains, for example, bioturbation (floralturbation

Figure 3.15 Thick sequence of crudely burnt remains with some faint lenses with intact hearth structure (arrow). Such remains are interpreted as hearth rake-out. MP, Pech de l'Azé IV, France.

Figure 3.16 Extensive accumulation of white heated limestone blocks related to stone hearth ovens. They are found inside dark brown sediment rich in burnt remains, shown better at the edges of the stone accumulation. Late Paleoindian, Wilson-Leonard, Texas, USA.

(a)

(b)

(c)

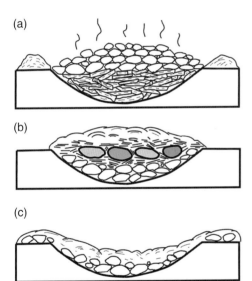

Figure 3.17 Organization and structure of a typical earth oven. (a) Firing stage showing wood fuel to heat cooking stone. (b) Baking stage with food packages in between green vegetation. (c) Abandonment stage after removing of the food package. After Thoms (2015).

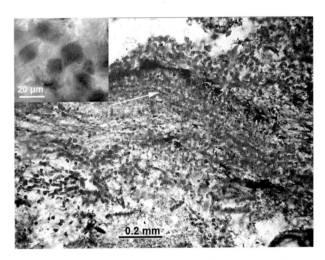

Figure 3.18 Ash crystals (arrow) produced by experimentally burnt wood. Note the arrays of dark rectangular ash crystals following the plant structure. Inset shows rhomboid and rectangular ash crystals consisting of micrite aggregates.

by roots and tree throws) can result in disarticulation of the ovens and movement of the stones to a depth of 40–70 cm (Thoms, 2007).

Microstructure Ashes are composed of aggregates of micrite (microcrystalline calcite). They may vary in colour from white, grey to yellow, and brown, depending on burning intensity (Mentzer, 2014; Wattez, 1988, 1992; Shahack-Gross and Ayalon, 2013). The micrite occurs in rhombic forms (Figures 2.90 and 3.18), which are derived from the transformation of calcium oxalate (pseudomorphs of

micrite after calcium oxalate, the POOC of Brochier and Thinon, 2003), which is found preferentially in bark and leaves (Brochier and Thinon, 2003; Canti, 2003b; Shahack-Gross and Ayalon, 2013). In addition to the rhombic types, triangular and lozenge shapes occur (Mentzer, 2014), but these are less common. The rhombic pseudomorphs consist of very fine nanometre-sized crystallites, which plainly differentiate them from geogenic calcite monocrystals (microsparites and sparites) (Figure 3.18). Thus, their interference colours are a palimpsest of coloured spots of second-order interference colours.

On the basis of experimental studies, it seems that ash pseudomorphs are preferably formed in low-temperatures fires (~500°C); at higher temperatures (above ~800°C) pseudomorphs start to decompose into small calcite particles, often appearing as fused droplets (Shahack-Gross and Ayalon, 2013). Note that a normal campfire reaches temperatures up to 800–900°C, but this peak temperature is not maintained during the whole burning event, and often fires do not reach such a high temperature for numerous reasons (wet fuel, strong winds, wrong fuel, and fire maintenance). High-temperature types of ashes have been reported from metallurgical and ceramic kilns (Courty et al., 1989: 110). Ashes of ceramic hearths have a heterogeneous fabric and a vesicular aspect, and are strongly disaggregated and disorganized. Indication of high-temperature burning at 1000–1200°C includes melted quartz in the form of 'dewdrops' or bubbled quartz (Figure 3.19).

As will be discussed in Chapter 5, the presence of ashes – while important in itself – is essentially just an indication that burning (usually wood) took place at the site. Perhaps more importantly, however, is the internal, microscopic organization of the ashes and other burnt components, which reflects on the original integrity of the combustion feature from which the ashes were derived. Articulated plant tissues (including articulated phytoliths and partially combusted organic material), as well as the presence of contiguous ash rhombs within a thin section, suggest that the material is intact, with minimal displacement/disturbance since the time it was burned (Figure 3.20a). Overall, the ash crystal arrangement forms a crudely bedded configuration (Figure 3.20b). In addition, the presence of a reddened, oxidized substrate

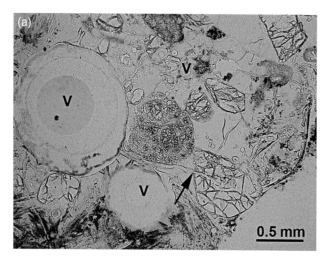
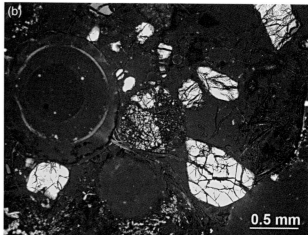

Figure 3.19 Photomicrographs of melted quartz in (a) PPL and (b) XPL. Quartz crystals appear densely fractured with fissures filled with melted isotropic material in XPL (arrow in PPL). Note somewhat fuzzy boundaries of quartz within a glassy isotropic matrix, and large and small vesicles (V). Viking Age, Hedeby, Northern Germany. Courtesy of Svetlana Khamnueva.

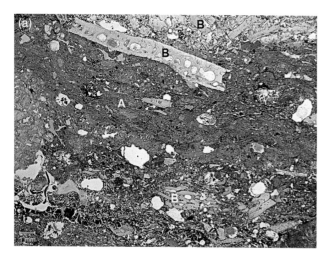
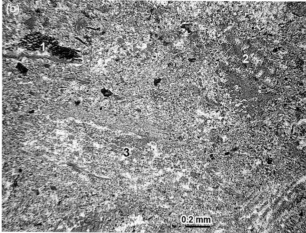

Figure 3.20 Ashes. (a) Microscopically bedded ashes (A) on top of pre-existing burnt remains. Bones (B) burnt at different temperatures are shown. MP, Roc de Marsal, France. (b) Photomicrograph of articulated ash crystals. Semi-ashed wood is shown in (1), well-organized ashes pseudomorphic after plant structure in (2), and slightly disorganized ashes in (3). Such ash structure suggests *in situ* hearth remains. MP, Klissoura Cave I, Greece.

associated with the burnt remains is an unmistakable sign (Figures 3.1, 3.2, and 3.4). Interestingly, although red heated substrates are easily identified in the field, under the microscope the change in colour almost passes unnoticed unless observed with oblique incident light (OIL) or incident dark field illumination (IDF). The reason is that neoformed hematite crystals are very fine, in very low quantities, and disseminated within the matrix, and therefore the colour does not change much under transmitted light.

On the other hand, loose aggregates of ash chaotically mixed with burned bone fragments or burned clay, for example, point to reworking of original combustion features, perhaps by burrowing, colluviation, trampling, dumping, or rake-out during hearth maintenance activities (Goldberg, 2003; Meignen et al., 2007; Mentzer, 2014; Sherwood, 2008). All these processes that alter the original structure of the burnt remains are discussed in detail in the relevant sections of the aforementioned processes. Nevertheless, generally, the degree of compaction decreases from trampled and raked-out hearths to very loose dumped hearth remains. Dumped hearth remains are very spongy, resembling air-fall aggregates (Figure 3.21) (Meignen et al., 2007). Raked out and trampled hearths are relatively 'in place', whereas dumped hearths are mostly completely disorganized without preservation of any fragile ash structure as viewed microscopically with IDF.

In certain cases under the microscope, it is possible to observe partially carbonized wood such that the original wood has not been completely transformed into ashes (calcite) (Mallol et al., 2007; Mentzer, 2014) and where plant tissues are still preserved (Figure 3.20); they would be destroyed if combustion were to be complete or reworked. Where combustion is incomplete (e.g. lack of oxygen or premature extinguishing of the fire) the residue is charcoal, which is composed of an organized phase, with graphite-like microcrystallites, and a non-organized phase. Whereas modern charcoal has a higher proportion of the former, the opposite is the case with fossil charcoal, which has a higher proportion of the non-organized phase (Cohen-Ofri et al., 2006). Charcoal, seed, and phytoliths can be studied for identifying the source and type of material originally burned. Such information could be used to identify possible fuels.

Excessive heating of phytoliths can partially or completely melt them and even turn them into a vesicular, silica slag, an amorphous and isotropic phase with large and small rounded vesicles (Cabanes et al., 2010; Gur-Arieh et al., 2014; Macphail and Goldberg, 2010; Mentzer, 2014) (Figure 3.22). In the latter case, excavations at the Bronze

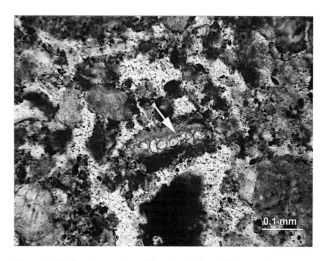

Figure 3.22 Burnt remains with melted phytolith (arrow) producing a vesicular glassy phase. Neolithic Makri, Thrace, Greece.

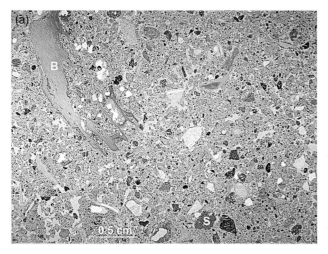 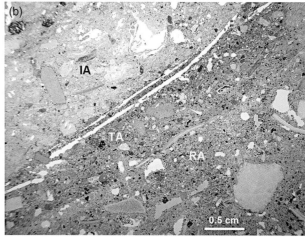

Figure 3.21 (a) Dumped loose aggregated ashes mixed with bone (B) having various orientations, burnt soil (S), and charcoal (with black). UP, Klissoura Cave I, Greece. (b) Compacted, trampled burnt remains (TA) over disorganized, slightly looser raked-out ashes (RA), overlain by *in situ* ashes (IA) with intact ash pseudomorphs after plant structure. UP, Klissoura Cave I, Greece.

Age site of Tel Yinam revealed a 2–3 cm thick layer of blue-green frothy glass that extended over an area of ~2 m^2. It exhibited the remains of siliceous plant fibres representing the original material from which the glass was produced (Folk and Hoops, 1982).

Sediments can take on a dark appearance for a number of reasons (Mallol et al., 2013; Mentzer, 2014). These include reduced iron produced by water saturation and the presence of organic staining from humates, for example. In addition, particulate matter will induce a dark colour, comprising aged (humified) and combusted/charred plant materials. Humified matter can be differentiated from charred material using OIL or IDF: charcoal appears black, opaque, and is often characterized by highly reflecting dots, whereas decayed matter appears transparent to semi-opaque and has dark brown colours (Figure 3.23). Some metal oxides are also black and opaque, but they do not have a cellular plant structure and normally are not elongated or platy in form; hematite appears red with IDF.

Most recently, the field of organic petrology has highlighted the presence of black grains of fat-derived char that are associated with former burning activities (Goldberg et al., 2009a, 2012; Mallol et al., 2013; Miller et al., 2010; Villagran et al., 2013). Char is gelified organic matter originating from the burning of fat from bones or meat. Due to the small size of the material, it is generally difficult to observe it directly in field, but it can be isolated by sieving and is visible in thin section. Microscopically, it can coat grains or be found as isolated fragments that are typically fissured/cracked and contain vesicles (Figure 3.24) (Villagran et al., 2013). It is more readily apparent when blocks of indurated sediment are polished and examined in reflected light, where it is strongly reflective.

Since char is related to cooking, it should normally be associated within hearth and combustion feature deposits. In the case of intact hearths, char occurs around grains as it was dripped from above, or as individual vesicular pieces mixed with other grains. The latter situation would suggest that it formed nearby and was displaced very locally, on the order of millimetres to centimetres. Both kinds of char are found within hearths from Castanet, an Upper Palaeolithic site in the Dordogne (southwest France). Similarly, at the nearby Middle Palaeolithic site of Pech de l'Azé IV, pieces of char are scattered throughout the organic-rich combustion features. Although intact hearths occur, many of the materials resulting from combustion there (e.g. ashes, burnt bone, rare charcoal) have been redistributed through the process of rake-out by spreading out the cooled materials of a former fire in preparation for the next one (Dibble et al., 2009; Goldberg

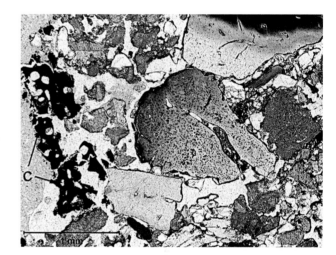

Figure 3.24 Black charred fat (C) with vesicles surrounding and coating a burnt bone. UP, Castanet, France.

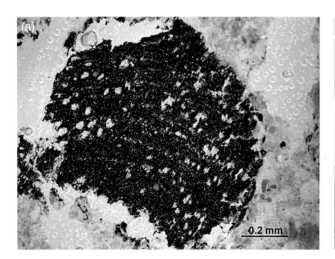

Figure 3.23 Photomicrographs under incident dark field (IDF) of (a) charcoal showing a metallic aspect and plant structure, and (b) decayed organic matter showing a brownish black amorphous appearance.

et al., 2012). Interestingly, at Pech de l'Azé IV, it is the presence of char and not charcoal that provides the darker appearance to the layer in the field.

The recognition of burnt peat and peat as a resource in general is beyond a curiosity. In Iceland, for example, peat as a combustible may be related to the associated function, such as industrial use vs. domestic use in hearths (Sawyer, 2016; Simpson et al., 2003). At the Viking site of Reynistaður in Iceland, light-coloured layers in the field were thought to represent the remains of wood ashes. In thin section, however, calcite was totally absent, and instead these layers were composed of phytoliths and diatoms, and were ascribed to remains of burnt peat rather than burnt wood, a feature not previously recognized on the island. Moreover, this

observation has ramifications in terms of resource preferences and fuel availability (Sawyer, 2016). In thin section, the constituents of fresh peat appear as decomposed stems, roots and leaf tissues, and commonly diatoms, with an amorphous fine fraction that at higher magnifications is composed of a gel-like mass of finely divided plants, phytoliths, and diatoms (Bal, 1973; Fitzpatrick, 1993; Fox and Tarnocai, 1990). When burned, peat can undergo a number of transformations with loss of organic matter, and enrichment of silica and some calcite (Figure 3.25). Simpson et al. (2003) experimentally burned peat at both 400°C and 800°C, and examined thin sections from the combusted material. At 400°C the mineral fraction was reddened as viewed in OIL, and at 800°C – after the organic material had been combusted – the fine fraction had a characteristic yellow colour in OIL. At 400°C, woody materials is associated with '…crystallitic groundmass b-fabric dominated by calcite and by the frequent occurrence of macro-sized black charcoal fragments' (Simpson et al., 2003: 1409). Moreover, siliceous elements (i.e. phytoliths and diatoms) become more dominant.

Burnt bone is identified relatively easily under the microscope. In plane-polarized light, unaltered bone is typically yellow to yellowish brown, but upon heating at low temperatures it becomes pale orange, often with brown edges; at 400–500°C it is completely brown and locally blackish brown due to carbon accumulation. The characteristic ropy histological structure of bone disappears. At even higher temperatures, bone becomes pale white and shows bright white low-order interference colours (Figure 3.26). Burning also produces cracking, some of it characteristically emanating from the Haversian canals and becoming prevalent and wider in higher temperatures (Hanson and Cain, 2007; Courty et al., 1989).

Figure 3.25 Burnt turf consisting of compact fine charred burnt remains well mixed with silty sand and occasionally some larger plant particles. Late Antique, Tarquimpol, France. Courtesy of Richard I. Macphail.

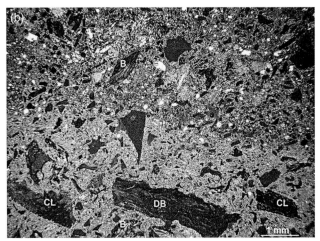

Figure 3.26 (a) Bones burnt at differing degrees, PPL. With increasing intensity of burning from beige (BE), to brown (BR), dark brown (DBR) and black (BL). UP, Castanet, France. (b) Photomicrograph of burnt bones in XPL. Brown bones (B) retain their ropy structure and have birefringence, dark brown bones (DB) have almost lost their birefringence, and calcined (CL) are practically isotropic. Archaic, Kalapodi, Greece.

Chemistry and Mineralogy As described above, ash is mainly composed of calcite, a common mineral found in various geological and archaeological materials as well. Regularly, different kinds of calcite present different crystallinity ranges (atomic order), size of crystals, and habits. Through the use of sophisticated Fourier transformed infrared (FTIR) analysis it was possible to differentiate among the various types of calcite, including calcitic ash (Regev et al., 2010). Calcitic ash is formed in two different ways. At around 500°C the calcium oxalates of plants (CaC_2O_4) are directly transformed to calcite ($CaCO_3$) by loss of monoxide (CO) (Huaqing, 1989). At higher temperatures (~750°C), the calcite undergoes decomposition to CaO and CO_2, absorbs water (humidity), and recarbonates to form calcite (Boynton, 1980). These two types of calcitic ash show different crystallinities, the high-temperature calcite being much more disordered at the atomic level and comparable to other pyrogenic forms of calcite such as lime plaster (Poduska et al., 2011; Regev et al., 2010; Shahack-Gross and Ayalon, 2013). In principle, these differences can work well in several cases but recrystallization of ash and certain disordered types of biogenic calcite can obscure identification of ash.

Experimental results indicate that the two types of ash formed at low and high temperature, respectively, show a significant difference in carbon and oxygen isotopic composition (Shahack-Gross and Ayalon, 2013). Indeed, the high-temperature ash shows the same isotopic composition as lime plaster, since both form in the same way (see section 3.4 on the formation of construction materials). It was also found that at low-temperature burning the carbon isotopic composition of ash may reflect the photosynthetic pathway of plants (C3 vs. C4 plants), opening up the possibility of differentiating between ashes of C3 and C4 vegetation.

The crystalline structure of bone apatite can be modified by heating, resulting in distinct changes in X-ray diffraction (XRD) and FTIR patterns (Shipman et al., 1984; Stiner et al., 1995). However, such mineralogical changes can result from weathering and interestingly Stiner et al. (1995) note that 'Our infra-red data on splitting factor and carbonate content did not reliably diagnose burning on prehistoric bones…' (p. 235). Nevertheless, FTIR is probably the most widespread analytical approach to document heating of bones (Berna and Goldberg, 2008; Surovell and Stiner, 2001; Thompson et al., 2009; Weiner, 2010), whereby heat alters the crystalline structure of the bone. At present, however, only ranges of temperatures can be inferred and not specific heating temperatures.

A sophisticated and experimental approach to evaluate manganese staining vs. blackening of bones by burning was taken by Shahack-Gross et al. (1997). Using FTIR and an electron microprobe to differentiate the two, they noted '…that it is very difficult, if not impossible, to differentiate reliably by eye between bones that are black due to staining, or bones that are black due to burning, or both' (Shahack-Gross et al., 1997: 443).

Relation to Other Archaeological Materials and Artefacts

Several other types of archaeological materials are affected accidentally or on purpose when burnt in a fire. Here we describe the alteration features of the most common materials other than those described already.

Chert and Other Lithic Materials The surface and physical properties of stone artefacts (observed in the field), particularly siliceous materials such as flint and ironstone, commonly show vestiges of burning (although we avoid here a discussion of natural vs. human fires). Modifications include, for example, 'breakage, spalling, crenulating, crazing, pot-lidding, microfracturing, pitting, bubbling, bloating, smudging, discoloration, adhesions, altered hydration,…and weight and density loss' (Deal, 2012: 98). In investigating Mesolithic hearths in the sandy northwest European plain, Sergant et al. (2006: 1000) found that weakly heated artefacts (flint and quartzite) showed little damage '…except for a weak reddish shine and a few isolated cracks'. Increased heating resulted in the formation of cracks, (non-specified) colour changes, and the formation of pot-lid fractures (centimetre-sized, circular, convex fragments without bulbs of percussion that become detached from flat surfaces); heavy burning resulted in white and grey colours resulting from total dehydration. Similar types of effects of fracturing, pot-lidding, and colour have been documented by Deal (2012), who also notes internal changes in lustre with heating; she also discusses comparable changes for other lithic material such as obsidian, which can bubble, melt, or bloat.

In the archaeological record, the presence of pot-lidding has been one piece of evidence to infer the presence of early fire used by hominins. It includes the presence of pot-lids developed on banded ironstone slabs from ~1 Mya layers at Wonderwerk (Northern Cape, South Africa), which occurred in the same layer as burned bone and possible calcitic ash crystals, and is perhaps one of the earliest records of fire (Berna et al., 2012). Similarly, researchers observed pot-lidded microdebitage from the slightly younger site of Gesher Benot Ya'akov (Jordan Valley, Israel) to document former repeated fires (Alperson-Afil et al., 2007; Goren-Inbar et al., 2004; Richter et al., 2011); this evidence has also been evaluated using thermoluminescence (TL) (Richter et al., 2011).

It should be noted, however, that other processes may produce small flakes and spalls that resemble pot-lids. These include salt weathering, for example, which is common both in desert environments (regolith soils; Dan and Yaalon, 1982), where there are many nights of dew, and in areas close to the sea. In addition, in desert environments diurnal contrasts can evoke insolation weathering. Thus, several lines of evidence and context are needed to demonstrate the production of pot-lids by past fires.

Limestone   Calcareous stone is often used to demarcate a fireplace or to form the base of a stone oven. Based on the experimental work of Gur-Arieh et al. (2013) it appears that a normal open fire is not capable of transforming limestone to lime even if the temperature of the fire exceeds the temperature of calcite decomposition (i.e. above ~700–800°C). The evident explanation is the duration of the fire temperature above this threshold, which must be in the range of several hours (Figure 3.27). Constant temperatures at this range and duration require continual fuelling and very good maintenance of fire intensity, something that is generally not expected in campfires or cooking fires. Nevertheless, duration depends on the size of the limestone clasts and the type of limestone (see Karkanas, 2007). Limestone pebbles are found cracked in the centre of the fire irrespective of their size (volume range 200–20 cm^3), but at the periphery of the hearth where the temperature was less than about 300°C, pebbles did not crack. Similarly, pebbles in the centre were oxidized (reddened) whereas at the periphery they were only blackened due to incomplete oxidation (burning) of the organic matter (Gur-Arieh et al., 2013).

Insights into the degree of movement of heated rocks around hearths can be gained through studies of rock magnetism of individual stones within a feature and can also offer similar insights into heating history and site formation. Gose (2000), for example, examined the magnetic

properties of the individual stones from burned rock features. Prior to burning in the oven, the magnetizations of rocks lying on the surface should be distributed randomly, displaying magnetizations inherited from the time of their formation. If the rocks are then heated above 580°C (the Curie temperature for magnetite (680°C for hematite)), they will acquire the same thermorenanent magnetization that is parallel to the existing field at the time of heating (Gose, 2000). Undisturbed and oriented samples collected from individual rocks and measured in the laboratory can reveal how many and which particular stones were moved after they were heated.

Sediment and Earthen Constructions   When heated, clay undergoes a series of temperature-dependent transformations. At about 450–600°C, clays lose their structurally bonded water (dehydroxylation). In kaolinite, this transformation occurs in lower temperatures whereas in other clays, such as illite and montmorillonite, transformation is more gradual and occurs at higher temperatures. At around 900°C new minerals are formed at the expense of clay and other silicate inclusions, with the formation of a glassy phase at the same time. XRD, FTIR, and differential thermal analysis (DTA)/TGA (thermogravimetric analysis) have been used to identify the temperature of heating of sediment in archaeological contexts (Karkanas et al., 2004; Berna et al., 2007; Gur-Arieh et al., 2013; Friesem et al., 2014a).

Substrates of hearths normally do not show such transformation, as the temperature even 1 cm beneath is below 500°C most of the time (Canti and Linford, 2000), although see recent experimental data of Aldeias et al. (2016). Burnt soil lumps also do not often show burning at temperatures high enough to be identified by FTIR, for example. Sediments altered at higher temperatures

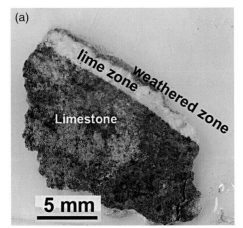  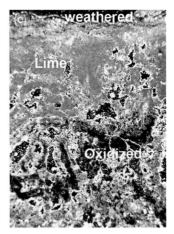

Figure 3.27 Limestone (travertine) masonry affected by an apparently prolonged conflagration event: (a) field sample, (b) thin section scan, and (c) photomicrograph in XPL. A few millimetres thick white outer rim in the sample turned out to be heat-altered to lime. Inside, a thicker oxidized zone is observed. Classical, Kalapodi, Greece.

are mostly associated with kilns and ovens. These are not necessarily reddened, as they are mixed with other burnt materials (charcoal and ash), giving an overall grey aspect to the sediment instead (Berna et al., 2007; Canti and Linford, 2000).

Clay constructions such as ovens and kilns are regularly affected by the high temperature of burning. Through time, the mudbrick or clay-rich earthen walls are transformed into fired brick and may become reddened and vitrified, and in some cases the walls may be coated with glass resulting from the reaction of silica in ashes (Nicholson, 2010). Large pottery kilns and installations decay similarly to other adobe structures (see section 3.4.2) as the construction materials are basically the same. However, in the case of kilns, the adobe does not decay in the same way as the bricks are consolidated by heating and the surface facing the fire can be vitrified or coated with silica. These aspects reduce the effects of breakdown of the brick components by slaking by rainwater, but more likely by gravitational deposition of whole pieces of adobe.

3.3 Organic Remains and Human Activities

3.3.1 Biological Constructions (Matting, Roofing)

Vegetal bedding is reported to be widespread among contemporary hunter gathers, and the use of more elaborated constructions such as carpets, rugs, and mats is not unknown. Several cases of thick mats of vegetal matter have been reported from Holocene and Palaeolithic sites, and have been interpreted as being remains of bedding (Miller and Sievers, 2012, and references therein). Covering a floor with fresh grass provides comfort, pleasure, and insulation, and bedding was used for a variety of purposes such as sleeping, sitting, and resting (Nadel et al., 2004). Most of the time, the evidence is indirect, as biological constructions consisting of organic matter, mainly reeds, sedges and grasses, readily undergo oxidation and decay, and thus leave very little, if anything, behind. Only under very special conditions, like complete absence of humidity (Nahal Hemar; Bar-Yosef and Alon, 1988) or anoxic environments (Ohalo II; Nadel et al., 2004), can organic matter be occasionally preserved for a long time. In most cases, only when bedding material is burnt can we possibly identify evidence of human practice (Goldberg et al., 2009a; Cabanes et al., 2010). The same is true also for other biological constructions like matting and roofing, although as these are associated with buildings they are mostly preserved when buried by volcanic ashes (Akrotiri; Shaw, 1977) or when they are

burnt, which leaves their ashy remains or imprints in their fired clay substrate. In rare instances, impressive mats are already preserved on floors inside buildings from Pre-Pottery Neolithic in Jericho (Kenyon, 1981). Recognition of such constructions – when not preserved under unique conditions – normally requires microscopic analysis for their identification because they leave only faint traces in the form of a very fine burnt layer. The very few well-studied cases of good preservation provide a hint of what may represent the traces of bedding and roofing in other cases.

The most impressive bedding remains come from the previously submerged 23,000-year-old Ohalo II site, at the shore of the Sea of Galilee in Israel (Nadel et al., 2004). The submerged anaerobic conditions resulted in excellent preservation of the organic matter. A discrete, ~1 cm thick horizontal layer of bedding was identified at the bottom of a dark anthropogenic layer that was identified as an occupational floor. The bedding layer was composed of bunches of partially charred grass stems and leaves, arranged in a repeated pattern around a central hearth inside the remnants of a brush hut.

An even older case (about 60 ka) is that of Sibudu Cave in South Africa, which preserves unprecedented details of bedding activities and their maintenance (Goldberg et al., 2009a). In the field, thick, fine laminated burnt sequences appear to represent a set of colourful palimpsests containing many interfingering hearths, white, grey, and orange ashy lenses, and black and brown organic-rich layers (Figure 3.28). Some of these combustion units seemed in the field to represent intact hearths, but their lenticular nature makes it difficult to trace them horizontally for more than a few tens of centimetres (Wadley, 2006). Detailed micromorphological analysis, though, revealed a somewhat different picture (Goldberg et al., 2009a). Three distinct microfacies were identified associated with burnt bedding (Figure 3.29). The lower one consists of laminated fibrous charcoal, derived from sedges. The middle consists of laminated fibrous charcoal and laminated sedge phytoliths and the capping one solely of laminated sedge phytoliths. The whole unit was the result of *in situ* burning. Sedges collected from the adjacent river were used to prepare bedding that was continuously trampled and compacted to produce the laminated microstructure. Microscopic trampled stringers of bone and chipped stones inside these burnt bedding layers suggest that sedges were repeatedly laid down to refresh the bedding surface; at times, the bedded was intentionally burnt, possibly for eradicating pests. Subsequent experiments conducted by Miller and Sievers (2012) confirmed that the laminated layers 'were most likely produced by human activity related to the construction, maintenance and burning of bedding'. Other associated microfacies suggest also sweeping,

Figure 3.28 Field photograph of colourful bedded burnt remains as a result of bedding activities and their maintenance. A micromorphological monolith sample is also shown. MSA, Sibudu Cave, South Africa.

raking out, and dumping activities as well adding to the maintenance practices of the Middle Stone people at Sibudu Cave (Goldberg et al., 2009a; Miller et al., 2010).

Evidence of matting remains is mostly identified in the form of millimetre-thick articulated phytoliths on top of floors inside buildings (Matthews, 2010). However, it is not always possible to securely associate this occurrence with matting unless relatively large areas are evenly covered with articulated phytoliths of plants that can be woven. Remains of other fibre-based construction (e.g. baskets, textile, netting) can produce similar configurations. Fodder bedding in stables is also characterized by layers of articulated phytoliths but the associated microfacies are totally different, as discussed in section 3.3.2. Other indirect evidence is derived from processes related to mat maintenance such as sweeping, which leaves very finely sorted dusty material on the top of floors (Figure 3.30) (see section 3.5).

Often roofs are constructed solely of vegetal material, or in some cases the vegetal matter provides the substrate upon which a mud ceiling is plastered. Thatching is a term often used for all forms of roofing consisting of vegetal matter such as reeds, sedges, grasses, heather, willow, straw, or even seaweed. Turf has been used as roofing and building material mostly in northern European and American environments. It is characterized by a dark colour, high organic content, and biogenic soil features related to a top soil: roots, root channels, and a high amount of excremental pedofeatures (Goldberg and Macphail, 2006: 279).

Very rarely burnt remains have been securely assigned to vegetal roofing. In the Pre-Pottery Neolithic site of Tell Qarassa in South Syria, occurring between a clear arrangement of burnt roof beams is a layer of ash and large fragments of charcoal – some of which are carbonized wooden elements – that is mixed with heavily burnt and reddened adobe (Balbo et al., 2012). Under the microscope, three sublayers were identified: the lower one contained a fine sediment groundmass with burnt wood from the roof beams, the middle one consisted of phytolith-rich ashes, and the top represented the fired daub. The phytolith sublayer contained two bands, an upper one with random orientations close to the fired daub and a lower one consisting of articulated and horizontally oriented phytoliths of leaves and culms of grasses. Inside the groundmass, sedge phytoliths were identified surrounding the burnt beams. It was suggested that the sedges might have been used to fill the gaps between the beams of the collapsed roof. Two layers of grasses in the form of hay or straw were laid to support the overlying daub without the need of twigs (Balbo et al., 2012).

A somewhat similar feature was identified in the Bronze Age site of Akrotiri, in Santorini, Greece, which was destroyed and buried under the volcanic tephra of the Theran eruption. A thin layer of vegetal matter was laid upon a grid of beams and poles in order to hold in place the poles and the earthen material that was placed above (Shaw, 1977). In another Bronze Age house at Palaikastro, Crete, fired collapsed roof remains have a first layer of clay with marks of reeds that formed the ceiling and then another layer of clay with imprints that imply mixing with seaweed, a tradition that was used up until recently (Figure 3.31) (Shaw, 2009).

3.3.2 Stabling

Enclosures for holding livestock appear early during the Neolithic as a by-product of the domestication of animals (Stiner et al., 2014). Terms for describing these enclosures that confine animals include pens, corals, and stables. From the Neolithic up to the present, sheep, goat, pigs, cattle, horses, and other animal pens are ubiquitously found inside sites or in special locations such as caves or isolated enclosures. The accumulating organic-rich deposits (the ubiquitous *fumiers* in the Mediterranean Basin) are the result of the patterns of

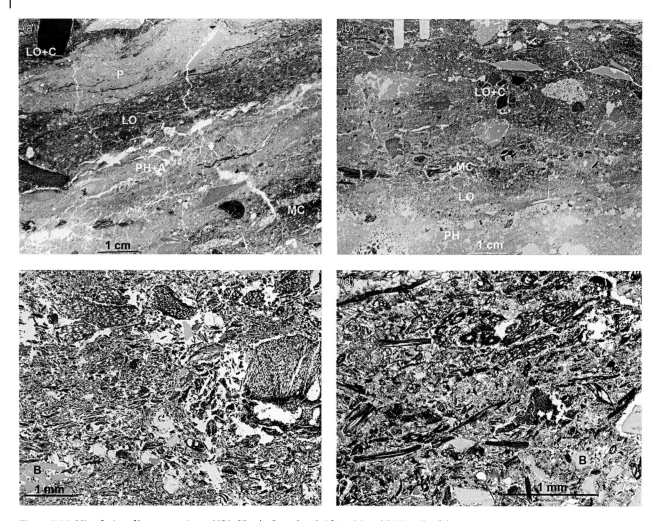

Figure 3.29 Microfacies of burnt remains at MSA, Sibudu Cave, South Africa. (a) and (b) Details of thin section scans showing sequence of thin layers of massive charcoal (MC), calcareous and phosphatized ashes (PH+A) (PH = phosphatized guano), laminated organic matter (LO), phytoliths (P), and laminated organic matter with charcoal (LO+C). (c) Photomicrograph of laminated organic matter with some burnt bone (B) overlain by massive charcoal. PPL. (d) Photomicrograph of laminated organic matter with charcoal and some burnt bone at the bottom (B). PPL.

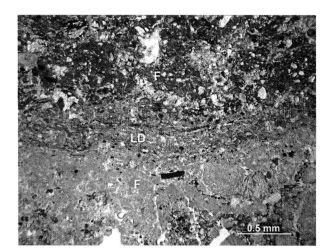

Figure 3.30 Photomicrograph of laminated dusty material (LD) trapped in between two constructed floors (F) related to mat maintenance such as sweeping. Neolithic Makri, Thrace, Greece.

Figure 3.31 Roof sequence containing seaweed from a recent abandoned house at Skyros Island, Greece. R = reeds, SW = seaweeds, CR = stony coarse sediment, C = clayey sediment.

human and animal relationships with the environment (Figure 3.32). The identification of stabling remains in archaeological sites can be used to 'reconstruct the subsistence base of the sites' inhabitants in the past' (Shahack-Gross, 2011; see also Macphail et al., 1997); therefore villages and early urban centres are universally characterized by penning activities. On the other hand, later urban centres – after having been gradually transformed to 'cities' after the second 'urban revolution' around the late sixth century BCE – often moved such activities to special places or outside to the country (Albert et al., 2008). Nonetheless, reappearance of animal husbandry in urban centres is observed during the post-Roman Period (Macphail, 2008).

Although special lots and structures have been used for enclosures, yards and abandoned or temporarily used buildings – including houserooms – were often exploited

as well (Rowley-Conwy, 1994). Ground levels in multi-storey houses and occasionally barns were also regularly used for keeping animals. Caves represent a special case as they have been intensively used as pens in the history of animal husbandry (Angelucci et al., 2009; Brochier et al., 1992).

Stabling and penning result in accumulations primarily of animal dung mixed with fodder. The shape and fabric of dung pellets or pats depends on the animal and on its diet (Courty et al., 1991). Herbivores can be browsers or grazers and consequently fodder includes straw, leaves, or even fine branches and bark depending on the animal diet, but also the feeding practices of the surrounding (Courty et al., 1991; Macphail et al., 1997; Akeret and Rentzel, 2001). Omnivores essentially fed on seeds, fruits, and vegetables. Thus, various enclosure deposits include not only different types of fodder but also dung. Amounts of clastic minerogenic sediment often find their way inside enclosures as they are attached to the fodder, which is often eaten by the animals as well (Macphail et al., 1997). Aeolian activity and rainwash are additional contributors of natural clastic sediments, as most enclosures are open or semi-open spaces.

Nevertheless, the primary processes for the final appearance of stabling deposits are animal trampling, post-depositional decay, and deliberate burning by humans, especially of the organic matter (Shahack-Gross, 2011). Trampling is very intensive and in herbivore enclosures it primarily affects fine plant material (Figure 3.33). In omnivore pens such as those for keeping pigs, clastic sediment is derived from the substrate and will be equally affected and intermixed with organic matter. Organic decay and chemical diagenesis takes place as the pen deposits are accumulating, but these processes act rapidly during abandonment phases, with rates depending on the climatic conditions of the

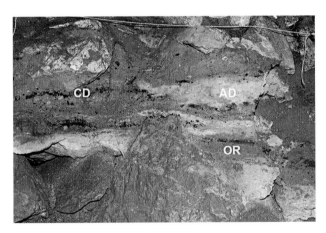

Figure 3.32 Example of *fumier*, layered burnt dung remains, consisting of organic-rich layers (OR), ashed dung (AD), and stringers of charred dung pellets (CD). Neolithic, Leontari Cave, Attica, Greece.

Figure 3.33 (a) Compacted dung (CD) in a recently abandoned pen at Franchthi Cave, Greece. From the surface view it is not apparent that there is dung underneath, but the cracking is typical of dried dung accumulations which reveal the dung content. (b) Detail of compacted dung from a recent pen composed of fibrous trampled dung and fodder. Levadia area, Greece.

area (e.g. Shahack-Gross et al., 2003; Shahack-Gross, 2011). Spontaneous burning of dung deposits has been observed in modern enclosures (Shahack-Gross, 2011). However, in several places around the world burning of dung pen deposits is a practice already known from the Neolithic, their deposits being among the best-studied sequences of their kind (Courty et al., 1991; Macphail et al., 1997; Akeret and Rentzel, 2001; Boschian and Montagniari-Kokelj, 2000; Karkanas, 2006; Angelucci et al., 2009).

Coprolites are mineralized fossil faeces, but often the term encompasses ancient herbivore dung as well. In the archaeological record coprolites are typically fossilized carnivore faeces, including human excreta. Carnivore coprolites are ubiquitously found in Palaeolithic caves where animals such hyenas competed with hominids for shelter (Macphail and Goldberg, 2012; Horwitz and Goldberg, 1989). As such, they cannot be considered anthropogenic sediment but they are presented here for the sake of completeness. Carnivore coprolites are also found in later periods associated with domesticated animals such as cats and dogs. Their study and analysis aim to determine their feeding habits and social life (Toker et al., 2005). Examinations of human coprolite material can shed light on prehistoric diets and reveal the presence of certain microorganisms, bacteria, and pathogens. Phytolith, pollen, bone, botanical, chemical, and aDNA analysis, among others, have been employed in the study of human coprolites (Bryant and Dean, 2006). Carnivore and human coprolites are often found dispersed into archaeological sediments and therefore rarely form archaeological deposits by themselves.

However, human coprolites when found in privies and latrine waste constitute the dominant component of these deposits (Macphail and Goldberg, 2010).

Recognition

Field The remains of stabling, when the organic content is still preserved, can be occasionally recognized by their likeness to dung pellets (Courty et al., 1989, 114), and the shape of the faeces can be used to identify the animal species (Chame, 2003). Ovicaprines and generally all Artyodactyles except Bovinae (cattle, buffalo, and bison) produce dark, small centimetre-sized, cylindrical or rounded pellets usually pointed at one end and concave in the other extremity (Figure 3.34). Cattle and generally all Bovinae produce dark green to brown relatively large flattened faeces that accumulate in circular piles, known as cowpats – rarely preserved as such in the archaeological record – and found mostly in waterlogged environments (Akeret and Rentzel, 2001). Faeces of horses and the family of Equidae are firm, single reniform (kidney-shaped) droppings usually several centimetres in diameter (Figure 3.34). Pig dung is usually amorphous and liquid.

Herbivore dung pellets consist mostly of plant fragments, and their size varies among species. Clauss et al. (2002) observed that browsers have a higher proportion of large plant particles in accordance with their lower fibre digestibility. Several millimetre-long particles are found in large quantities in cattle dung but less so in sheep, for example. On the other hand, fine amorphous dark brown organic matter constitutes a large proportion of the ovicaprine matrix (Courty et al., 1989: 114; 1991; Macphail and Goldberg, 2010).

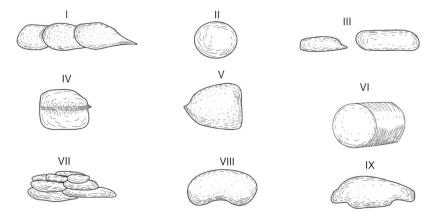

Figure 3.34 Faecal morphometry for conspicuous species of terrestrial mammals: I, cylindrical (sausage-shaped) = carnivore order; II, well-round little and single pellets in small or large accumulations = Lagomorpha (hares and rabbits) and some ungulates as the hyrax and aardvark; III, single and cylindrical pellets with one extremity slightly tapered = rodents; IV, single cylindrical pellets, inflected, and with characteristic furrow along the length (coffee bean shape) = African rodents (great canerats); V, cylindrical or rounded pellets usually pointed at one end and concave in the other extremity = artiodactyls, except Bovinae; VI, big and cylindrical like bars = elephants, hippopotamus, rhinoceros and South American anteaters; VII, flattened faeces that accumulate in circular piles = Bovinae (cattle, buffalo, and bison); VIII, single reniform (kidney-shaped) = Equinidae and warthog (Suidae); IX, mixed shape and size, amorphously cylindrical or rounded = opossums, primates, armadillos, and insectivores. Adapted from Chame (2003).

Pig dung is rich in both amorphous organic matter and an amorphous fine mineralized matrix. It may also contain remnants of partially decomposed plant material and amounts of fine clastic mineral particles (Macphail and Goldberg, 2010). Normally, pig dung is found intermixed with the sediment substrate, forming dark brown, deformed and fragmented crust-like features produced by heavy trampling.

Carnivore faeces are cylindrical and sausage-shaped, with segments that have a tapered end at one of the extremities (Chame, 2003). They are characterized by their animal and fish bone content and their compact beige fine mineralized matrix. Human coprolites are rather amorphously cylindrical but without any common and specific characteristic are often misidentified with other carnivore or even herbivore dung (Bryant and Dean, 2006; Goldberg et al., 2009b). However, human coprolites have a yellowish colour and do not normally contain coarse bone and much clastic mineral material. In latrine wastes and cesspits, coprolites occur in the form of yellowish cemented crusts (Macphail and Goldberg, 2010; Macphail, 2017).

Highly trampled dung remains may not be preserved as intact pellets or pats. In general, stabling organic-rich remains are in the form of layered accumulations of compact, finely bedded plant remains with a fibrous structure. The overall geometry of the remains can be flat or in the form of a shallow heap. Boundaries in organic-rich dung deposits are always clear, although the upper one can be gradational, a result of natural and biological processes during the abandonment phase. Often, in dry environments and when stabled inside structures, the surface of the dung accumulation can be smooth and polished due to severe trampling and foot traffic

(Figure 3.33). Since these accumulations include fodder and bedding, the sizes and distribution of plant particles are not so helpful for determining the animals kept in the enclosure. In decayed and thus organic-poor dung remains, field observations alone can very rarely provide evidence of penning. In some cases, diagenetic features such as horizons of phosphate or gypsum nodules and filaments, along with other concentrations, may indicate the presence of decayed dung (Figure 3.35). In caves, stone-paved surfaces are almost ubiquitously associated with penning activities. Polishing of bedrock walls and blocks through the action of fine particles within the fleece of the animals (Brochier et al., 1992) is a subtle indication of penning. Restructuring and maintenance of these dry surfaces can produce a depositional sequence with scattered but clustered horizontal concentrations of relatively flat stones (Figure 3.36a).

Sequences of intentionally burnt dung remains in caves have a characteristic layered and colourful appearance (*fumiers*) (Figures 3.32 and 3.36) (Boschian and Miracle, 2007), and these stratified sequences normally comprise couplets. The lower basal unit consists of dark brown organic-rich, partially charred sediment that may still preserve dung remains and pellets, albeit in various stages of decomposition (Figure 3.36b). This part represents the lower, urine-stained, and relatively unburnt part of the sequence but its presence depends on the frequency of burning and the humidity inside the cave (Macphail et al., 1997). Plant remains can still be recognized in this part of the sequence, such as small twigs and leaves. However, in other cases blackish layers represent the mostly charred basal part of the sequence.

The upper part, overlying the organic-rich base, is a layered brown grey, bluish grey, pinkish to whitish layer

Figure 3.35 Banded accumulation of white nodular gypsum (arrow) related to animal penning. Classical, Drakaina Cave, Greece. Tools in the background for scale.

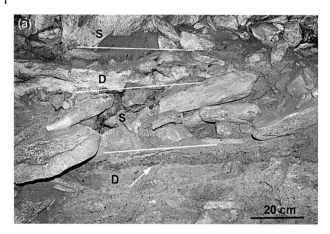
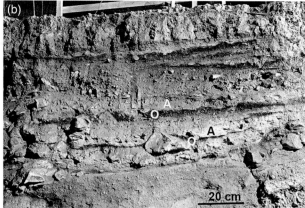

Figure 3.36 (a) Layers of burnt dung (D) separated by layers of predominately flat stone structures (S). Neolithic, Leontari Cave, Attica, Greece. (b) Bedded burnt dung remains from Medieval, recent Theopetra Cave, Greece. Each layer consists of a dark organic-rich base (O) and light-coloured ashed dung at the top (A).

of ash (Figure 3.36). Charred and ashed coprolites may be still recognizable, and the sediment can be punctuated by black charred dung pellets. The grain-size of all sediment is in the silt to sand range with generally few dispersed charcoal particles. Overall, the deposits are organized in planar, centimetre- to decimetre-thick bodies that display quite a lateral continuation that cannot be the result of household fireplaces. Concentrations and layers of flat stones may interfinger with the burnt dung remains. The layering is subhorizontal, but often follows the topography of the cave floor and can form heap-like or flat features.

Microstructure The macroscopic shapes of the different dung faeces are expressed microscopically as well (see also section 2.6.1). Ovicaprine pellets, for example, consist of convoluted fine, relatively decayed (fermented) plant remains in a porous microstructure dominated by brown amorphous organic matter (Figure 2.65b). Cattle dung consists of longer and fresher plant fragments that have a microlaminated smoothly folded appearance in a porous microstructure without strong convoluted structures (Figure 2.65a and 2.66b). Plant remains are surrounded by brownish to yellowish isotropic organic matrix, rich in phosphates; plant structures are still discernible. Microcharcoals, burnt and unburnt small bone fragments, diatoms and sponge spicules, and silt to sand-sized mineral particles are often found in all dung faeces that were included within the food and water of the animal (Macphail et al., 1997; Akeret and Rentzel, 2001; Shahack-Gross, 2011; Macphail and Goldberg, 2010).

Characteristic inorganic components associated with the plant materials are several microscopic features that comprise calcium oxalates, opal phytoliths, and dung spherulites (Figures 2.66a, 2.67a and 3.37) (see also section 2.6.1). The first two are inherited from the vegetal fodder whereas the last is produced inside the animal digestive system (Brochier et al., 1992; Canti, 1997, 1998, 1999). Calcium oxalates are organominerals commonly found in leaf material in many species of dicotyledonous plants. They appear in crystal druses and have characteristic high birefringence colours. Phytoliths are most prevalent in grasses and can be used to reconstruct the animal diet (Albert et al., 2008; Tsartsidou et al., 2009; Lancelotti and Madella, 2012). Dung spherulites are one of the most commonly used traces for identifying dung in archaeological sites (Canti, 1997, 1998, 1999; Shahack-Gross and Finkelstein, 2008; Albert et al., 2008; Lancelotti and Madella, 2012). Spherulites (Figures 2.67a and 3.37b), however, are not always present because animals do not always produce them and they are prone to chemical dissolution in even slightly acidic environments (Canti, 1999).

Pig dung contains larger amounts of amorphous organic matter embedded in an amorphous, isotropic phosphate matrix (Figure 3.38). It is often intermixed with clastic mineral sediment making for intercalations with vegetal remains (Macphail and Goldberg, 2010). As described in section 2.6.1 carnivore coprolites are characterized by a beige-grey to dark dotted grey phosphate matrix, which after fossilization matures to apatite mineral phases (Figure 2.68a). The groundmass thus has an undifferentiated isotropic appearance. Carnivore coprolites are often found in the sediment as yellow to beige, angular to subrounded fragments and hence can be easily differentiated from phosphate nodules (Figure 3.39). Bone content is partially dissolved, often becoming amorphous. Voids and vesicles – characteristic features of hyena coprolites – are probably produced by gastric gases, whereas elongated pores are presumably imprints of decomposed hair and fur. In addition, plant remains, pollen, and phytoliths are occasionally found

Figure 3.37 (a) Photomicrograph of dung spherulite-rich layer (between lines) with characteristic colourful dotted appearance (S) and some phytoliths in between (P). XPL. Late Bronze Age, Mitrou, Greece. (b) Photomicrograph mainly of large oxalate druses (OX) and small spherulites in between (arrows). Neolithic, Makri, Thrace, Greece.

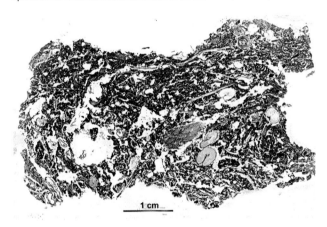

Figure 3.38 Pig dung from a recent pen in the recently abandoned village of Kranionas, Greece. Dark isotropic, organic-rich matrix with dispersed fibrous plant matter in a relatively open, porous structure.

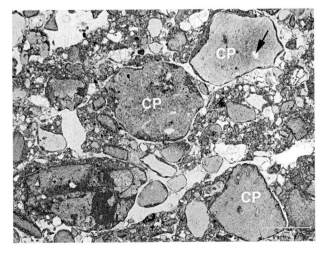

Figure 3.39 Dusty beige, rounded carnivore coprolites (CP) from Roc de Marsal (MP), France. Note the vesicle (arrow) in the one from the upper right, a feature that is common in such coprolites, which are isotropic in XPL (not shown here).

(Horwitz and Goldberg, 1989; Fernández Rodrigues et al., 1995; Karkanas and Goldberg, 2010b; Macphail and Goldberg, 2010).

Human coprolites have a phosphatic matrix similar to those of carnivore coprolites but usually have a stronger yellowish colour and are often heavily stained and impregnated with iron-manganese oxides due to the oxidation of urobilinogen, a metabolic product of the breakdown of red blood cells (Figure 2.68b) (Bryant and Dean, 2006). Clastic material content is low and bone remains are few and smaller. If cellulose is present it is much fresher than in herbivore dung. Parasite analysis of nematode eggs is often helpful in identifying human coprolites (Macphail and Goldberg, 2010).

Trampled modern and archaeological organic-rich dung deposits as well as archaeological organic-poor dung deposits appear microlaminated (Shahack-Gross, 2011). Organic-rich dung sequences show a dense platy structure separated by planar voids (Figure 3.40). They consist of horizontally bedded plant fragments, occasionally including seeds and roots. Linear arrangements of articulated phytoliths and spherulites are observed after the ingested organic grass fibres have degraded, and are often associated with laminae of amorphous organic matter (Shahack-Gross et al., 2003). Burnt dung remains are composed of similar laminae of articulated phytoliths. Ash derived from vegetal material like straw and faecal spherulites may be widespread. Layers of wood ash are often found on the top of burnt sequences, probably representing burning of wood material for igniting the fire.

Nevertheless, these sequences often appear truncated and mixed with other types of deposits, such as dumped materials or domestic phases (Macphail et al., 1997). They are often associated with homogeneous deposits, mostly made up of dung spherulites, phytoliths, and other

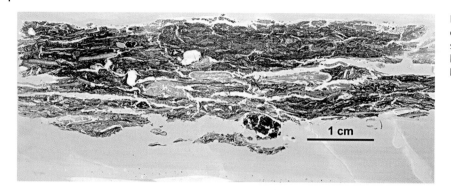

Figure 3.40 Banded appearance of dense compacted dung remains and fodder separated by planar voids. Basement of house, recent abandoned village of Kranionas, Greece.

components that are randomly dispersed in the groundmass, which can be enriched in fine clay material; it also displays a microgranular aggregation that is probably the result of weathering and trampling of a mixture of unburnt and partially burnt dung deposits (Angelucci et al., 2009).

Decay of the organic matter is part of the diagenetic process that results in the formation of new, 'authigenic' minerals whose composition depends upon the chemical variables of the sediments (see section 2.7.2; Karkanas et al., 2000; Shahack-Gross et al., 2004b). Apatite, vivianite, and other phosphate minerals are the most prevalent mineral phases that form where dung has degraded (e.g. Brochier et al., 1992; Macphail et al., 2004; Shahack-Gross et al., 2004b; Karkanas and Goldberg, 2010b). Hydromorphic features such as manganese mottling are also prevalent due to water and urine circulation through the organic matter. Authigenic gypsum is often found in decayed dung deposits in relatively dry environments and together with dung spherulites are often the only indications of the former presence of dung (Karkanas, 2017). Elevated $\delta15N$ values of preserved organic matter are a good indicator of the presence of dung (Shahack-Gross et al., 2008a; Shahack-Gross and Finkelstein, 2008).

Relation to Other Archaeological Materials and Artefacts and Processes

The average stabling remains contain relatively few other cultural vestiges, such as lithics and pottery, and often there are only very few faunal relics (Angelucci et al., 2009). Nevertheless, remnants of stabling are often interbedded with other occupational deposits, suggesting rotational use of space. Note that burials are often found in dung enclosures, which indicates the high value some societies put on animal dung (Hodder, 1987).

Dung has been widely used as fuel, fertilizer, and as construction material. Dung in the form of 'cakes' has a compact but porous structure that burns steadily, making it ideal fuel for cooking as well as for pottery firing (Sillar, 2000; Miller, 1984). Dung is widely used as filler in clay structures, giving extra strength and reducing cracking when dry. It is also currently used as modern material

flooring in a South African church, where Berna (2017) showed that particle size and orientation of the plant fibres were distinctive in comparison to those found in livestock enclosures. Therefore, identifying dung in archaeological sediments has enormous implications for the interpretation of the economic domain of ancient societies. The same criteria for identifying intact and stabling dung remains can be used to identify dung utilized in other activities.

3.4 Formation of Construction Materials

Knowledge of the formation of construction materials is fundamental for understanding site formation processes in architectural sites. Often, the overwhelming amount of deposits in such sites consists of decayed construction material such as mudbrick, mortar, plaster, and daub. Hence, the study of the nature and fabric of these materials can help in the identification of their decayed and weathered products that create the archaeological sediment of several sites. Sometimes the end product is similar to the raw material, the soil of the construction material, and therefore it is difficult to identify its anthropogenic origin. Nevertheless, identification of still intact parts can serve as a reference for the microscopic recognition of decayed parts (Goldberg, 1979b; Rosen, 1986; Friesem et al., 2011, 2014b; Goodman-Elgar, 2008; Macphail and Goldberg, 2010).

In the following sections we do not describe construction materials that are durable and accordingly easy to identify, such as ceramics, mosaic, stoneware, hydraulic mortars, etc.

3.4.1 Living and Constructed Floors

The study of occupational surfaces can provide the most impressive information for reconstructing everyday activities. In non-architectural sites the identification and definition of occupational surfaces, or 'living floors' is much disputed. The term was first used in ethnographic studies and implies a surface on which activities are contemporaneous. According to Villa (1976) a 'living floor' should be a thin layer of medium artefact density that represents more or less contemporaneous human activities. More recently Malinsky-Buller et al. (2011: 90) further constrained the term 'living floor' as a temporal construct, the defining criterion being that of a single episode of site use. It follows that a 'living floor' can be recognized by the preservation of objects and features. These include single, intact fireplaces, matted surfaces, and discrete bone and lithic patterns (Figure 3.41) where certain combinations of lithic categories coexist in the same location and are a product of certain activities such as knapping, butchering, etc. (Malinsky-Buller et al., 2011 and references therein).

Thus, 'living floors', if well preserved, should represent discrete, thin sedimentary horizons with certain characteristics that will differentiate them from overlying and underlying sediments. Their natural content will not be different from the substrate on which it has been formed, but it will be more compact due to trampling and more enriched in anthropogenic remains.

In practice, such occupational surfaces can be preserved and recognized only if they are quickly buried by natural sedimentary processes or specific human activities such as massive dumping. Indeed, the most conspicuous examples of 'living floors' are found in inundated coastal flats (Boxgrove: Pope and Roberts, 2005) and loess sequences (Händel et al., 2009). If buried by sediments derived from similar occupational processes, it will be difficult to isolate occupation episodes during excavation, although in some cases they can be separated microscopically, but again only locally.

Figure 3.41 Single intact hearth representing several fire episodes perhaps during a seasonal visit to the site; this suggestion is based on the thickness of the red substrate and ash accumulation (arrow). Such features can be associated with a 'living floor'. MSA, Site PP5-6, South Africa.

In more recent cultural periods, the term floor is often used to designate any occupational or walking surface, including open areas. Sometimes any interface that is thought to represent a depositional break is uncritically considered an 'occupational surface'. However, this is not often the case because any depositional break that forms a *de facto* surface for some time is not necessarily one used by people, as, for example, in the case of abandoned areas subjected to intermittent natural deposition. Occupational surfaces should have sedimentary characteristics of exposure and weathering, trampling, and increased accumulation of artefacts and other anthropogenic remains on them.

In sites with architectural remains, floors constitute an important component of house stratigraphy, for example. Floors are normally found in between walls and often demarcate habitation spaces. In some Neolithic sites where walls and postholes are totally destroyed or decayed, only floors can demarcate former houses. Floors can be also found in some open spaces such a yards and plazas where they define prepared or constructed surfaces.

Floors can be intentionally constructed sedimentary bodies or the natural substrate that may have been cleaned and levelled for occupation. The term 'beaten floors' has been used for some non-constructed floors (Macphail et al., 2004) essentially implying trampled occupational debris formed 'through gradual deposition of debris on a sequence of accruing surfaces' (Milek and French, 2007: 340). On the other hand, the term 'mud floors' has been used for a local substrate that has been intentionally prepared or modified by compaction, sweeping, or trampling (Courty et al., 1989: 124). Constructed floors can be composed of local soils and sediment, and have been called clay floors (Macphail and Goldberg, 2010), plaster floors (Matthews and Postage, 1994), or mud plaster floors (Moorey, 1994; Matthews et al., 1996). So, although any of the above names can be used, it is essential to indicate when a floor is intentionally constructed in order to clearly differentiate it from non-constructed floor sediments. Natural substrates prepared solely by cleaning and levelling are not included in constructed floors.

Constructed floors may have a preparation layer made of material different from the capping plaster floor (Figure 3.42a). The former can be composed of matrix-supported gravel (Moore, 2000; Karkanas and Van de Moortel, 2014), plant-tempered adobe (Macphail and Crowther, 2007), or simply levelled occupational debris (Matthews et al., 2005). The plaster surface is often finer-grained and can be clayey, or composed of gypsum or lime plaster (Moorey, 1994; Karkanas, 2007). Millimetre-thin lime wash, coloured clay, or ochre finishing coats are often observed (Figure 3.42b) (Karkanas and Efstratiou, 2009;

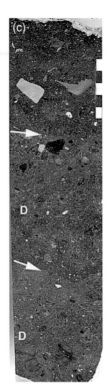

Figure 3.42 Scans of resin impregnated slabs of constructed floors. (a) Lime plaster caps (arrow and L) above preparation layers (PL), the upper one consisting of rounded gravel. LBA, Mitrou, Greece. (b) Sequence of plastered clay and lime floors with some red clay finishing coats (R). Neolithic, Makri, Thrace, Greece. (c) Thick constructed floors with greenish colour (rich in serpentinite clasts) in between dark occupational debris. The upper contacts of the floors are relatively sharp (arrows) and their lower ones diffuse (D). LBA, Mitrou, Greece.

Matthews et al., 1996). Although it is useful to separate the construction layer from the capping plaster floor itself, in practice it is not always possible. The coarse construction layer – often called the mortar floor – may have been used as the floor surface without any finely prepared capping plaster or finishing coat. Benches and platforms are often capped with a floor surface that is in some cases similar to the floor (*sensu stricto*) of the room or from a slightly different material, but the same logic and technique are employed.

Cases have been reported where floors are constructed by single and multiple quarried slabs of fossiliferous tabular formed tufa (Macphail and Crowther, 2007) or, more often, pure white calcareous sediment, which both resemble lime floors (Matthews and Postage, 1994). Fresh wetland turf has been used for constructing floors in northern cooler and damp environments (Milek, 2012), and dung is a basic component for relatively drier environments (Boivin, 2000). Floors can be constructed with pure natural soil, sediment, or gypsum and lime plaster, but they can be also mixed with 'occupational debris' in all kinds of combinations (Karkanas and Van de Moortel, 2014). Addition of straw, chaff, dung, wool, or more exotic components like bitumen provides extra strength, durability, and waterproofing capacities. Indeed, lime and gypsum plaster have to be combined with temper in order to become hard and durable, which is also the case with pure clay, which cracks when not tempered.

Based on ethnographic studies, the processes of making construction floors are rather simple and straightforward (Milek, 2012; Boivin, 2000; Matthews and Postage, 1994; Tsartsidou et al., 2008; PK, personal interviews). Preparation of the surface may include simple cleaning but may also encompass digging, wetting, non-mechanical compaction (e.g. using beaters), and levelling. These actions are followed by mixing sediment or soil and other additives (e.g. straw, dung) with water and stirring, kneading, and pugging in order to prepare a viscous muddy material to be applied on the ground surface. The material may be used immediately, but it can be left to stand overnight before being applied. Application is made mainly by hand or with leafy branches, but other tools such as shovels and plaster floats can be used depending on the local traditions (e.g. Shaw, 2009: 145). Stone tools may have been used for burnishing coats. In all cases, what characterizes the application is deformation and compaction of a semi-plastic, mostly clay-rich material under wet conditions. Thus, in several aspects, clay floors resemble natural mudflows in terms of their internal fabric.

Constructed floors can also made without adding water by simply covering the surface with a layer of sediment brought from somewhere outdoors. Even spreading and compaction is necessarily made by treading or using tools such as floor beaters. However, to our knowledge there are no ethnographic analogues of 'dry' constructed floors, perhaps because such a floor would not have the durability and evenness of a mud floor and would not be any different from the natural soil substrate.

The technology of production of lime and gypsum plasters is more complicated. They are prepared by heating calcareous and gypsum rock, respectively, at temperatures sufficient to decompose calcium carbonate (calcite) and calcium sulfate dehydrate (gypsum). For the production of lime (*sensu* quicklime), temperatures in the range of 800–900°C have to be maintained for a long time (several hours to overnight), but certain calcareous materials like tufa, calcrete, or calcarenite, which have been used in antiquity for lime production, need shorter times (a few hours) (Karkanas, 2007; Shahack-Gross et al., 2005). For gypsum (*sensu* plaster-of-Paris) much lower temperatures are needed, in the range of 100–200°C. Both materials are mixed with water, with lime requiring preparation in

advance for an even putty to be produced. However, hot mixing is often observed, that is, unslaked quicklime is mixed directly with aggregate (temper) and with enough water to work the material *in situ* without the laborious and time-consuming stage of putty (slaked lime) preparation (Karkanas, 2007 and references therein). The final stage of application of lime and gypsum plaster is similar to that of clay floor preparation. However, because of the highly caustic nature of lime, it could not be applied with bare hands.

As discussed in the introductory chapters, mosaics, tiles, and building stone floors are outside the remit of this book, and animal stable floors are treated in section 3.3.2.

Recognition

Field In the field, 'living floors' are very difficult to recognize, and only their overall content and geometry have been used to infer their presence. The existence of a thin cultural horizon sharply underlying natural sediments devoid of cultural material is a good candidate for further analysis. Non-disturbed 'living floors' have a sharp upper contact following the irregularities of their artefact content and have a gradational lower one. Originally, they develop horizontally, so to some extent they reflect the activities of the human occupants. The sedimentary matrix of the 'living floor' can be compacted due to trampling, particularly under wet conditions, but its short lifetime precludes deep and severe compaction. The organic cultural content of the sediment will often impart a darker colour to the layer depending on the nature of the activities at the site and the space where the floor is situated.

Since 'living floors' constitute part of a non-constructed, non-protected natural environment, they

Figure 3.43 Sequence of constructed floors. Note the one marked with an arrow having a planar shape, a sharp upper contact, and a wavy lower one. EBA, Mitrou, Greece.

Figure 3.44 Remnants of a plastered lime floor (arrow) preserved close to a wall. Late Bronze Age, Mitrou, Greece.

are prone to post-depositional alteration, which further hampers their recognition. Therefore, their identification will mostly rely on the patterning of lithics and other cultural material or the existence of single hearth constructions (Figure 3.41) (Machado et al., 2013).

Constructed floors are ~1–5 cm thick planar bodies of sediment with a sharp, knife-edge upper surface and a gradational lower one (Figures 3.42c and 3.43). They are often recognized by their compacted or even indurated nature relative to the overlying and underlying looser occupation debris or levelled construction fill, respectively. Normally, they are expected to be found immediately above the foundation part of a wall and are often better preserved close to walls or other constructions (Figure 3.44). If they are well constructed, upon excavation they separate easily at the contact between sediments above and below the floor. Most conspicuous are floors made of exotic material not found in the geological substrate of the site. Materials that have been often used are white marly sediment and red soils mixed with coarse components derived from the soil substrate such as schist, limestone, and quartz rock fragments (Figure 3.45).

Very commonly, floors are parts of multiple sequences in which they are interleaved with loose occupational debris that may have been levelled for building a foundation level for the overlying floor. In such cases, floor sequences appear as relatively thick horizontally bedded sediment (Figure 3.46). If they are not so well bedded – because of frequent destruction and repair – they can form variegated sequences consisting of mixed planar and lensoid bodies, as well as very thin streaks representing finishing coats (Figure 3.47) (see also Matthews et al., 1997).

Figure 3.46 Sequence of floors (F) appearing as crudely finely bedded sediment (between lines). Wind deflation and washing of the section by rain has enhanced bedding produced by horizontally, parallel-oriented schist fragments used for the construction of floors. These are not natural deposits because of the angular schist fragments, which are not native to the bedrock underneath this hill, and also the dense organization of the layers (see also Figure 3.45). EBA, Palamari, Skyros Island, Greece.

Figure 3.45 Constructed floor made of crushed greenschist (arrow), a raw material not found in the bedrock of the site. MBA, Palamari, Skyros Island, Greece.

Figure 3.47 Constructed floors. (a) Complex sequence of floors interbedded with thin occupational debris, burnt layers, and red clay constructed surfaces. Scale is 30 cm. (b) Corresponding scan of resin-impregnated slab showing the fine bedding of the sequence (lines connect some marker layers in the field). EBA, Mitrou, Greece.

Indeed, in relatively massive indoor sequences, the existence of horizontal, discontinuous parallel streaks of conspicuous sediment such as red clay can be a diagnostic element of amalgamated floor sequences (Figure 3.48).

Although floors have a characteristic internal fabric, it is not easy to identify it in the field, and so microscopic analysis is needed as described in the following section. A sequence of floors may also be separated by a thick construction fill that actually demarcates a new phase of house construction. Floors made of lime and gypsum are indurated and form readily identifiable cemented planar bodies (Figure 3.44). Nevertheless, in prehistory, lime was mixed with clay and ash-rich sediment that yielded inferior results in terms of the consistency and strength of the floor (Goren and Goldberg, 1991; Karkanas, 2007).

Although floor construction is horizontal, post-depositional deformation due to differential subsidence of its organic content can distort its geometry to form relatively smooth undulating surfaces (Shahack-Gross et al., 2005). Additionally, however, subsidence must have affected walls and other constructions as shown by bowed foundation walls, for example.

Microstructure The only probably universal microscopic feature of 'living floors', as defined above, is the occurrence of a thin layer characterized by compaction and trampling, which is presented in detail in section 3.5.3. However, suffice it to say here that the most important changes will be reduction in the amount and size of packing pores, as well as the formation of horizontal tapering cracks and voids close and parallel to the surface. Organic or organomineral (Fe and Mn oxides) staining may be present in the form of impregnative nodules and coatings due to the decay of organic matter in a damp environment. Under the microscope, parallel orientation of some fine elongated particles and artefacts

may be evident. However, overall the modification of the natural substrate is not that obvious and the evidence is often meagre and ill defined.

On the other hand, constructed floors are characterized by a dense homogeneous microstructure that exhibits few pores that are mainly vesicles and elongated vughs with smooth surfaces. Characteristic is embedded porphyric coarse/fine (c/f) related distribution (coarse particles are floating in a finer matrix) (Figure 3.49a) (Gé et al., 1993; Matthews et al., 1996; Shahack-Gross et al., 2005), as well as welded aggregates that individually are often of different composition (Figure 3.49b). Although porphyric c/f related distributions appear quite often in natural soils, in well pugged floors all sizes of clasts are rather equally distributed in space, forming a homogeneous appearance at the mesoscale of observation. This 'mosaic' structure resembles in many ways the mosaics of polished tiles and concrete floors prepared in houses until recently.

If dung or straw is added, pseudomorphic voids after plants often dominate the pore system (Matthews et al., 1996). Elongated particles normally show orientations parallel to the surface during plastering of the floor (Karkanas and Van de Moortel, 2014). Vesicles and smooth vughs are produced by water or air escape during application of the plaster layer and suggest that the floor was at least partially saturated with water (Figure 3.50).

In addition, diligent observation reveals deformation structures such as rotation (galaxy) structures, folded slaking bands, and shear zones marked by parallel striated fabric; *boudinage* (stretched deformation) features can be observed mainly when a mixture of immiscible material occurs (Figures 3.49 and 3.51). In some cases, fine planar vughs, parallel arrays of vesicles, or parallel fissile parting can be identified (Figure 3.50). A matrix-supported

Figure 3.48 Constructed clay surfaces. (a) Thin red clay lens (with arrow) overlain by a white ash lens demarcating the occurrence of a sequence of floors above (F) not clearly shown in the field. Sequence is 1 m thick. MBA, Palamari, Skyros Island, Greece. (b) Sequence of red clay constructed surfaces (arrows) interbedded with white ashes and burnt remains. Some of the clay surfaces pinch out laterally, appearing as thin discontinuous lenses. The ashes typically rest on the red surfaces with a sharp contact. Late UP, Yuchanyan Cave, Hunan Province, China.

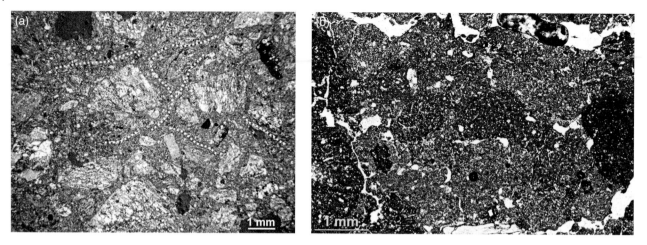

Figure 3.49 Photomicrographs of fabric and microstructure of floors. (a) Embedded 'porphyric' related distribution of mica-quartz schist coarse particles in a fine matrix with evidence of deformation as a result of preparation (kneading) of the floor. Major flow lines are marked with dotted lines. MBA, Palamari, Skyros Island, Greece. (b) Welded clay aggregates within clay surface similar to that in Figure 3.48b. The different compositions and welded nature of the individual aggregates demonstrate that they are constructed. Late UP, Yuchanyan Cave, Hunan Province, China.

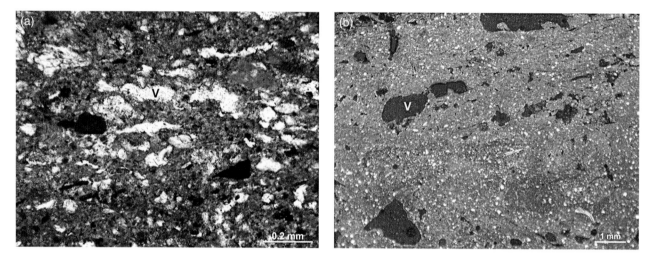

Figure 3.50 Void patterns in constructed floors showing parallel oriented arrays of voids (V), some having lenticular shapes. (a) Occupational debris made to a floor. EBA, Mitrou, Greece. (b) Floor made of loess. Neolithic, Szeghalom-Kovácshalom, Körös Region, Hungary.

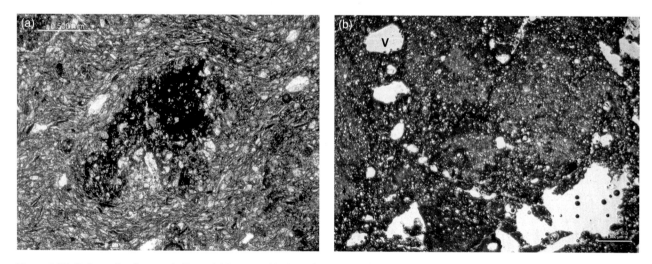

Figure 3.51 Deformation features in floors. (a) Rotational (galaxy) feature with fine matrix rapping around an organometallic concentration. Neolithic, Szeghalom-Kovácshalom, Körös Region, Hungary. (b) Array of vesicles (V) in a water escape feature. LBA, Pylos, Greece.

tiled structure (imbrication) of coarse and elongated particles has been also observed. The features described above are reflected in the indurated appearance in the field and in the process of formation of the floor by kneading and pugging wetted sediment and plastering it on the surface.

Floors are often associated with trampling, with most conspicuous features observed in some lime floors and generally floors that are very well cemented. Generally, an array of parallel cracks is formed on the surface of these floors (Figure 3.52).

Orientation Analysis

Constructed floors have their characteristic fabric as described above, but archaeological materials found on the surface of the floors must show orientation fabrics similar to those of the 'living floors': planar orientations but random distribution of artefacts on this planar surface (Petraglia, 1993; Henry, 2012). When objects are discarded they generally lack any systematic or specific orientation on the surface (Bernatchez, 2010). Clustered concentrations of certain types of artefacts are often associated with craft-related activities characteristic of areas of the room or the living space.

Chemistry

Chemical analysis of floor surfaces has been conducted in several cases to identify former functional areas (Cook et al., 2005; Knudson et al., 2004; Middleton and Price, 1996; Terry et al., 2004; Middleton, 2004; Wilson et al., 2008, 2009; Canuto et al., 2010; Milek and Roberts, 2013). The presence or absence of certain elements on floors has helped reveal the location and the nature of human activities. Usually higher concentrations of barium (Ba), calcium (Ca), copper (Cu), phosphorus (P), lead (Pb), strontium

Figure 3.52 Planar cracks in a lime floor. These are curing cracks formed during carbonation process. Neolithic, Drakaina Cave, Greece.

(Sr), and zinc (Zn) are found on floor sediments. Peak concentrations of all elements are often associated with the hearth and kitchen areas (Wilson et al., 2008, 2009), whereas phosphorus is particularly related to food preparation, organic refuse, and the presence of dung (Middleton, 2004; Macphail et al., 2004; Canuto et al., 2010; Wilson et al., 2008, 2009). Wood ash is particularly rich in phosphorus, potassium, and calcium (Middleton, 2004). Low concentrations of trace elements imply meticulous cleaning of the area, high-traffic areas, or low-utilization areas such as sleeping and meeting rooms. High concentrations of heavy metals like copper, iron, manganese, mercury, lead, and zinc have been reported in areas where metallic-based pigments and dyes were used (Parnell et al., 2002). However, it appears that charcoal and bone are the main trapping materials for most of these elements (Wilson et al., 2008, 2009). Enhanced magnetic susceptibility is also associated with these burnt materials on floors surfaces (Macphail et al., 2004). Consequently, in cases where ashed organic matter is a major component of occupational debris on floor surfaces or where ash has been used to dry damp floor surfaces, different activities do not produce unique chemical fingerprints and multi-element analysis alone cannot be used with confidence (Milek and Roberts, 2013).

Relation to Other Archaeological Materials and Artefacts

'Living floors' are by definition a locus of concentration of artefacts and archaeological remains in general, unless the floor has been swept clean (Matarazzo et al., 2010; Matarazzo, 2015). As already stated, the formation processes of a living floor are characterized by trampling and therefore some fragmentation by foot traffic might be expected (see section 3.5.3). Concentration of organic matter and chemical compounds – such as phosphates due to intensive human activities – could result in diagenetic changes whose intensity will depend on the local hydrological regime affecting the particular layer. Nevertheless, few specific studies are available to further evaluate post-depositional processes affecting 'living floors' (although see Malinsky-Buller et al., 2011).

Constructed floors in architectural sites are not expected to directly affect archaeological materials, yet in several cases floor construction materials contain large amounts of previously deposited fine archaeological objects. In general, this reality goes unnoticed, and floors are considered to be perfect materials for doing microartefact analysis (Rosen, 1989), wet sieving, or for other recovery procedures of fine and microscopic materials. However, this approach has to be evaluated on a site-by site basis, as even wall plasters can contain large amounts of microartefacts presumably unrelated to *in situ* activities, and much of the material in floor assemblages is background noise (Hodder and Cessford, 2004).

Floor sequences, on the other hand, are perfect refuse traps for everyday activities, and as a result, archaeological objects *on* former surfaces and *within* these sequences are one of the best testimonies of human activities and use of space (LaMotta and Schiffer, 1999; Hodder and Cessford, 2004).

3.4.2 Mudbricks, Daub and Other Mud Construction Materials

Mudbricks and daub constructions prevail from the early Neolithic to modern times in all parts of the world and their extensive use has culminated in the formation of tells, which are found from the Balkans in the northwest to India in the southeast (Rosen, 1986). Due to its astonishing mechanical qualities, clay was – and still is – the essential building material in several parts of the world where wood is not so abundant. Clay combined with wood to form timber-framed structures (wattle and daub) provides an alternative solution to stone constructions. Clay when wetted is plastic and malleable but when dry becomes a solid, hard substance. Pure clay, particularly a certain kind of clay, referred to as 'fat clay', cracks on drying and therefore has to be mixed with sand and silt to become less plastic and less liable to shrinkage and cracking. Many factors influence the plasticity of clays, including mineralogy, grain-size, soluble salts, organic matter, and other non-clay mineral content. Expansive (swelling) clays such as those belonging to the smectite group (e.g. montmorillonite and bentonite) are usually the so-called fat clays. However, exceptionally fine-grained clays composed of non-expansive clays, such as kaolinite and illite, can also behave as fat clays.

Wall-making daub is prepared the same way whether it is used in baulk or in the form of mudbricks and is not much different from the preparation of floor construction material. Daub is mixed with water and pugged by being trodden upon. At the same time, tempering agents such as sand or fine gravel, straw, dried grass, cereal chaff, dung, dung ash, seaweed, grog, or any kind of fine refuse can be added (Courty et al., 1989; Adam, 1994; Shaw, 2009; Nodarou et al., 2008; Love, 2012). Rammed earth (*pisé* or taipa) involves compressing daub into an externally supported frame or mould, which determines the width of the wall. Mudbricks (adobe) are also moulded in wooden frames without a base, but cases have been reported where the bricks are handmade, particularly during the Neolithic period. Clay walls are manufactured with lean clay with only fine temper. However, daub in general can be quite coarse and gravelly even when used for adobe.

All earthen architectural constructions are prone to weathering and decay readily if not protected with plaster. Therefore, understanding the degradation process is important for recognizing the former presence of mudbricks and for identifying and distinguishing them from various archaeological infill sediments. Often, it is exceedingly difficult to differentiate degraded mudbrick material from local soils (Goodman-Elgar, 2008). Indeed, it is often the case that mudbrick is not reported during excavations, possibly because it is relatively easy to dig through mudbrick walls. Nevertheless, detailed field and microscopic observation often reveals the existence of decayed mudbrick and other daub constructions (Goldberg, 1979b; Karkanas, 2015).

In situ degradation occurs even while the bricks are still emplaced in the wall. The rate of the process depends on the annual precipitation in each area. Exfoliation due to salt precipitation in arid environments (Courty et al., 1989: 239), rainsplash and erosion (McIntosh, 1974), and freeze-thaw activity also contributes to daub and adobe decay. Accelerating decay is observed upon abandonment, once brick and daub walls are not protected from rain by a roof (Friesem et al., 2014a). This stage of degradation involves water and gravity flow processes.

Recognition

Field Identification of adobe and daub material is mostly based on brick shape and the presence of elongated voids attributed to additions of straw temper during mudbrick production (Goldberg, 1979b; Rosen, 1986). Remnants of wattle and daub structures are readily identified by the presence of impressions or wood branches on the daub (Figure 3.53). However, although (unheated) mudbricks are often not preserved in their original shape, it is possible to recognize their original structure and fabric by using accidentally fired mudbricks as controls (Nodarou et al., 2008). Although degradation processes result in the loss of elongated voids (Friesem et al., 2011), daub and adobe are dense and relatively indurated, and they can thus be generally recognized in

Figure 3.53 Burnt fragment of daub with impressions of wood branches. Neolithic, Plastira Lake, Greece.

archaeological deposits in the form of conspicuous aggregates (Figure 3.54).

Ethnoarchaeological observations and studies show that mud wall structures degrade into small mounds, as shown by the formation of a talus on both sides of degrading walls (Koulidou, 1998; McIntosh, 1974, 1977; Friesem et al., 2014a,b). Consequently, slope processes dominate the formation of these miniature talus forms, which initiate a variety of flows that may operate simultaneously (Figure 3.55). Apart from wet flows, slope processes also encompass dry gravitational movements. Gravitational collapse of intact wall segments following cracking and fissuring involves rolling of single bricks. Furthermore, the moderate sorting and the weak bedding of clasts with the inclination along the talus slope

suggests alternating wet and/or dry colluvial grain flow processes (Friesem et al., 2014b; see also section 2.3).

Observations of talus morphology reveal that talus formation next to degrading walls results in a post-depositional U-shaped basin between adjacent opposing walls. With time, the area between the walls becomes flat and horizontal. However, the deposits within the U-shaped basin between the walls is still preserved in the form of crudely bedded, lensoid mudbrick-derived sediment, which commonly alternates with other types of sedimentary layers, such as wind-blown sand in arid environments (Friesem et al., 2011, 2014b). Such talus profiles attributed to wall degradation have recently been identified in post-Roman ancient Corinth based on field and microscopic observations, suggesting that careful observations can reveal fills formed by degraded walls (Figure 3.56).

Microstructure Remnants of the original mudbrick material have to be identified in order to establish the different microscopic decay facies. Often, the process of kneading the wet soil material during daub preparation is not intensive enough to break up and eliminate microscopic features derived from the original parent material; hence, pedofeatures such as clay coatings or fragments of alluvial deposits are still recognizable (Matthews et al., 1996). The kneading process is also commonly evident from the well-developed porphyric-related distribution of the coarse and fine fraction, as well as an occasionally homogeneous and dense fabric (Figure 3.57c). Additional features include (a) slaking and separation of the silt and clay fractions during settling of a watery slurry, (b) directional flow of the mainly silty matrix along bands with flame-shaped margins (water escape features), (c) dusty silty and clay coatings, and (d) rotational

Figure 3.54 Decay of mudbrick wall with conspicuous reddish brown mudbrick remnants (M) that can be still recognized along with some light-coloured pieces of mortar (arrows). Scale is 30 cm. LBA, Mitrou, Greece.

Figure 3.55 Talus of decayed mudbrick (DW) on the sides of a remnant of still standing mudbrick wall (MW) above a stone foundation wall (SF). Soil-formation processes have begun acting on the talus (S) homogenizing the sediment. Scale is 20 cm. Abandoned village of Kranionas, Greece.

Figure 3.56 A conspicuous reddish brown talus (T) is observed abutting two robbed walls (RW), presumably formed by decay of mud constructed materials from the superstructure of the related building. Scale is 3 m. Post-Roman, ancient Corinth, Greece.

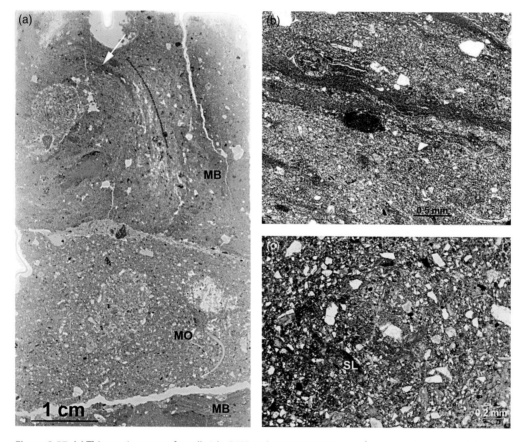

Figure 3.57 (a) Thin section scan of mudbricks (MB) and mortar (MO). Note separation of clay (arrow) in water escape (red dotted line) along large rotational feature. (b) Photomicrograph detail of (a) showing separation of clay and silt during settling of water saturated parts of the slurry. Middle Formative Period, Río Viejo, Mexico. (c) Embedded, 'porphyric', related distribution of coarse to fine, matrix fractions. Dusty clay slaking features are also shown (SL). Minoan, Mochlos, Greece. Courtesy of Marta Lorenzon.

features; the last indicate ductile deformation due to moulding (Figure 3.57a,b) (Goldberg, 1979; Courty et al., 1989: 239–242; Friesem et al., 2014b).

Koulidou (1998) identified a reduction in grain-sizes from the wall towards the centre of the studied rooms during the degradation of mudbrick walls and Goldberg (1979) observed an increase in coarse sand in mudbricks probably reworked by water flow. Lower energy water and gravity flows seem to be prominent processes acting on bricks degrading in place, as well as on infill sediments. Water flows are evident in the form of laminated sediment, with clay-rich and sandy silt-rich couplets, some of which show moderate sorting. Good indications of debris flow processes are coarse grains floating within a fine matrix associated with vughs and vesicles, and the presence of recognizable mudbrick fragments embedded within the talus matrix (Figure 3.58). Also indicative is the presence of continuous clay coatings around clasts that are embedded in the sediment matrix, also known as 'rolling pedofeatures' (Angelucci and Zilhão, 2009; Boschian, 1997). On the other hand, prevalence of packing voids, lateral normal grading, and reverse grading can be indications of dry grain flow (Bertran and Texier, 1999; Postma, 1986; see also section 2.3).

Finally, mudbrick and its decay products can also be affected by frost activity. Micromorphological pedofeatures include localized platy structures with deformation features that are often embedded in a matrix with single-grain sand and silt (see also section 2.7) (Friesem et al., 2014b; Bertran and Texier, 1999; van Vliet-Lanoë et al., 1984; van Vliet-Lanoë, 2010). Nevertheless, in all the above cases mudbrick remnants have to be identified in order to be sure that these processes are a result of mudbrick decay.

Relation to Other Archaeological Materials and Artefacts

Observations of microartefacts in floor sequences are generally valid for mudbricks as well. Mudbricks can contain artefacts and botanical materials (charcoal, phytoliths, seed, and macrobotanical remains) that were derived from previously deposited sediment and that had been incorporated into the matrix during their manufacture (Seymour and Schiffer, 1987). When mudbrick decays, these artefacts and organic materials are released into the room fill that is commonly found in between the walls; it is clear that such objects are unrelated to the activities performed in this area prior to the decay. In addition, collapsed mudbrick buildings that are still standing up to a certain level are often used as trash places, thus the decayed mudbrick accumulations can be gradually enriched in objects and anthropogenic remains that are not related to the original use of the space (Schiffer, 1985; Seymour and Schiffer, 1987). On the other hand, decay and collapse of mudbrick walls result in rapid burial of floors and thus may preserve remains of former activities, which, if identified, can be good stratigraphic markers for locating floor surfaces and for revealing 'clean' activity areas (Friesem et al., 2014a).

3.4.3 Mortar, Wall Plaster

'Mortar is defined as any material in a plastic state which can be trowelled, becomes hard in place, and is utilized for bedding and jointing' (Cowper, 1927 in Holmes and Wingate, 2002: 60). According to this definition mortar is defined by its physical characteristics and utilization, rather than the composition of the material. In practice, though, plasticity in wet conditions can be achieved with

Figure 3.58 Sheetwash processes affecting mudbrick degradation. Note crude laminae of unsorted sand (S), silt (SI), and sand-sized rounded mudbrick aggregates (M) and vesicles (V): (a) PPL and (b) XPL. Abandoned village of Kranionas, Greece. From Friesem et al. (2014b).

relatively few materials, such as clay, lime, and gypsum. White, fine calcareous silty clay has also been used in antiquity (Matthews et al., 1996). Mortar materials are not much different from those in plaster, but their requirements are different. Plaster describes any coating of a building surface with plastic material that eventually hardens (Holmes and Wingate, 2002). Plastering of exterior surfaces is also called rendering. Plaster tends to be finer grained, less porous and more homogeneous than mortar, which is characterized by its coarse component. Indeed, mortar is a Latin word meaning crushing. However, frequently the first coat of plaster wall in many ways can resemble mortar. Moreover, in some cases the constructed foundation layer of floors on which the plaster floor is laid is also called mortar; hence, the terms mortar and plaster are often used interchangeably.

Plaster walls appear to be as old as building constructions, as, for example, in the famous sixth millennium BCE plaster walls of Çatal Höyük. However, the use of mortar for bonding stones with the aid of a cementing agent occurs with the use of gypsum in Egypt during the third millennium (Carran et al., 2011). Nevertheless, bonding stone with simple mud mortar cannot be readily differentiated from post-burial natural mud infill between the stones of a construction. Plaster is also used to cover wooden piles, platforms, ceilings, and roofs (Shaw, 2009). In the Old World, the use of lime and gypsum mortars was not widespread outside Anatolia and the surrounding areas. It was only after the Hellenistic period and particularly the Roman period that lime replaced clay, where it was widely used for the manufacture of mortar to bond rubble masonry (Adam, 1994).

Different substances, called aggregates, are used as tempering agents in mortars, and these are more or less the same materials as those of mudbricks, although sand and gravel – and later in Roman times, crushed pottery – are the dominant aggregate types for manufacturing mortar. The use of coarse-grained aggregate reduces the need for plant inputs (e.g. straw, dung) since they accommodate shrinkage during the setting of lime. Furthermore, the presence of lime is reported for mortar and plaster remains but not for mudbricks or daub.

Decay of buildings after abandonment will also yield mortar and wall plaster remains, in addition to rubble and decayed mudbrick deposits. Indeed, the same processes that act on mudbrick walls also affect mortars and plaster.

Recognition

Field Fragments of mortar and wall plaster can be identified by their denser and relatively indurated nature, particularly if they are made of lime and other cementing materials (Figure 3.54). Therefore, as in the case of mudbrick, they are generally differentiated by being conspicuous aggregates in the archaeological deposits (Figure 3.59). They are more indurated than mudbrick and have a whitish colour when made of lime-related cements. Cemented construction materials maintain their hardness for some time even when immersed in water, whereas unburnt clay and calcareous muds become soft and 'collapse' (slake).

Mortars often contain colourful coarse sand to fine gravel sized inclusions inside an indurated whitish matrix. Wall plaster, on the other hand, is more homogeneous and often identified by its layered to laminated structure representing repeated coats during the initial application or during maintenance procedures. Obviously remnants of painting and other decorative

Figure 3.59 Lime mortar aggregates in sediment fill. Porous, crumbly white aggregates (black arrow) that when degraded become distributed throughout the layer as conspicuous finer white particles (white arrow). Seventeenth century ancient Corinth, Greece.

features can be very helpful in identifying small pieces of decayed wall plaster. Nonetheless, most of the time fragments of wall plaster cannot be differentiated from floor plaster if they are part of a fill.

Microstructure Already from the early Neolithic, wall plaster presents evidence of remarkable maintenance activities. At Çatal Höyük, for example, as many as 80 couplets of calcareous clayey plaster have been microscopically identified in samples of wall plaster (Matthews et al., 1996). Couplets consist of a millimetre-thick preparation plaster, often with small amounts of straw stabilizer, and <1 mm-thick finishing coats with very smooth and compacted surfaces. Elongated platy voids represent decayed vegetal stabilizers and horizontally aligned elongated vughs, and vesicles are produced during the application process as the wet plaster dries. They are often more compacted, fine grained, and better sorted towards their surface (Figure 3.60). Microscopically, mortar shows fabrics and microstructure similar to those of plaster floors (see section 3.4.1).

Lime is generally easily identified when pure and in large amounts. However, it is often mixed with clay and household debris, thus hampering its secure identification. Nevertheless, lime produces a 'reacted' appearance, whereby individual calcite crystallites are welded together; in contrast, individual micrite crystals in natural calcareous clay can still appear isolated. In addition, cryptocrystalline areas of ill-reacted lime and transitional textures of partially carbonized slaked lime can be observed within the lime lumps (Karkanas, 2007); shrinkage fractures, vesicles, smooth vughs, and occasionally colloidal forms also occur (Figures 3.52 and 3.61).

Relation to Other Archaeological Materials and Artefacts

Mortars and wall plaster can also contain fine artefacts, and vegetal and charcoal material. Thus, they add materials to construction or destruction fills that are unrelated to the use of the specific area in which the plaster occurs.

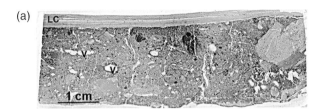

Figure 3.60 (a) First coat of coarse plaster and fine laminated finishing coats (LC). Note vesicles and arrays of vughs (V). Abandoned mudbrick house in modern Mycenae. (b) Mudbrick mainly composed of sand of varying mineral and rock types (R) – with clay coatings inherited from a B-horizon of the nearby soil – showing outer surface smoothed by infilling of lime plaster (C), which is overlain by fine gypsum coat (G). Officers' Quarters (1776), Presidio, California.

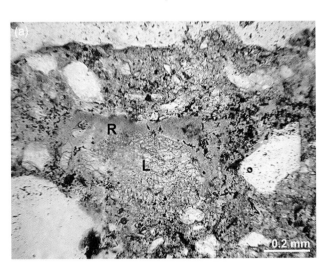

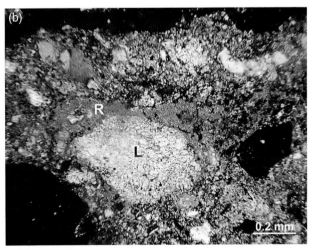

Figure 3.61 Photomicrographs of (a) PPL and (b) XPL of lime fabric showing half-burnt limestone fragment (L) with a beige very low relief and low birefringence rim of slaked lime (R). This ill-crystallized calcite gives way to high birefringence calcite in the rest of the matrix. Neolithic Makri, Thrace, Greece.

3.5 Maintenance and Discard Processes

Maintenance and discard processes have been a matter of fervent discussion in the archaeological literature. They are considered very important issues in studying activity areas and the use of space at a site, since different activities generate debris during procurement, processing, consumption, or manufacturing (Brooks and Yellen, 1987). These processes lead to dropping, tossing, resting, positioning, and dumping of items on the surface that potentially can enter the archaeological record (Binford, 1978). Nonetheless, only a very small number of these items are found in functional articulation as a result of abandonment behaviours and natural post-depositional processes (Newell, 1987; Schiffer, 1972, 1985). Schiffer (1972, 1987: 58–98) defines two main categories of refuse: primary and secondary. The former reflects spatial patterns originating by original activities and the latter artefact disposal away from the use location.

In most case studies, described formation processes focus on objects: their nature, value, and patterning that are found – or are expected to be found – in relation to certain activities (Schiffer, 1985; Newell, 1987; Oswald, 1987; Binford, 1987; Brooks and Yellen, 1987; Wilson, 1994; Tani, 1995; Hutson et al., 2007). Considered only rarely are the fabric and the characteristics of the deposit that includes the discarded item. The majority of the discard activities as mentioned above are not responsible for the final emplacement of the archaeological objects, nor the way they are found in the excavated sediment. Actions like tossing, dropping, loss, abandonment, and what constitutes *de facto* refuse according to Schiffer (1985), that is, refuse left behind when the structure was abandoned, are not always responsible for the incorporation of the objects into a body of sediment. Foot traffic and other anthropogenic or natural burying processes are overall accountable for the final pattern that is produced by primary refuse. As a result, primary refuse will be reoriented and redistributed by trampling and *de facto* refuse by post-abandonment processes, such as wind, rain, plant colonization, fauna activity, wall and roof collapse, or more 'distracting', natural sedimentary processes, such as water flow (Schiffer, 1985). It is the subsequent burying processes that are actually responsible for the appearance of primary or *de facto* refuse in the archaeological record; otherwise we are still seeking sites like Pompeii.

Most of these processes do not necessarily rework the discarded objects substantially so as to totally distort the general location of their original emplacement. However, they do disturb their original position, that is, orientation, inclination, and, in some cases, distribution patterns.

This is not to deny that analysis of the objects found on a surface cannot give important information about activity areas and some clues to the discarding processes, but studying the sedimentary processes that buried and incorporated the objects can supplement and assess better the produced patterns. Indeed, in constructed sites in general, where analysis of objects alone has provided conclusive data, floor assemblages have been clearly differentiated from the overlying secondary room fills. However, in the majority of sites this is actually the issue that is at stake, and thus the inquiry of 'refuse structure' (Tani, 1995) should be actually transformed into the investigation of 'sediment structure'.

Nevertheless, there are discard activities that produce characteristic bodies of sediment, and these are mostly secondary refuse processes, *sensu* Schiffer (1976). These activities can produce large amounts of debris that bury themselves and the included archaeological objects. Although the term 'secondary refuse processes' was originally coined to mean processes that 'distort' the archaeological record, it has become apparent that they are actually part of the systemic context and important elements of human behaviour (Brooks and Yellen, 1987; Binford, 1981, 1987; Tani, 1995; Wilson, 1994; Shillito, 2015; Villagran, 2014). Dumping is probably the most important discarding process and certainly the most volumetrically significant one. Maintenance processes such as sweeping and raking produce negligible amounts of sediment that can be resolved only microscopically. Trampling is a very important depositional process and together with dumping is perhaps the most significant anthropogenic means of burying. It can not only fragment and 'retouch' objects but also displace them into underlying, older strata.

3.5.1 Sweeping and Raking

Sweeping and raking activities are part of regular maintenance processes to keep occupational areas clean and devoid of dangerous items such as lithics, bone, pottery, and metal debris. They have been recognized in all archaeological periods from deep prehistory onwards (Figure 3.15). The act of sweeping and raking can be carried out by hand or with feet, or with implements such as wood or leafy branches, straw or hay, or any kind of ready-made broom or rake. The objective is to drag loose debris over an occupational surface and move it to the side or to transport it using containers such as a basket or dustpan in order to dump it. The act of sweeping normally affects both the remaining surface and also the material that is swept away. Sweeping earthen occupational surfaces normally does not entirely remove all occupational debris, and thus artefacts large and small can be left on the surface. Any material left on the sides is a direct product of the sweeping process and therefore the sediment will be imprinted by the characteristics of this action. Dumped materials are described in the following section.

Since the act of sweeping involves particle movement by the force of sweeping, dry gravity flows and falls best describe the process (cf. Miller et al., 2010; Matthews, 1995). Thus, sorting and grading may be encountered in swept sediment. However, very few studies or experiments have been undertaken and consequently sweeping activities have been mostly inferred from the lack of debris on the occupation surfaces. The same holds for raking processes. The latter is probably better described by natural, fast-moving, dry mass waste processes such as debris avalanches, which also include a gravity component. The aforementioned slope processes are gravity driven, whereas in sweeping and raking the driving forces are bodily actions made predominately by hand movement. In the case of sweeping earthen floors and occupational surfaces, the movements are abrupt and of relatively high energy. Thus, the affected particles attain high kinetic energies such as those observed in free-falling grains on a slope. Sweeping over planked floors results in free-fall of fine particles through spaces in between the planks. In raking, there is a kind of incremental steady movement that drags, turns, and mixes particles in a chaotic way, which can approach debris avalanches in natural slopes.

Recognition
Sedimentary bodies formed by sweeping on an earthen floor are normally not identified with the naked eye.

They are millimetre-sized thin layers covering the floor surface and following the geometry of the latter. Compacted lenses of silty clay with wavy boundaries have been assigned to material percolated through mats and rugs, whereas thin layers of unorientated sand-sized particles may have formed when mats were lifted for shaking (Figures 3.30 and 3.62) (Matthews and Postage, 1994). Beneath planked floors thicker layers of fine dusty, mainly organic-rich sediment accumulation can occur (Gé et al., 1993), although they are very rarely preserved in the archaeological record. If they manage to be preserved, they appear as thin planar and often laminated bodies of sorted organic dust.

Sweeping deposits consist of microscopic layers with finely comminuted and subrounded particles (Figure 3.62). They show crude orientation and layering and an enaulic (intergrain aggregated) related distribution where aggregates of fine material partially fill the voids between coarser particles (Matthews, 1995). Experimental sweeping produced an open, spongy, structure of different materials very loosely organized. Such a structure will have very few chances of surviving in the archaeological record (Miller et al., 2010). Swept sediment will be fine grained and locally sorted, consisting of mostly silt-sized mineral particles and sediment aggregates (dust-sized) rich in light material, such as charcoal, bone, straw, phytoliths, and similar anthropogenic fine debris (Goldberg et al., 2009a). Aggregates and anthropogenic components will be subangular to subrounded due to repeated sweeping and previous trampling (Figure 3.63). Millimetre-thick platy layers on top of constructed floors with a general microscopic

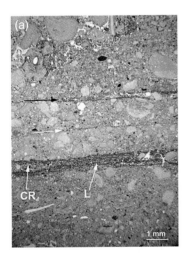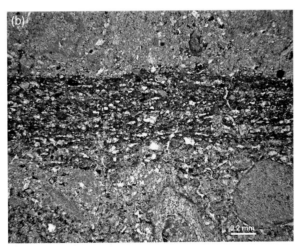

Figure 3.62 Fine occupation debris trapped between constructed lime floors. (a) Finely laminated parts (L) are probably related to dirt percolating through mats and rugs or due to their maintenance such as shaking and sweeping, particularly in the case of the coarser lenses (CR). Note that the middle floor (white horizontal arrow) is practically clean – probably due to meticulous sweeping – whereas for the upper one (black horizontal arrow) some fine debris has been trapped on it. From Karkanas and Goldberg (2017a). (b) Bedded fine organic-rich occupation debris, detail of area of L in (a). The dense and compacted nature of this material suggests that it was covered with a mat or rug or that the material was swept after being dampened. PPL. Neolithic, Makri, Thrace, Greece.

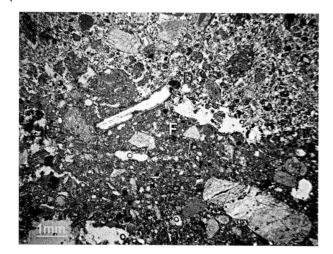

Figure 3.63 Loose debris most likely swept over a worn-out floor. The debris consists of dust-sized occupation debris (D) and rounded to subrounded sand-sized aggregates mostly derived from the underlying floor (F). PPL, MBA, Mitrou, Greece.

stratified appearance can be assigned to sweeping activities, probably after the material has been dampened (Karkanas and Goldberg, 2017a). They are made of anastomosing or interfingered couplets of thin clayey, and thicker and coarser silty laminae (Figure 3.62).

Swept up or raked sediment along the edges of activity areas will appear as crescent, scalloped, or doughnut shapes (Tani, 1995). If moderately disturbed, they retain a generally upraised lenticular form with a relatively flat, lower surface. The sediment is characterized by a fine dusty matrix supporting coarse particles of anthropogenic

origin. The structure is loose, and the fabric chaotic, including vertically oriented elongated particles. In general, they are fine grained, matrix supported, massive, and structureless (Figure 3.64). In several cases, they cannot be differentiated from fine-grained, single-event dumped materials. In general, the grain-size distribution of swept material is more homogeneous compared to that of dumped material (Miller et al., 2010).

3.5.2 Dumping and Filling

The most pronounced discard process in the archaeological record is undoubtedly the act of dumping. In sites with no construction, dumping normally occurs in the periphery of activity areas (Figure 3.14) (Binford, 1978). In general, these areas are characterized by repeated dumping actions, which in the archaeological record are identified as a midden, trash deposit, and domestic waste or discard deposit (Newell, 1987; Shillito, 2015). Shell middens, for example, are one of the most studied types of refuse features (Villagran, 2014). The same behaviour is observed in constructed sites, although dumping does not normally occur inside constructions but often not far away, as in the case of 'door dumps' (Binford, 1981). Provisional dumping is normally close to the constructed space (Tani, 1995). On the other hand, post-abandonment dumping is observed inside houses and can be found intermixed with collapsed roof and wall material. Some locations of preferential dumping include exterior parts of walls, paths, and roadsides, periphery of yards or fences, and abandoned constructions. Large-scale dumping in

Figure 3.64 Raked out massive, structureless hearth remains containing large amounts of burnt bone showing various degree of burning. The horizontal orientation of the coarse particles suggests that some trampling have been involved as well. (a) MP, Pech de l'Azé IV, France. (b) MP, Roc de Marsal, France.

ancient cities can be located in nearby dry torrent banks, cliffs, and valley slopes. Abandoned pits, wells, and privies are also characterized by secondary refuse fills (Wilson, 1994; Macphail et al., 2008).

Dumping sediment, soil, or construction and demolition material for purposely filling pits or for levelling flat areas involves more or less the same bodily activity as dumping refuse, although the composition of the filling material is normally radically different. Backfilling is ubiquitously observed in architectural sites and often forms the foundation layer upon which a new building is constructed, hence the often-called construction or foundation fill, or construction packing (Matthews and Postage, 1994; Karkanas and Efstratiou, 2009; Macphail and Goldberg, 2010; Shillito et al., 2011a). Such fill also comprises the building material of mounds, and several types of tombs (Macphail and Goldberg, 2010; Sherwood and Kidder, 2011; Karkanas et al., 2012).

The act of dumping involves throwing or emptying material generally from a small height such as a container and consequently includes a gravity component. Dumping for potential future use (provisional dumping) is also possible. Hands, baskets, animal skins, shovels, pans or similar equipment are normally used. Dumping from a greater height produces a pile of debris whereby some of the material rolls down or avalanches along temporally bounded slopes of the pile (Karkanas et al., 2012). In this sense, this type of dumping is probably the anthropogenic activity that most resembles the dry gravity sedimentary processes observed along natural slopes (see section 2.3.2).

Recognition

Field Refuse dumps may take on any shape. Small ones are often in the form of piles or fans, with convex transverse surfaces abutting a construction feature, such as a wall, or against pre-existing natural morphologies, as, for example, a slope, cave wall, or depression (Figure 3.65); therefore their morphology is often characteristic, as is their content. Their lower boundary is rather horizontal and follows the morphology of the previous surface. Their morphology might be modified by subsequent human activities or natural processes, often leading to truncated dump sequences. They appear as pockets and festoons in an archaeological sequence.

Larger scale dumps that represent long periods of dumping activities often fill small depressions or large pits and ditches, and they partially even out pre-existing topography. Their overall geometry is often sloping (Figure 3.66), or they appear to sag in the middle of depressions, ditches, or pits (Figure 2.16). Indeed, what often characterize pit fills and middens are deformation phenomena due to self-compaction (Figure 3.67). Streaks,

Figure 3.65 Massive loose dumped ashes filling a pit (above line) at the UP of Klissoura Cave I, Greece. Grid is 1 m wide.

layers, or features that were originally horizontal appear to form festoons (Cornwall, 1958: 60). This topography is considered as evidence of deliberate, relatively fast filling, and it offers a plausible explanation for the greater compaction in the centre relative to the sides because of differential loading and pore water circulation. In any case, the fact is that there is not sufficient time for the sediment to become spatially equilibrated when loosely dumped sediment accumulates rapidly. In addition, dump fills often consist of soft compressible material such as ash and organic matter, which at the same time are prone to secondary dissolution and decay; the latter further contributes to the loss of volume most pronounced at the centre of the fill (for more detail see section 2.7.2).

Large-scale dumps can be massive or layered depending on the type of dumped material and the rhythm of dumping activities. Large objects tend to produce a clast-supported body, but dumping the same material along a pre-existing slope produces clustering of coarser clasts with some finer tailing. In general, large objects are angular and certain materials such as ceramic sherds tend to be tightly clustered because they interlock (Figure 3.68). Probably, some lateral sorting of materials (e.g. flakes and body sherds) can occur at the periphery of the dumped concentration because they roll and slide down temporary piles or fans. Finer material will later infiltrate along the open spaces of the dumped objects, but in all cases the whole sedimentary body is loose and porous. In general, finer dumped material will tend to form aggregates that are produced when the material was originally detached by cleaning at its primary location.

Dumping ash-rich deposits produces different structures (Shillito et al., 2011a; Shillito and Matthews, 2013; Villagran et al., 2009). Massive ashy midden deposits are composed of homogeneously mixed, redeposited

Figure 3.66 (a) Larger scale sloping dumps that represent long periods of dumping activities. They consist mainly of hearth remains and some construction materials. Their bedded nature probably suggests distinct dumping phases. Neolithic, Imvrou Pigadi, Thessaly, Greece. Scale is 50 cm. (b) Bronze Age rampart construction dump with individual layers sloping in both sides. The construction was later truncated and covered by modern debris. Width of trench is 1 m. Fort Harrouard, France.

Figure 3.67 Prehistoric shell midden accumulation along the shoreline at British Camp, part of the San Juan Island National Historical Park, Washington St., USA. The midden is generally composed of cumulative interbedded shell layers and finer sediments rich in ash and organic matter with some intrusive cooking pits. Note the dumped, inclined shell unit (arrow) and the general wavy appearance of the lower part due to self-compaction.

material that mainly consists of loose, aggregated ashy sediment with an overall grey aspect and a chaotic fabric. Coarser elements, such as charcoal, bone, flint, daub, and pottery 'float' inside the fine matrix in all kinds of positions and orientations (Figure 3.21a; see also section 3.2). Finely stratified ash and plant-rich midden deposits appear as pure, fine, whitish, or grey layers associated with bone, charcoal, decayed plants, and dung remains (Figure 3.67). These ashy layers, however, do not represent *in situ* fires since they do not show the typical layering of *in situ* combustion features (see section 3.2).

Other layers are often very rich in organic materials derived from the burning and decay of plants used for covering, or as matting, basketry, crop processing, or stabling. However, there are cases where *in situ* activities such as burning have been identified in midden areas, and therefore caution has to be taken when interpreting such sequences. Organic refuse in the form of orange inclusions identified as coprolitic material is often found in middens. On the basis of organic residue analysis, it is possible to differentiate human from animal coprolites (Shillito et al., 2011b).

Figure 3.68 Dump construction material (D) with clast-supported structure filling a pit. Scale is 2 m. Medieval, ancient Corinth, Greece.

Construction or packing fills are not so porous mainly because they are often deliberately levelled. Construction fills constitute planar bodies of aggregated construction materials, or any other type of sedimentary deposits, including open-work rock fragments. Their upper contact is horizontal and sharp, but their lower contact can be undulating but still sharp while following the former substrate (Figure 3.69). They usually do not show any festoon structure because they normally do not contain large amounts of compressible materials such as organic matter and ashes, and are more tightly packed. These bodies are ordinarily constrained by walls or similar architectural features but they can also be found in open areas filling urban terraces, abutting retaining walls, or filling entire areas to elevate and level the substrate. They often contain different types of material, with some construction elements showing clear signs of having been overturned. Larger fragments display all kinds of orientations, and the structure is chaotic (Figure 3.70). When composed of open-frame coarse rocks, the fills are often porous, and some open spaces remain unfilled even after secondary infiltration by natural processes. Indeed, collapse of sediment into empty spaces during excavation is a good sign of rapid dumping (see Figure 3.69).

Sedimentary features characteristic of natural formation processes such as bedding, sorting, grading, and imbrications are normally not observed. However, if dumping happens from some height, crude sorting and grading may occur very locally in some cases (Figure 3.71). These features will crosscut the overall geometry of the filling packets (Karkanas et al., 2012).

Another type of fill found in monumental mounds is characterized by 'basket-loading' (Cremeens, 2005; Ortmann and Kidder 2013), which represents visible single dumping events that together are combined within one fill event. In pre-Columbian mounds, individual basket fills have highly variable texture and fabric that represent

Figure 3.69 Example of levelled construction (packing) fills A and B that consist of unsorted material with chaotic internal structure. Note their planar geometry and that their contact is being hollowed (arrow) an indication of rapid filling upon a stable compacted surface. Ottoman, ancient Corinth, Greece.

Figure 3.70 Jumbled construction fill (DF) in between two sequences of floors. The construction fill consists of floor fragments and stones floating in a finer matrix. Neolithic, Makri, Thrace, Greece.

Figure 3.71 Sedimentary structures in a sloping dump. (a) Layers of relatively sorted gravel-sized particles coarsening downwards (arrow). Hellenistic, Laona, Palaipaphos, Cyprus. (b) Detail of a fill from a scan of resin-impregnated slab showing inclined beds with reverse grading (arrows point to fining direction) as a result of dry flow processes during dumping activity. LBA, Agia Sotira cemetery, Nemea, Greece.

mixed material mined from different places around the mounds. Basket loads are often made into dome-shaped piles. Loads in each pile dip and appear highly compressed, although their interfaces are often sharp and clearly seen in the field. Overall, piles of basked loads appear as overlapping colourful lentoid features. In other cases, the individual events are no longer evident, creating a larger more homogenized fill. Such fills can mantle extensive parts of the mound, whereas in other cases alternating light and dark layers form what is called zoned fill (Sherwood and Kidder, 2011), which probably represents broadcasted sediment in between basket loaded piles (Kidder et al., 2009).

20 cm

Figure 3.72 Packets of fills truncating each other and producing a wedged appearance in a sloping fill of a corridor of a Mycenean chamber tomb. LBA, Agia Sotira cemetery, Nemea, Greece.

Certain types of tomb, tumulus, and some mound fills can be characterized by more pronounced dry gravity processes, since material is often thrown from greater heights and accumulates along temporal slopes as the fill is built up. As a consequence, crudely sorted pockets of coarse sediment with inverse grading and crude imbrication have been observed in corridors of chamber tombs and earthen mounds (Figure 3.71) (Karkanas et al., 2012). Packets of sediment follow the slope of the corridor but are unrelated to the slope of the surface (Figure 3.72). Differentiating natural slope processes from anthropogenic filling is normally straightforward, since in human constructions, fabrics and structures are obviously related to a planned building process. Some examples that demonstrate intentionally dumped material include (a) distinct sedimentary packets that do not necessarily follow the predominant slope but are oriented so that they better fill the space, (b) monotonous repetition of the same sedimentary structure without signs of evolution of processes, as is commonly observed in slope processes (i.e. repeated sequences of alternating free-fall, grain flow, and debris flow deposits), (c) sharp boundaries between packets of fills of slightly different composition that are not erosional although they occur in sloping areas, (d) filling of tombs whose walls are fresh, not affected by weathering, and do not show signs of collapse, and (e) lack of water flow or debris flow processes.

Microstructure Dumping is identified under the microscope by complex packing voids, in some cases with spongy microstructure (interconnected voids) and different coarse/fine related distributions: intergrain aggregates (enaulic: fine material occurs as microaggregates between coarser components) or linked (gefuric: coarse particles are linked by bridges or braces of fine particles) (Matthews, 1995). These structures reflect disaggregation of the original material after cleaning and reworking during transport and deposition (Figure 3.73). Coarse components float within a finer matrix displaying random orientations, including vertical ones (Courty et al., 1989; Simpson and Barrett, 1996; Matthews et al., 1997; Villagran et al., 2009; Goldberg and Whitbread, 1993; Banerjea et al., 2015b). Dumping can be inferred from an open structure with little to no compaction and the lack of any *in situ* crushed softer materials (e.g. bone) that could be attributed to trampling (Schiegl et al., 2003). Mixtures of burnt and unburnt shell or bone with random distribution is additional evidence of dumping (Villagran, 2014). Dumped deposits show more chaotic structures and a greater range of size classes than swept sediments (Miller et al., 2010). Note that even after physical compaction and consolidation, the porous microstructure is retained. Overall, dumped materials are microscopically characterized by their chaotic fabric, their variable aggregated content, and their complex porous nature.

During loading of construction fills complex packing voids and generally loose intergrain related distributions are produced by disaggregation of deposits during extraction and redeposition, followed by separation of coarse and fine fraction as they fall to the ground (Figure 3.74) (Matthews and Postage, 1994; Matthews, 1995). The open voids can be later filled with dusty to impure clay under damp or wet conditions. In such cases, textural intercalations are formed due to slaking of the disturbed soil (Macphail and Goldberg, 2010). Artefacts, charcoal, bone, shell, and construction material are rather angular and the sediment overall is highly unsorted.

Chemical weathering that can result in the formation of phosphate nodules (including vivianite), iron and manganese staining, and decomposed organic matter are common features of middens (Figure 2.91) (Macphail and Goldberg, 2010). Basket loads in mound fills are often rimmed by iron oxide, implying weathering in place to form a band around loads that were still dump when emplaced (Kidder et al., 2009).

Figure 3.74 Photomicrograph of intergrain related distribution and incomplete dark dusty clay coatings related to dumping activity in an open space. Classical, Molivoti, Thrace, Greece.

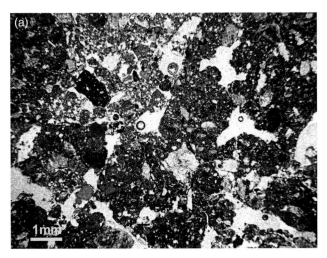

Figure 3.73 Photomicrographs of dump fills. (a) Chaotic arrangement of large aggregates in an open structure. BA, Mitrou, Greece. From Karkanas and Van de Moortel (2014). (b) Loose, porous aggregated material with some finer infiltrated particles. Classical, Molivoti, Thrace, Greece.

Relation to Other Archaeological Materials and Artefacts

Clearly, dumped deposits contain archaeological materials that are out of primary context. Nevertheless, they are very important for a number of reasons. First, their often-extraordinary archaeological richness can provide valuable information about material culture and, second, they themselves are evidence of maintenance and cleaning activities that were undoubtedly important cultural aspects of the society. Ritual aspects have also been associated with dumping activities (Hodder, 1987), further underlying the importance of their study and analysis. Moreover, dumped deposits often contain large archaeological objects, which provide the possibility of identifying individual dumping events.

Finally, construction fills may also contain large artefacts, including parts of statues and construction parts, such as frescos and mosaics. These objects can be directly related to the surrounding structure or may be totally unrelated, having been derived from somewhere else. In tombs, fills may contain remains of burial ritual practices albeit in a secondary context.

3.5.3 Trampling

Trampling is considered an important process in ancient hominid sites and generally in all hunter-gatherer sites (Gifford, 1978; Brooks and Yellen, 1987; Behrensmeyer et al., 1986; Gifford and Behrensmeyer, 1977; Domínguez-Rodrigo et al., 2014). It has been also identified in earthen floors and is considered an important building element in the formation of 'beaten floors' (Macphail et al., 2004; see also section 3.4.1). Together with dumping, trampling constitutes the main burying and penecontemporaneous modification process produced by human activities (Schiffer, 1985). As already mentioned in Chapter 1 trampling may be superficially linked to post-depositional processes; however, in reality it acts to form a new body of sediment on the surface and thus from the depositional point of view it is a syn-sedimentary process.

Both human and animal traffic occur in a site, and normally – but not exclusively – the latter is restricted to outdoor areas. Human traffic is a reflection of intensive use of space. Trampling will not be recorded when the area is vegetated (particularly with grass; Weaver and Dale, 1978) and is covered by materials such as a carpet, straw, or reeds. In open-air sites and caves, micro-refuse trapped in the substrate by trampling can be used to reveal use of space. Although trampling normally changes the original position of the artefacts, it does not radically change the general area of the initial emplacement, even during heavy animal traffic (Eren et al., 2010). To a large degree, the nature of the substrate dictates the resulting pattern of the trampled artefacts, and substrate permeability is probably a major factor in primary refuse preservation (Gifford, 1978). Therefore, damp substrates behave differently from dry ones, and sandy substrates are affected more than clay-rich or stony ones. Rentzel and Narten (2000) have shown experimentally that in dry substrates the effects of trampling are limited to a few millimetres below the activity surface, whereas in wet sediments trampling phenomena can be observed up to a depth of 3 cm. Heavy animal traffic can distort the substrate more deeply (Eren et al., 2010). Sandy substrates are not significantly compacted, but artefacts are buried in it, and one has to be wary of mixed assemblages in this material (Brooks and Yellen, 1987). Trampling effects are greater on slopes than on level ground (Weaver and Dale, 1978).

Recognition

Field Depending on the nature of the substrate, trampled deposits are planar to lenticular bodies a few centimetres thick. The characteristics of trampling will be more evident on the surface and gradually diminish downwards. Thus, trampled bodies will have a relatively sharp horizontal upper contact and a gradational lower one. In general, trampling produces horizontal distribution of features including minerogenic and anthropogenic particles. Most of the trampled deposits appear compacted and hard on their surface, particularly when they are damp or clay rich. Under very dry conditions, however, trampled areas consist of a loosely sorted and aggregated, millimetre-thin horizon overlying a hard, compacted substrate (Figure 3.75).

Continuous trampling leads to rounding of some relatively soft artefacts such as pottery, charcoal, and bone due mainly to attrition as the objects crash into each other (Rentzel and Narten, 2000). More durable and brittle

Figure 3.75 Modern trampling of hearth remains. Because of continuous reworking under hot and dry conditions, the sediment has been turned into a loose aggregated material with rounded charcoals. Note that there is lateral sorting of the coarser charcoal particles away from the viewer. Diros Bay, Greece.

artefacts such as flint will break and become blunt (Villa and Courtin, 1983). In shell middens, trampled shells are subhorizontally distributed, compacted, and highly fragmented *in situ* (Balbo et al., 2010).

Microstructure   Under the microscope, trampled surfaces are characterized by parallel, horizontal thin cracks (Figure 3.76). However, often this uppermost horizon is not preserved. Furthermore, in very dry environments trampling produces thin, millimetre-sized horizons composed of rounded silt-sized microgranules (Rentzel and Narten, 2000; Miller et al., 2013).

The main body of trampled sediment appears aggregated, often rounded to subrounded and compacted, with very few pores other than horizontal cracks (Figure 3.21b) (Gé et al., 1993; Courty et al., 1989: 124–125; Milek and French, 2007). Overall, trampled sediments show an

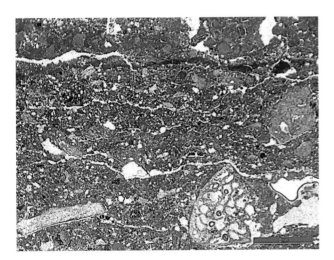

Figure 3.76 Photomicrograph of parallel oriented fissures produced by trampling of a constructed floor that contains fragments of clay. Archaic, Kalapodi, Greece.

embedded, porphyric-related distribution (Matthews, 1995), but enaulic-related distributions have also been observed. On prepared plaster floors, dislodged plaster aggregates are incorporated into the overlying trampled sediment (Matthews and Postage, 1994; Mathews, 1995). In most cases, the orientation of artefacts is horizontal or undulating, but other orientations have been reported in wet plastic substrates, rich in clay, where deformation of the sediment is also observed (Eren et al., 2010; Huisman et al., 2009; Rentzel and Nartern, 2000; Gebhardt and Langohr, 1999). Therefore, plastic folding, rotational features and microfaulting are evident in such cases; however, they should be pronounced close to the surface of the body, not evenly spread, and mainly affect aggregates and not the matrix, as in the case of constructed floors, for example. In addition, prepared floors are more homogeneous, often have planar lower boundaries, and exhibit aggregates that tend to exhibit a high degree of welding within the matrix due to more careful kneading.

Compaction seems to be more pronounced in damp conditions. Compressed clods collected on the soles of feet form lenticular features in relatively dry substrates; the superposition of these lenses can result in thin laminated bedding structures that are also associated with subhorizontal fissural porosity (Gé et al., 1993; Goldberg and Macphail, 2006: 221; Banerjea et al., 2015a,b). Dusty impure clay coatings have been linked to trampled doorway deposits under relatively damp conditions, whereas microlaminated silty clay coatings have been observed in trampled sediment under wetter conditions in probably semi-open places (Banerjea et al., 2015a). Additional evidence of trampling is *in situ* crushed bone, shell, pottery and other artefacts, and the general lateral continuity of *in situ* broken fragments (Figure 3.77) (Balbo et al., 2010; Goldberg et al., 2009a; Miller et al., 2010; Mallol et al., 2010; Aldeias et al., 2014a). In damp environments,

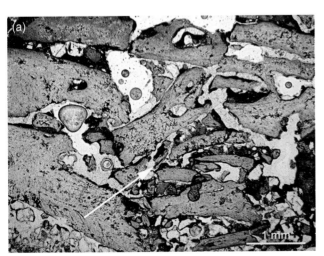
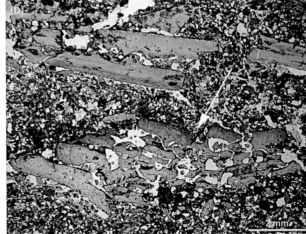

Figure 3.77 Photomicrographs of *in situ* crushed bone (arrows) due to trampling. MP, Pech de l'Azé IV, France.

chemical weathering is also evident and generally resembles the features described in midden deposits (Gebhardt and Langohr, 1999; see also section 3.5.2).

Relation to Other Archaeological Materials and Artefacts

During trampling, displacement and substantial damage of artefacts may occur (Nielsen, 1991; Villa and Courtin, 1983; Gifford-Gonzalez et al., 1985; Eren et al., 2010). Based on experimental studies, horizontal displacement is in the range of a few tens of centimetres up to about 1 m in some cases. Vertical displacement is constrained to a few centimetres, with maxima up to about 20 cm in wet substrates modified by heavy animal traffic (Eren et al., 2010).

Trampling produces modifications in anthropogenic materials such as bone and lithic artefacts that in some cases produce false 'cut marks' and 'use-wear polish', or bone fragments that resemble tools (Reynard, 2014 and references therein; Sala, 1986). In shell middens, shells are found often severely broken, although they are still articulated. Trampled street deposits appear to contain small sizes of sherds due to continuous breakage of potsherds into ever-smaller sizes (Figure 3.78) (Rowley-Conway, 1994). Human trampling arranges archaeological assemblages in a planar or linear manner according to surface geometry (Benito-Calvo et al., 2011).

3.6 Concluding Remarks

To summarize, archaeological sites are comprised almost universally of a depositional 'glue' that holds all the other components of the archaeological record together, including architectural elements, ceramics, lithics, and so on. Moreover, as in Chapter 2 with natural sediments and processes in sites, we tried in this chapter to demonstrate the richness of archaeological deposits as significant records of human history, including their mode of formation and what they can tell us about very short-term (almost instantaneous in 'archaeological time') human actions and longer-term activities involved in their production. In a way, anthropogenic sediments can be thought of as being a DVD with scratches, where parts of the video are faithfully recorded (e.g. individual dumping or burning events) along with parts of the story having been skipped, pixelated, or blurred. Such gaps can be caused by 'missing time' due to erosion, non-occupation

Figure 3.78 Scans of resin-impregnated slabs of street deposits. Due to continuous trampling on the street surface, pottery has been broken to produce carpets of potsherds. Hellenistic – Classical, Molivoti, Thrace, Greece.

(equivalent to watching paint dry), or by post-depositional effects such as bioturbation and diagenesis (see sections 2.6.2 and 2.7.2). Nevertheless, we hope that the reader has come away with the concept that the detailed study of anthropogenic deposit should be an integral part of doing archaeology.

4

Site Stratigraphy

4.1   Introduction

In Chapter 1 we introduced the essentials of the concept of stratigraphy and its definition. In this chapter we present a more thorough analysis of stratigraphy and its essential role in doing archaeology.

The analysis of the relative chronological sequence of a site is the foundation for the interpretation of its archaeology. This analysis is normally based on three major constituents of the site: the objects, the architectural remains when they exist, and the deposits that encompass all the above (Warburton, 2003). More rarely, other features like pits or burials are fundamental elements of a site and hence constitute a major analytical component. The chronological sequence of objects is based on their typology and technology, and those of architectural remains on their construction phases, which are assigned to a chronological position that is also based on their typology (Warburton, 2003). However, both objects and architectural remains are integral to the sedimentary deposits and thus their ultimate context is encapsulated within by the stratigraphy of the deposits.

In turn, the stratigraphy of the deposits is defined by the sequence of layers. Architectural features such as walls help us to differentiate sequences of layers and therefore reveal their spatial relationships, and we must remember that the establishment of contemporaneity of buildings, for instance, is based on correlation with their associated deposits. Although commonly the excavation of the deposit and the exposure of the objects that are contained in them are simultaneous, digging ultimately precedes the final exposure and observation of the objects and constructions. It is therefore strange not to base the primary chronological succession on the sequence of the layers that contain the archaeological objects. As will be shown below, the process of filling a space with a deposit is not related to its content. Here, the notion of filling has no specific genetic connotation and refers to how the material is placed into the empty space on the earth's surface and then occupies part of it. Obviously this process may be only marginally linked to the content and not necessarily defined by it. However, only by deciphering this connection can we safely move forward and correlate the objects and deposits, and thus define their chronological succession. Indeed, the whole excavation endeavour is about revealing this connection.

Another problem with the study of stratigraphy based on the objects and structures and not on the deposits is the subliminal perception that the deposits do not carry cultural meaning. Indeed, this is the fundamental misconception and shortfall of just about all theoretical treatises on site formation processes. We have already stressed in the Introduction that the excavated sediment is not 'noise' that we have to get rid of in order to understand the archaeology of the site. All kinds of pure anthropogenic deposits (e.g. hearths, pits, occupational surfaces, mud and plastered floors) are archaeological objects *per se*, and the only reason that we do not treat them as such is essentially a problem of education (or lack thereof) and technology. If we look at recent archaeological excavations, we see today that more and more fractions of these deposits are treated as archaeological objects (Matthews et al., 1996, 1997; Macphail et al., 1997; Goldberg and Macphail, 2006; Mallol, 2006; Goldberg, 2008; Goldberg et al., 2009a; Goldberg and Berna, 2010; Karkanas et al., 2004; Shahack-Gross et al., 2005; Milek and French, 2007; Karkanas and Goldberg, 2008; Karkanas, 2007; Maureille et al., 2010; Shillito and Matthews, 2013; Ismail-Meyer et al., 2013; Miller et al., 2013; Mentzer, 2014). However, the geogenic deposits or the anthropogenic sediments deposited or affected by geogenic processes also carry cultural meaning that is related not only to their archaeological content but also to the process of their formation as a three-dimensional body. Their formation is governed by the anthropogenic substrate, that is, all the activities that have shaped the site up to that point. Their content may also provide proxies for environmental reconstruction, but still, if they do not, they participate in the constitution of a site and its mere existence. *There is no such thing as a deposit that is irrelevant to the archaeology and history of a site.*

Reconstructing Archaeological Sites: Understanding the Geoarchaeological Matrix, First Edition. Panagiotis (Takis) Karkanas and Paul Goldberg.
© 2019 John Wiley & Sons Ltd. Published 2019 by John Wiley & Sons Ltd.

4.2 Historical Overview

The constituents that define the chronological succession, that is, objects, constructions, and deposits, actually define the major schools of stratigraphic analysis (Warburton, 2003). For Palaeolithic excavations, the stratigraphic record is usually dominated by geogenic processes as based on the study of their deposits. This method was one of the first used in archaeology in Europe (Lyell, 1863; see also van Riper, 1993; Rapp and Hill, 2006). Depositional stratigraphy was rarely used in historical sites because it was believed that such sites are dominated by anthropogenic processes – including architecture – and therefore they cannot be interpreted using natural sedimentary paradigms; in historic archaeological timeframes, archaeologists also rely on texts, and thus believe that there is little need for other 'external' sources of information. Moreover, in such sites the stratigraphy is often understood in terms of construction phases that are based on the analysis of architectural remains (e.g. rooms, monuments, plazas, walls), and in fact this strategy is indeed very successful in defining the history of erected constructions. All deposits in between the architectural remains are linked to the construction phases, assuming a direct relationship between the constructed phases and their deposits. The archaeological objects of the deposits are further analysed in terms of their typology and therefore a chronological succession is produced. Often this typological approach leads to defining stratigraphic units based on the statistical homogeneity of their assemblages. Hence, in practice, a few younger or older objects in an otherwise chronologically consistent group of layers are treated as 'intrusions' and not further evaluated, despite the fact that logically major mistakes can be made in interpretation. Furthermore, the architectural and typological approach does not take into consideration the different origins of deposits inside buildings (see Courty, 2001). Construction fills, floor sequences, primary and secondary use occupational sequences, and phases that are related to temporary abandonment cannot always be linked to construction phases without their sediment being studied and interpreted.

During excavation, archaeologists tend to interpret such deposits based on their experience, often attained by traditional and orally disseminated knowledge. Archaeologists learn empirically the types of deposits of a site as they excavate them, which results in almost site-specific knowledge, or in the best of cases, a knowledge applicable only in certain cultural periods and specific regions; this knowledge base is always being questioned when an unknown feature appears. This situation stems partly from the widely held – and mistaken – belief that there are no universal characteristics that can be used to differentiate such types of anthropogenic deposits. However, in spite of the fact that the repertoire of human activities is infinite – as well as the issue of equifinality where different activities can produce similar deposits – there are also some fundamental differences among major types of anthropogenic deposits. These differences are based on the different dynamics of the activities that produce them and are analysed in detail in Chapter 3 on anthropogenic sediments.

Today, the majority of historical excavations make use of the Harris Matrix to take care of stratigraphic recording (Harris, 1989; Harris et al., 1993). No doubt, the Harris Matrix represents a successful tracking tool of the stratigraphy of a site. In a way, the Harris Matrix can be considered as a version of the geological approach as it considers stratum as the basic unit to be reconstructed and adopts some of the geological stratigraphic nomenclature, albeit in a revised manner. However, as discussed in the Introduction, its major flaw is the assumption that interfaces can be defined during excavation without knowledge of the processes that deposited the layers and therefore produced these interfaces. But, as very cogently pointed by Warburton (2003: 16–17) discussing the problem of recording and identifying the relationship of walls and associated floors, '...only the analysis of the character of the deposits and interfaces can confirm whether they are floors and whether they abut the wall or are cut by it'. Moreover, the Harris Matrix approach to stratigraphy considers drawing of sections as being non-representative and unnecessary. Although it is true that deposits exposed in a corner of a building do not necessarily have to be similar to those of the entire building, they are still informative. Indeed, since layers normally cannot be followed along an entire room, it is expected, for example, that if a space is delimited by walls, most major contacts that define the change from construction fill to floor sequence and to destruction phases should extend along the whole area. Consequently, a section of the deposits will probably not reveal details of the geometry of a particular major contact or include all variations of the occupational deposits; nevertheless, it can provide the basic information on the nature of major lithological changes.

Finally, we stress that a section drawing is actually a test whether the succession of layers as depicted has an internal consistency and portrays a meaningful stratigraphic sequence. Anecdotally, one of the authors remembers his experiences in geological field camp decades ago, where the instructor 'encouraged' us to draw or sketch geological structures and deposits, saying that doing so forces you to see if your story makes sense. It is mirrored more recently by Stow (2010: 18): 'Field sketches and logs are as important or more important than many words of description'. (see section 4.11).

More recently, the need for a theoretical framework for interpreting stratigraphy was raised in a series of discussion articles by McAnany and Hodder (2009a,b), as well as by Berggren (2009), Helwing (2009), Maca (2009), and Mills (2009). What is interesting in these articles is that for the first time archaeologists coming from different schools of thought discuss the possibility of regarding layers and deposits themselves as finds and admitting at the same time that stratigraphy 'defines archaeological context' and is 'the jugular vein of archaeological practice' (as previously quoted). Indeed, irrespective of whether someone will agree on the need for a social interpretation of stratigraphy – or its scientific validity – one of the take-home messages of this discussion is that basic stratigraphic data is the starting point and a prerequisite for doing archaeology. Period.

4.3 The Definition of Stratigraphic Units in an Excavation

The vast majority of stratigraphic entities in an excavation are described as 'layers' or 'features', and in some excavations 'context', 'locus', or stratigraphic unit is also used for describing all of them. Following the definitions of archaeological deposits presented in the previous chapter we will not consider and discuss here stratigraphic entities such as standing constructions (walls, etc.) or other lasting constructions that normally are not parts of the excavated deposits *sensu stricto*. We wish only to remind the reader that the stratigraphy – if it can be called this – of architectural levels is based on typological and architectural characteristics and usually cannot be analysed on the basis of stratigraphic principles of deposition. We also believe that the existent methodology of defining construction levels has been proven successful, and there is no need for revising it. However, we *do* need to consider the stratigraphic relationship of standing walls and their associated deposits.

In geological nomenclature, a *bed* is the basic stratigraphic unit and is defined as a body of sediment that has been deposited by similar physicochemical conditions with physical properties that differ from those above it and below it (Collinson and Thompson, 1989). Note though, that in archaeological sites the term 'bed' is not normally used (except perhaps as a short form of 'bedding'), but instead the terms 'layers', 'strata', or 'stratigraphic units' are used (Stein, 1990). In sedimentary geology, a new bed is formed whenever the source of the materials or the conditions of deposition change. If the new source has a different mineralogical composition it will supply a different content to the new bed. The conditions are described by the medium of transport and deposition (air, fluid, and gravity) and the energy conditions prevalent at the time of deposition. For instance, a decrease in the velocity of fluid will lead to deposition of the coarser particles that cannot be carried farther. This change will lead to the formation of a bed with a characteristic grain-size distribution, which is different from the bed above it or below. Thus, the distinction of a bed is based on sedimentary characteristics, such as colour, grain-size, mineralogical and organic content, as well as shape, roundness, and geometric relationships (fabric) of the particles. In geology, sedimentation units thicker than 1 cm are referred to informally as layers or strata and below 1 cm as laminae (Figure 2.7) (Collinson and Thompson, 1989).

Furthermore, a geological bed is defined by its upper and lower bounding surfaces, which reflect the changes in the depositional conditions. These surfaces are known as bedding contacts or bedding planes. A contact can be defined by the change in the characteristics of the whole body of the bed and therefore demarcates this change. However, there are often cases where similar beds are separated from each other only by the presence of well-defined contacts. This situation implies that all these beds were actually deposited under the same depositional conditions but a recurrent change in these conditions produced the contacts themselves. Such changes could be only a variation in the rate of deposition, which leads to a short period of stasis and consequently to enhanced compaction during the time of depositional starvation; it can also lead to deposition of undetectable microscopic laminae of dust or clay at the bed contact, all of which lead to the development of bedding.

Certain depositional environments are characterized by rhythmic alternation of some of their conditions, which leads to the deposition of rhythmically alternating layers or a set of layers that define discrete beds. This occurs, for example, when ripples are formed by wave or current oscillations, and as they move in space and time they are deposited at an angle to the main depositional surface and construct cross beds (Figure 2.14). Another example of bedding formation occurs when there is rhythmic alternation of quiet conditions and current deposition (e.g. slackwater deposits), which will lead to alternating deposition of fine- and coarse-grained layers that overall build a single bed (Figure 2.10c). A new bed may be formed if the rhythm is changed or there is a different supply of material. Although someone might be tempted to subdivide the bed into their smaller depositional units, this cannot be done in practice because such temporal rhythmic changes do not produce laterally persistent contacts nor clearly separated three-dimensional bodies of sediment. Nevertheless, these obscure changes can be studied microscopically in thin section, which will reveal details of the depositional processes.

Figure 4.1 Lateral variations in layers. (a) Burnt layers with varying degrees of lateral reworking. Unit 1 contains several intact lenticular hearth features surviving inside raked-out hearth remains. Unit 2 consists of homogenized raked-out and trampled ash remains that laterally fade into burnt remains of units 2 and 3; the latter probably represents mostly raked-out burnt remains. Such burnt sequences are difficult to separate into distinct, laterally continuous layers. MSA, PP5-6, South Africa. (b) The grey layer delineated with dotted lines contains several distinct features, such as some hearth features in the upper contact (H) and the red clay lens (R) in the right side as well as some diffuse ones like the brown concentration in the left side (B). Furthermore, the matrix of the layer is quite heterogeneous with lateral and vertical changes. Overall, however, the layer makes a very conspicuous body of sediment with distinct, clear boundaries between the underlying and overlying deposits. Such a deposit implies a gradual accumulation resulting from a series of background activities interrupted by short-term specific activities. Scale is 30 cm. MBA, Kolona, Aegina Island, Greece.

The somewhat simple definition of bed given above is valid for sediments deposited in large basins such as oceans and lakes. However, in small terrestrial depositional basins, such as riverbeds or hillslopes, the situation can be quite complex. Nevertheless, the preceding general concept of what defines a sedimentary unit is valid. All changes in depositional conditions are recorded in the formation of new beds even in the most complicated cases. Because depositional conditions also vary laterally, beds often change laterally. If the change is enough to produce distinct bounding surfaces then a new bed will be formed (Figure 4.1).

Practically, though, the stratigraphy of such complex sequences is based on a hierarchical approach, which is described below in more detail. Here, suffice it to say that once the depositional processes of a site or sequence is understood, one should not be overly concerned with the drawing of each and every fine lamination, say, in a sequence of laminated sheetwash deposits; grouping them into a layer of 'finely laminated interbedded sands and clays' would be sufficient (Figure 2.40). Over the years, both authors have experienced archaeologists who insisted on drawing every detail in an excavated profile, largely because they did not understand the sedimentary dynamics operating at the site. Obviously, it would be preferable to record as many stratigraphic units as possible during the excavation and then later lump them together and draw the profiles accordingly, rather than realize later that important contacts are missing.

The definitions described in the previous paragraphs cannot be directly applied to the realm of the archaeological site, particularly when it comes to purely anthropogenic deposits, which are not laid down by geogenic processes. Obviously, however, discrete stratigraphic units are formed by anthropogenic activities affecting a certain type of material when (a) the activity that produces the material of the layer or the feature does not change during the time period of the filling of the three-dimensional space by this layer or feature, or (b) the activity is repeated with a rhythm that overall produces the same material. For example, the discard of burnt remains from a domestic kitchen each morning on the side of the yard – assuming that they are not affected by natural geogenic processes – will lead to the formation of a distinct layer of burnt remains. If the overall composition of the burnt material produced every day is generally the same, then one massive layer will be formed. If, on the other hand, the material varies then several layers will be formed with rather diffuse contacts (Figure 3.14). Some layers will be more distinct than others, depending on the differences of the discarded material. If there is no discard for a considerable time, surface alteration will occur to produce a contact separating two similar layers (other things being equal and if the process of discard continues after the break). A sequence of floor layers will be produced by the subtle variation of the construction material used or by the occupational debris produced on the floor surfaces. Note that as in purely geogenic sequences, where rhythmic alternations of a set of layers can represent a distinct bed, the same can happen with anthropogenic deposits like the ones produced by constructed clay floor sequences (Figure 4.2).

Similarly, a construction fill will be represented by a massive layer if it was filled at once; several layers can be represented if the same type of construction material accrued as a result of repeated deposition (Figure 4.3). In the latter case, bedding can be produced by changes in

the material used for the fill or by the exposure of the surface of each phase to atmospheric conditions of alteration.

Distinct layering can also be produced by random activities such as foot traffic. As we noted in Chapter 1, although many researchers would consider trampling a post-depositional process, it should really be considered as an activity that builds the body of sediment as it accumulates. For instance, the area around a Palaeolithic hearth will entrap burnt remains, and trampling will gradually homogenize sediments to produce mixtures of these burnt remnants, other human-derived residues, and geogenic sediments derived from the substrate. This trampled material will show up as a distinct layer that is laterally continuous with the *in situ* hearth remains; most often, it will include the undisturbed hearth vestiges as a separate lens (see also section 5.2.1).

The reason for the latter is that normally a Palaeolithic hearth represents a short-term (e.g. seasonal or similar duration) occupational feature. When people return, and if they choose a different locality to build their fire, they will often trample the burnt remains from previous visits. The extent of trampling will depend on the quantity of geogenic sediments that were deposited and which then sealed the burnt remains in the intervening time period. Therefore, in cases where there is some geogenic deposition, the sequence of sediments in the vicinity of fireplaces will consist of trampled, burnt material with islands of well-preserved hearths. This is exactly the case at Kebara and Hayonim caves (Figure 4.4) (Meignen et al., 2001, 2007; Goldberg and Bar-Yosef, 1998). In cases where there is not much geogenic deposition, trampling may produce thick massive layers of trampled hearth remains without preservation of intact hearths. It thus appears that the interplay of geogenic and anthropogenic sedimentation rates is crucial in the production

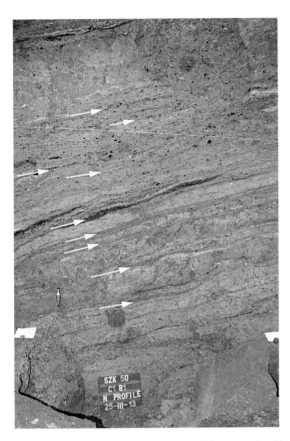

Figure 4.2 Thick sequence of constructed floors made of loess with interbedded occupational debris (some shown with arrows). Neolithic, Szeghalom-Kovácshalom, Körös Region, Hungary.

Figure 4.3 Types of construction fills: (a) massive homogeneous fill probably representing one episode and (b) sequence of fills formed in different times, the lower one (L) below the wall (W) and the upper one (U) after the destruction of the wall. MBA, Palamari, Skyros Island, Greece.

Figure 4.4 Plan view of intact (I) and trampled (T) parts of hearths. MP, Hayonim Cave, Israel.

of the final stratigraphic situation that we observe during excavation.

Following these previous definitions, each layer is characterized by its content and fabric. The fabric is the result of the depositional process (e.g. trampling, discarding, laying, applying) and it concerns the spatial arrangement of all constituents (particles and pores), their shape, size, and frequency. The content includes minerals and rock fragments derived from natural sources (e.g. the substrate of the site, a quarry, or water, and air-transported material) or newly formed minerals that have formed by anthropogenic activities (e.g. calcitic ash and burned clay from pyrotechnological activities). Content also includes organic matter (charcoal, plant remnants, phytoliths, starches, etc.), bone, shell, and other biological materials, as well as products and by-products of craft activities (e.g. pottery, metals, mortar). As a consequence, a different layer will be produced not only when the activity or the set of activities change but also when the content (and consequently the source or the material affected by the activity) changes as well. Note that post-depositional processes (e.g. cementation, oxidation, etc.), including soil formation processes, do not produce a new layer.

4.4 Nature of Contacts

As described above, stratigraphic contacts are a direct manifestation of changes in the depositional regime. Thus, they are of crucial importance for understanding the stratigraphy of the site and consequently the depositional processes responsible for its formation (Mallol and Mentzer, 2015). The nature of boundaries is at least as important as the characteristics of the layers themselves. The bounding surface or contact between two successive stratigraphic units can be abrupt or gradational (Boggs, 2005) (Figure 2.9a,b). A gradational contact is a boundary that cannot be precisely defined in the section but it is a diffuse zone between two units that can be measured. A gradational contact implies a gradual change in the conditions of formation, if the blurring effects of bioturbation can be eliminated. Abrupt changes produce sharp contacts that may further be defined as non-depositional or erosional. Erosional boundaries are of special interest because they imply loss of material and information, and often a temporal hiatus. In this respect, the geometry of the boundary is very helpful. Erosional contacts are abrupt and normally irregular, making a relief of protrusions and troughs (Figure 4.5a). They are often inclined but an inclined boundary, which truncates more than one contact, is by definition erosional (Figure 4.5b). In the limited area of the excavation trenches, however, some of these irregularities in the contact might not be clear.

Erosion is a natural process whereby water or wind removes previously deposited material. Channelized water flow is one of the most destructive erosional processes, since depending on the size and scale, entire layers or parts of sites can be removed, as in the case of Shiqmim in Israel (Figure 4.6) (Goldberg, 1987). On the other hand, in less energetic settings, surface erosion by wind (deflation) or by sheetwash might be also important erosional processes.

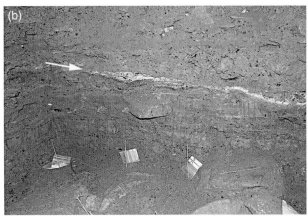

Figure 4.5 Erosional features. (a) Troughs and cuts (arrows) in a water erosional feature filled with stone lag. MSA, PP6-6, South Africa. (b) Inclined contact (arrow) truncating different parts of the underlying layer. Neolithic, Panos Cave, Attica, Greece. See also Figures 1.5b, 2.56b, and 2.79.

Figure 4.6 Erosion of Chalcolithic site of Shiqmim, Negev, Israel. (a) The architecture of the site is clear in this aerial view, but disappears in the right-hand part of the photograph (dashed line) because of erosion by the Nahal (Wadi) Be'er Sheva (NBS) and subsequent deposition during Byzantine times. (b) Cross-section of (a) showing alluvial silts of Byzantine age (B) eroded into Chalcolithic gravels (C) which intergrade into the occupation levels of the site in (a).

Anthropogenic activities leading to removal of previously deposited sediment are not normally called erosional, but activities such as digging lead to truncation of the deposits and formation of similar to geogenic erosional surfaces called 'cuts' (MoLAS, 1994).

Erosional contacts of whatever origin attest to temporal stratigraphic gaps. However, not all erosional contacts are temporal hiatuses simply because the time represented by the eroded material might be negligible (of course 100 years during the Middle Palaeolithic is different from the same interval during the Iron Age, for example) or the erosion occurred only locally. Well-developed soils can also be considered to represent temporal gaps because, as discussed in Chapter 1, they imply depositional stasis (Figure 4.7). Indeed, in several geological environments, erosional contacts and palaeosols are thought to represent the overwhelming part of

geological time, and only a tiny fraction of time is represented by the deposits themselves (Ager, 1993). In archaeological sites most contacts are normally gradational, with few of them clearly erosional. Gradational contacts are to be expected in heavily occupied sites because more than one activity each time is recorded in the final sedimentary product, and any change in the amount of input by one type of activity will produce a gradual effect. In other words, if one activity becomes more dominant the sediment will gradually change because there will always be the background of the other activities.

A floor surface, for example, accumulates material resulting from several activities, such as foot traffic, food consumption, or food production; through time these will be homogenized to become what will be called 'occupational debris.' However, the presence of a hearth will progressively enrich this debris with burnt remains in the vicinity of the hearth (Figure 4.8). Therefore, since many regular types of occupation are expected to produce a range of sedimentary layers with similar appearance and gradational contacts, any

Figure 4.7 Buried red soil (S) with prismatic structure under colluvium (C). Soil formation processes have affected the underlying clay construction feature by enhanced weathering along cracks. Neolithic, Koutroulou Magoula, Greece.

Figure 4.8 Burnt material spread laterally to the left of the main hearth features (H). White tag is 10 cm long. Neolithic, Vésztő-Mágor, Körös Region, Hungary.

major change that will include destruction, abandonment, or a major change in the type of occupation will be recorded as an erosional contact. Abandonment will be characterized by the prevalence of natural deflation and other erosional processes producing an uneven relief; if erosion is relatively minor, pedogenesis will occur.

Destruction is a complex process that includes anthropogenically induced natural processes such as conflagration; it may also include post-destruction natural processes if local abandonment occurs (e.g. van Keuren and Roos, 2013). All of these processes may lead to a new occupational phase that most often will need a major reorganization of the space in the area. During such a process, demolition, truncation, packing, and levelling are expected to happen. As a consequence, one will normally observe such a distinct contact that truncates all underlying features and layers (Figure 4.9).

In Palaeolithic sites where geogenic processes are much more prevalent, a broader spectrum of contacts has to be envisioned. In these types of sites, sedimentary stasis and natural erosion is the most important agent in producing major stratigraphic changes. Sedimentary stasis will not only enhance pedogenic processes, but also other physical and chemical alterations. Dissolution in particular may lead to differential subsistence and inclination of layers (e.g. Kebara Cave, (Schiegl et al., 1996), Theopetra Cave (Karkanas, 2001), Tel Dor (Shahack-Gross et al., 2005) (see also Chapter 5)). When sedimentation resumes, the new layers will be deposited unconformably on top of the altered sequence, thus recording an important temporal break (Figure 2.82a). However, such dissolution features may also be observed in quite recent sites because of the solubility of materials such as wood ash, which is found in large quantities in all pre-industrial sites. Yet, stratigraphic variability in old, non-architectural sites is produced by the presence or absence of occupation that introduces into the system a distinct anthropogenic component.

Figure 4.9 Dark ashy destruction layer (D) clearly truncating the underlying layer and overlain by a wall (W), which demarcates a new building phase of the settlement. MBA, Palamari, Skyros Island, Greece.

4.5   Time and Stratigraphy

In the above narrative, a stratigraphic unit represents a geometric body with certain characteristics that fill a space. However, the crucial aspect of stratigraphy is time. The ultimate use of stratigraphy is to reconstruct the history of the deposits, and in this sense deposits are proxy indicators for not only process, but time as well. The formation of a single stratigraphic unit implies that a certain minimum amount of time passed in order for this deposit to be formed. Thick layers *may* represent large periods of time, but at the other extreme landslides occur virtually instantaneously.

Obviously, it is impossible to subdivide further a homogeneously thick unit and reconstruct its internal time history. Therefore, the time resolution of a stratigraphic section is the resolution of the depositional characteristics that define the different individual stratigraphic units. In an area of occupation, the resolution of a single activity will be defined by the amount of time that the final product was exposed at the surface, which in turn is related to the sedimentation rate of this particular activity. It is expected that a single activity will produce sediments that overall have distinct and relatively uniform characteristics (e.g. colour, type, shape and form of particles, and fabric), and thus will produce a body of sediment that will be different from the one above and below it. Since more than one activity normally happens in an area, subsequent activities will tend to obliterate the characteristics of the sediments produced by the previous one.

Nevertheless, the chance of intact preservation of at least a fraction of each activity will depend on the amount of sediment produced by it, which in turn depends not only on the nature of the activity, but also on its duration. As we have already noted above, the activity can produce only one stratigraphic unit if it is repeated several times for a long period of time and produces the same final product. Since we are interested only in the final product that is left on the ground, more than one activity might be responsible for this product. We want to stress the point once more that discard and trampling are also distinct human activities that produce sediments; they also have duration and probably temporal attributes such as a rhythm. They are not post-depositional processes.

Geogenic processes can disturb previously deposited anthropogenic sediments, or seal them and thus protect them. When a natural process redistributes archaeological sediments, it produces a new layer with different attributes, and hence a new stratigraphic unit is formed that fills this particular space and thus records a new event at a new time. The same happens with anthropogenic activities that redistribute previously deposited materials. For example, producing a pile of sediment by digging a pit will result in the formation of a new

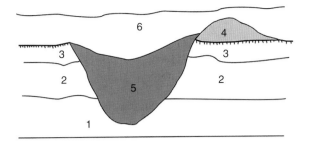

Figure 4.10 Sketch of the stratigraphy of a typical pit fill. The pit has cut into layers 1, 2, and 3. Layer 4 represents remains of the excavated pit sediment thrown at the side. Layer 5 represents the backfill of the pit that in this case partially covers the material that was dug out of the pit and placed to the side (layer 4). In principle, layer 5 is younger than layer 4, although often they are penecontemporaneous.

stratigraphic unit, above and to the side of the pit. Its age is the time of the formation of the pit. The fill within the pit will be another stratigraphic unit that post-dates the pile, although it is lower in elevation than the pile. Note that this arrangement does not violate any stratigraphic principle, as will be shown below (Figure 4.10).

4.6   Massive Thick Layers

A massive thick layer constitutes rather an undesirable puzzle in the archaeological record. There are cases where such layers could have been formed at once by a natural disaster such as a flood (Figure 4.11) (e.g. Boker Tachtit, Israel; Goldberg, 1983) or by human activities such as demolition and the levelling of an area (Figure 4.12) (e.g. Karkanas and Efstratiou, 2009; Shillito, 2011). Since these processes are well constrained in time, they do not really produce problems in the interpretation of the archaeology. However, there are cases where thick layers appear to have been formed over long periods of time that did not produce distinct features; this type of accumulation results in an absence of contacts that would enable us to further subdivide the deposit. Most of these layers have been deposited predominately by natural, undifferentiated processes that act in a similar way for a long time. Such processes are usually gravity related and do not yield well-defined sedimentary structures but rather chaotic and massive deposits, called colluvium (see section 2.3). Since each new phase of this kind of accumulation is not much different from the previous one – and generally the same types of materials are being continuously reworked downslope by gravity – a thick, homogeneous sequence is produced (Figure 4.13). In reality, if we look carefully, subtle changes can be observed and slight differences in lithologies can be recorded, although a secure stratigraphic division cannot be constructed. Nevertheless, the general formation processes can be established in all cases. This apparent

Figure 4.11 Marly alluvium containing Middle Palaeolithic and Upper Palaeolithic layers truncated by massive alluvium rich in cobbles and boulders. Boker Tachtit, Negev, Israel.

Figure 4.12 Sequence of massive, compact layers of packing fill containing demolition debris. White tag is 10 cm long. Neolithic, Vésztő-Mágor, Körös Region, Hungary.

homogeneity conceals the danger of extrapolating general processes from one part of the layer, which might not be applicable to all parts of the layer. Although a layer might appear to be formed by high-energy processes that have extensively reworked archaeological objects, there may be also hidden lenses where material was scarcely moved or even is *in situ* (see section 2.3.4).

4.7 Basic Stratigraphic Principles

Based on geological sedimentary stratigraphic principles and the above discussion we can formulate some basic archaeological stratigraphic principles that ought to be used to understand the history of archaeological

sequences and also correlate non-physically connected parts of a site.

4.7.1 The Principle of Superposition of Beds

As is well known in the Law of Superposition, younger archaeological layers are deposited on top of older layers in a sequence. Exceptions are only subterranean animal burrows or underground tunnels (Figure 4.14). Although walls are not considered deposits in this treatise, they do not violate this principle, contrary to what is widely believed (Figure 4.15). A wall is younger than the layer that it is built upon. It is also younger than all layers in the foundation trench that it cross-cuts as predicted by the principle of cross-cutting relationships

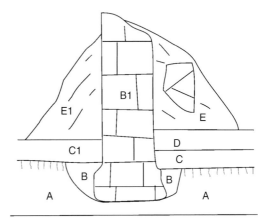

Figure 4.15 Stratigraphic sketch of a wall and its foundation trench. Layer A represents the substrate into which the foundation trench, layer B, was dug. Wall B1, from the stratigraphic point of view, is younger than its foundation trench, but in practice they are contemporaneous. Layers C and D represent occupational debris and floors in the interior area of the wall; C1 is the material that was accumulating outside the wall at the same time. E and E1 have been formed from the decay of the wall and superstructure.

Figure 4.13 Thick unsorted colluvium (CO) deposited by gravity flow processes (see section 2.3) burying an anthropogenic layer of snail shells (S) at Franchthi Cave (Mesolithic), Greece.

Figure 4.14 Horizontal burnt layers dated to the MP cut by a burrow filled with Neolithic pottery (arrows) some 3 m below the Neolithic strata of the site. Tag is 5 cm. Theopetra Cave, Greece.

(see next section). All other layers above this foundation layer are clearly younger than the wall; they bank up against it. In the same sense, a roof when it is part of the depositional sequence does not violate the principle of superposition: the time of collapse and actual deposition of the roof remains on the ground surface is younger than the layer immediately underlying it, which could be, for example, the last occupational floor, or even a soil developed on it. From an architectural point of view the time of construction of the roof might be younger than the ground surface, but from the depositional aspect this is irrelevant. If the subsequent analysis can reveal the exact time of the roof construction by relating it to a phase of the wall construction, the dating of the roof will be achieved by placing the roof into its real depositional stratigraphic context. All additional attributes that could relate a wall to a non-physically connected fallen roof are not stratigraphic but architectural. Stratigraphy is about strata and not about architectural elements.

4.7.2 The Principle of Cross-Cutting Relationships

A feature that cuts across a layer is younger than the layer. A pit, a foundation trench, or a ditch – and their fills – is younger than the layers it cuts (Figure 4.16). Erected structures such as walls do not cut across the fill in between them. Such features appear as exceptions to this principle, but as already discussed walls and similar constructions are not considered deposits *per se*. Note that in geology some formations also follow this rule and seem to violate the principle of superposition, for example volcanic intrusions.

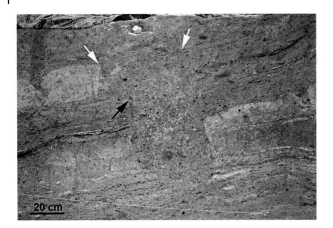

Figure 4.16 A pit cutting through stratified deposits. The uppermost possible rims of the pit are marked with white arrows at the point where clearly the stratigraphy is cut by the pit. However, the step-like cut on the left side of the pit (black arrow) might indicate two phases of digging of the pit. Bronze Age cutting through Neolithic, Szeghalom-Kovácshalom, Körös Region, Hungary.

4.7.3 The Principle of Original Continuity of Layers

Layers normally maintain their continuity for some distance and terminate in a regular manner, such as the end of a baulk, against a structure or wall of a cave. This termination can result from convergence and merging of the upper and lower contact or by lateral gradation of the layer so the contacts simply die away. It follows that when a layer terminates laterally in an abrupt manner, it does so by meeting a secondary cross-cutting feature, such as an erosional surface, a pit, or a previously erected construction (Figure 4.16). In all cases, except termination against a construction, the continuation of the layer should be sought beyond the cross-cutting feature.

The content of a layer can vary laterally because of subtle changes in the depositional environment or human activities, for example different uses of a room or plaza (see section 4.10 on facies). In most cases where the deposit is bounded by discrete contacts, we can safely assume that the overall process of infilling the space by the layer was the result of a persistent process with variations that did not result in new boundaries. This is the basis of the principle of continuity, which additionally states that a layer has the same age across its entire extent.

However, there are rare cases where a layer has been formed in different times along its lateral extent. In geology, this is often the case in certain depositional environments like river deltas, where the same formation continues to grow laterally in the coastal area over a considerable time. In archaeological excavations a similar diachronic type of layer can be formed when a depositional body is constrained by the same distinct boundaries, although laterally it exhibits markedly different attributes. The reason for the latter is that the content of this particular layer might be so distinctive and different from the surrounding material that although it was deposited by different processes at different time intervals along its extension it appears as one continuous body. Such is the case of the PP5-6 cave in South Africa, where a series of gravity flow deposits have moved burnt material from previously deposited layers and produced a new amalgamated layer. This layer has a distinct appearance that contrasts with the surrounding aeolian sands, but it has been formed in different stages (Figure 2.33b) (Karkanas et al., 2015a).

4.7.4 The Principle of Original Horizontality of Layers

All sediments deposited on the earth's surface whether water-laid, windblown, or accumulated along a slope are governed by gravity. Anthropogenic processes such as trampling, dumping, pugging, sweeping, brushing, discarding, and placing are also controlled by gravity. Therefore, when layers originally are laid down they were either parallel to the earth's horizontal or at angles generally lower than 30°; the latter is the angle of repose of most sediments, that is, the steepest angle on a slope on which granular material will remain stable and not slide. In other words, dipping beds can be a result of depositional dips associated with a sedimentary process (e.g. talus cone), or the result of purposely built constructions (e.g. ramparts and glacis; Bullard 1970), or they could have acquired this inclination from some post-depositional process, such as slumping, subsidence, or faulting (sedimentary features that are related to slope processes are described in section 2.3). Inclined layers are often observed in ashy sequences (Figure 4.17). However, it is generally expected that fireplaces were originally constructed on flat horizontal areas. Therefore, post-depositional processes are the most parsimonious explanation for the appearance of inclined layers rich in burnt remains. The same holds for organic-rich deposits, as their material is quickly degraded and decayed.

4.7.5 The Principle of Included Fragments

This critical principle is much underestimated and misunderstood in archaeological excavations. It states that material from older layers can be included in a younger layer but not *vice versa*. Indeed, strictly speaking, the majority of material within a layer is older than the layer that contains them. This seems contradictory, but the time of deposition of a layer is the time of the filling of a particular space with sediments. Inevitably, the content

Figure 4.17 Inclined intact stratified burnt remains. Since fireplaces are originally constructed in flat horizontal areas inclination should be the result of post-depositional natural processes such as slumping or differential dissolution of ashes. MSA, Site PP5-6, South Africa.

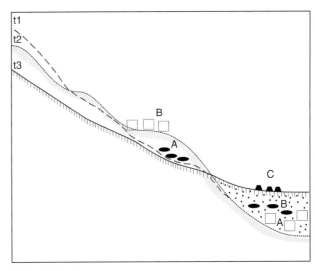

Figure 4.18 Stratigraphic sketch of "inverted" stratigraphy in slope deposits. At time t1, a flat area A of the slope is occupied. At time t2, erosion uphill of the slope buries site A. Site B is formed on the new surface thus produced. At time t3, the slope is eroded again, first moving downwards the materials of site B and then the ones of the buried site A. Site C is formed on top of the colluvial fill at the foot of the hill.

of the layer will be overloaded by materials that pre-existed and are reorganized as inclusions in the layer. All layers contain a geogenic component (e.g. sand grains), which is obviously much older, having been derived from the old bedrock and sediments of the area. But even the archaeological objects in most cases were not part of the activity system at the same time the deposit was laid down; it is possible that the objects are penecontemporaneous with the sediment and in general not really older than the enclosing deposits. However, it is not that old sediments and their archaeological objects are included in the formation of new layers. This inclusion might be attributable to a geogenic process, as for example rainwash eroding an area of older archaeological layers and depositing a new layer downslope (Figure 4.18). Nonetheless, anthropogenic activities commonly tend to mix older archaeological sediments, as in the case of using older anthropogenic-rich clay deposits for constructing floors, walls, etc.

In relatively rare instances, objects (e.g. lithics or ceramics) resting on an exposed surface or deposit may be incorporated into the upper part of the underlying strata, say, via trampling. In light of the above arguments, this upper part of the layer should be considered as a new layer. However, the incorporation of these pieces is normally so difficult to detect in the field (although they may be spotted in the laboratory during processing) that from an operational standpoint it is not possible to differentiate this layer as something different from the sediment that lacks the incorporated material.

The Principle of Included Fragments has also serious implications in dating. Charcoal pieces are always picked from the deposits of layer in order to date them. However, only by knowing the depositional processes responsible for the formation of a layer we can decide if the charcoal

piece was formed at the same time the layer was deposited (Goldberg and Berna, 2010; Goldberg et al., 2009c). Indeed, only intact burnt remains can securely date a layer. In all other cases, the dated charcoal piece gives a minimum age or what is called a *terminus ante quem* age because according to the principle of included fragments younger material cannot be included in a layer. Violation of this rule includes intrusion of younger bioturbated material and, in general, all disturbances that are included in post-depositional processes or the inclusion of materials trampled in from above, although they are not recognized as such; normally, trampled charcoal will likely not survive such abuse. We should stress once more that post-depositional processes are those that change the integrity of a layer. They are not processes that participate in the building of the original deposit as it was laid down on the earth before being buried by a new layer.

4.8 What is *'In Situ'*?

'In situ', literally in position, in its broader definition refers to archaeological objects not removed from their original place of deposition or use. 'Original place of deposition' implies that the deposition of the object in this place records a moment of human activity ('de facto refuse' according to Schiffer, 1983). It is not much different from the use of *'in situ'* in palaeontology whereby a fossil is found in living position. However, in this way the concept of *'in situ'* is not related to the concept of 'intact

deposit', although by definition an intact deposit is a closed system containing objects placed in this position and not disturbed afterwards. Intact (i.e. structurally complete) fireplaces, undisturbed burials, standing walls, whole pots, and debris of tool production are examples of '*in situ*' archaeological objects and features. Nevertheless, a broken pot found with its pieces fitting together is also considered *in situ*, although the act of breakage may not relate to a human activity but to the collapse of a wall after abandonment of the site, for example. As a consequence, 'grades' of *in situ* are commonly implied.

Indeed, as we move back in time where the results of a human activity cannot always be interpreted in a straightforward way, *in situ* might start out coinciding with an intact deposit. In this way, '*in situ*' refers to an archaeological object that can correctly date the layer in which it was found. We often read about layers with lithic tools that refit, hence they are *in situ*, but a specific human moment of activity cannot be demonstrated, since displacement of the refitted pieces over the site area could have taken place days or years after they were fabricated. Therefore, in these cases, the temporal dimension increases from an instant in time to the time interval of deposition of the layer. For even much older archaeological sites, it is sometimes sufficient to find out that the objects are 'geologically *in situ*' (*sic*). This term usually implies that although the artefacts might have been reworked and moved from their original depositional layer, they are enclosed in deposits that generally date to the time of their production and use. The Lower Palaeolithic site of 'Ubeidiya (Jordan Valley, Israel) is such an example (Mallol, 2006). The natural habitats of the site's hominids were mudflats, floodplains, and shallow subaqueous lacustrine environments located on a fluctuating lake shoreline. Frequent penecontemporaneous reworking of the anthropogenic remains was being carried out by low energy fluvial and lacustrine processes. *In situ* (*sensu stricto*) archaeological remains, therefore, are not encountered, but these vestiges – and the site as a whole – are considered to contain well-preserved and stratified successions of layers with anthropogenic remains dated to the deposition of this lake-margin sedimentary sequence, some 1.4 million years ago.

4.9 Human Constructions and Depositional Stratigraphy

As discussed above, construction phases are defined on the basis of architectural and typological features (see also Introduction). However, they have to be correlated with the overall stratigraphy of deposits, which includes constructions. A simple rule that is based on the above stratigraphic principles is as follows: a construction (or a phase of it) post-dates the layers that it overlies and crosscuts, but it predates the layers that bank up against it (Figure 4.15).

A construction is usually erected on a shallow foundation trench that is filled with sediment to stabilize the structure. This 'fill' is the first layer that banks up against the construction, although in real time it post-dates the construction – or at least its foundation. In fact, the fill dates the construction because it can be considered contemporaneous with the construction. Note that, although the age of the foundation fill generally dates the related construction, according to the principle of the included fragments its content can be much older. All other layers that overly the foundation-fill are *younger* than the construction. The layers into which the foundation trench was dug predate the construction (Figure 4.15). Nevertheless, defining which layers are floors or deposits that post-date the wall and which layers are part of the foundation fill that separate the wall from the pre-existing layers is not only a matter of the geometric relationships of the contacts, but also of the characteristics (composition and fabric) of the deposits themselves.

Postholes record the same types of sequence as walls, although in some wet soils posts may have been pounded into the substrate without first digging a hole. Once the post starts to decay, occupation deposits that accumulate during the lifetime of the site can fall into the posthole and eventually silt it up (Goldberg and Macphail, 2006: 255–257). As a result, filled postholes are younger than the construction but generally penecontemporaneous with the floor deposits or in general with the occupational phase related to this construction.

4.10 The Concept of Facies

Both small vertical and horizontal variations that produce ill-defined contacts as well as the rhythmic or gradational internal structure of layers are not easy to record during excavation. In addition, they cannot easily be resolved or unambiguously designated in a stratigraphic section or be assigned to a layer. This problem can be partly overcome by applying the concept of facies. Facies is a collective noun (never 'facie') for categorizing sediments with specified characteristics, and is a concept that is borrowed from geology (Walther, 1894; Middleton, 1973). A facies represents a body of sediment whose characteristics (e.g. composition, grain-size, and sedimentary structures) are recognizably different from the surrounding sediments.

A facies is the product of certain environmental conditions or activities that overall produce a distinct type of

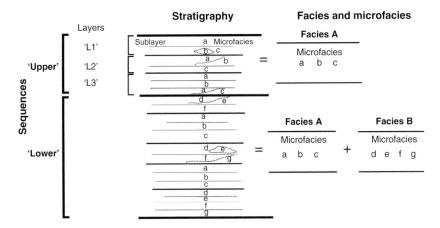

Figure 4.19 Schematic relationship between stratigraphic units, facies, and microfacies. Stratigraphic units (sublayers), layers, and sequences represent the smaller, intermediate, and higher hierarchical stratigraphic units, respectively. A single macroscopically identified sublayer consists of one or more microfacies (small letters) that produce the thinnest homogenous deposit identified in the field. A group of microfacies showing a distinct lateral and vertical distribution produces a higher order of homogeneity in the field, thus defining a facies (capital letters). A facies corresponds to higher hierarchical stratigraphic entities, a layer or a sequence. Sequences may consist of more than one facies. Stratigraphy and facies are different ways of describing a depositional sequence. Stratigraphy is based on vertical relationships of depositional units, whereas facies is about grouping depositional units based on their depositional characteristics and their vertical and lateral interrelationships, most commonly implying a genetic correlation between facies within the same group as well. Adapted from Karkanas et al. (2015a).

deposit. Geogenic deposits are usually grouped into facies based on grain-size distributions and internal bedding features that produce a sedimentary structure. As examples, we can have laminated sand facies, or structureless sandy gravel, or normal graded interbedded sands and fine gravels. A layer is by definition a unique body and therefore it receives a certain name on a stratigraphic section. Grouping can be as detailed as the particular study requires. Thus, in some cases facies can include several subfacies, or in fact subfacies can become facies in a different study. The study of sediments at the microscopic level can assist in further dividing facies into several *micro*facies (Figure 4.19).

An important aspect of facies is that it implies a relatively constant environment where the same activities happen regularly, and thus all contacts are normally not erosional or truncated. Often, sequences of facies are characterized by gradational contacts, which imply a gradual change within the same environment. In natural sedimentary environments, the Law of Facies predicts that a vertical sequence of facies will be the product of a series of depositional environments that lay laterally adjacent to each other (Middleton, 1973). Sedimentary layers essentially preserve the environment of their deposition. These depositional environments change through time and old depositional layers shift laterally and may become superimposed on surrounding deposits. For example, water flowing in a stream will deposit coarse sediments inside the channel and at the same time deposit fine-grained sediments on the overbank areas. Migration of the channel will lead to deposition of coarse

sediments on top of the overbank deposits and the latter will fill the abandoned channel.

The Law of Facies can occasionally be applied to archaeological sites, as certain sequences of facies can be considered characteristic of certain areas, environments, or activities. Hence, in the interior of a house the association of constructed clay floors, occupational debris (including lateral microfacies variation such as hearth-, food-, or craft-related debris), and foundation layers of floors often alternate and produce characteristic indoor sequences. Debris from floor wear, occupation, and maintenance practices produces a series of deposits that lie laterally adjacent to each other. Eventually, they form a vertical sequence, as these activities occur repeatedly and alternately in the same location (Karkanas and van de Moortel, 2014). In a similar way, at Sibudu Cave in South Africa, bedding and hearth-related activities and their maintenance practices (including sweeping and dumping) were happening contemporaneously and repeatedly in different locations during each occupation. Continuous occupation over time produced a thick laminated sequence consisting of all these various microfacies, with each representing different activities (Goldberg et al., 2009a).

Facies are different from layers in two fundamental ways (Figure 4.19). Facies is a way of grouping deposits and consequently the same facies can occur several times in a stratigraphic sequence. A layer, on the other hand, has a unique position in the stratigraphy of a site. Furthermore, a facies can contain more than one layer. Indeed, groups of layers that are characterized by similar characteristics are considered one facies. Moreover, facies have more of a conceptual flavour than a layer. In other words, a certain layer is generally site specific and restricted to that place. On the other hand, a certain facies type, for example a grey ashy sand, can exist at many sites of various types, especially in coastal areas of South Africa where combustion features and dunes are relatively common.

Application of the facies concept to archaeology is not so straightforward, but it is a very effective means to understand and record depositional processes, site formation, and structure. Grouping deposits into facies in archaeological sites is based on certain attributes that are most important for analysing the deposits of a site from a genetic point of view (Courty, 2001). For instance, in Palaeolithic sites the amount, distribution, and structure of burnt remains are often used to define and group occupational facies (Karkanas et al., 2015a). In sites dominated by architectural remains, the reorganization of sediments for clearance, levelling, and construction purposes produces a certain array of facies that is based on the amount, fabric, and overall organization of construction elements, such as broken pottery, charcoal, stones, etc.

The importance of the Law of Facies in archaeological sites – as also in geological contexts – is that we may predict and expect to find a certain facies sequence each time and exclude other associations of facies if the sequence has not been interrupted. This 'prediction' can hold with anthropogenic sediments because it is the *spectrum of sedimentary facies and their presence in certain combinations* that are indicative of distinct conditions during their formation. Thus they accurately define how a particular area has been formed. Consequently, it is particularly important to interpret each facies in its spatial association with other facies. Inside houses, for example, a normal vertical succession is as shown in Figure 4.20. Each combination will define a certain history of occupation and activities (see Karkanas and van

de Moortel, 2014; Regev et al., 2015). As a consequence, it appears that activity areas can be categorized by a limited number of depositional configurations that lead to the construction of a facies model such as the one produced by Macphail et al. (1997) for the Neolithic stabling in Mediterranean caves.

The application of facies and facies models in geology is a major advance in modern, sophisticated stratigraphy (Miall, 2016). This approach entails the development of understanding details of the physics of deposition and the accuracy and precision of absolute dating. It is probably time to move forward in archaeological stratigraphy and make a new sophisticated synthesis by using the powerful tool of facies analysis to better understand how sites are structured, and to better enable the correlation of deposits and stratigraphies across sites. Finally, absolute dating has to be part of the documentation of every stratigraphy, including those from historical periods. It is the only means to check the reliability of stratigraphic correlations.

4.11 Practicing Stratigraphy

This section does not give detailed instructions of doing and recording stratigraphy of certain features and cases. The reader can refer to manuals such as Barker (1993), MoLAS (1994), and Burke and Smith (2004). On the other hand, we provide here some general principles for practicing stratigraphy using the depositional perspective of analysing stratigraphy as discussed so far.

Most archaeologists tend to think that the content of a layer is the fundamental attribute that defines a layer. However, the change in the distribution of a constituent is equally important. In several archaeological sites, the general content does not vary much, but the amount – and more importantly the distribution – of objects within the layers varies markedly. Indeed, the distribution of the

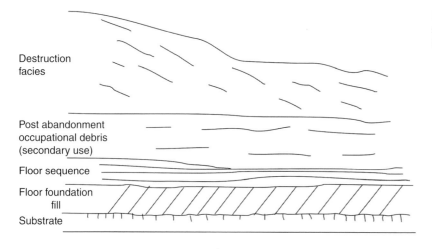

Figure 4.20 Vertical succession of sedimentary facies often occurring inside houses.

Destruction facies

Post abandonment occupational debris (secondary use)

Floor sequence

Floor foundation fill

Substrate

archaeological objects, which most likely constitute the coarser component of the deposits, is what readily makes a layer different from the others. Furthermore, in order to successfully define and record a layer, the contacts have to be mapped. Obviously, there is no layer without its contacts. So, in a drawn profile (or excavation sections), the lines that define the contacts themselves produce the history of deposition of the layers. This is a fundamental issue that has to be understood by individuals who draw the stratigraphy in profiles.

The ways the contacts appear, disappear, intersect, diverge, or converge make a sequence of events that is governed by the few stratigraphic principles described above. Hence, by looking at a finished drawing of a stratigraphic section someone has to portray a meaningful history of the sequence of the filling of the space with deposits. If there is not a straightforward unequivocal succession of the layers, then the drawing is simply wrong; there is no 'sort of correct'. For instance, an elliptical or lenticular feature floating inside a layer, without sharing contacts with other layers, is stratigraphically and logically meaningless (Figures I.2 and 4.21).

Three possibilities have to be figured out before such a stratigraphic drawing can be finalized and made correct:

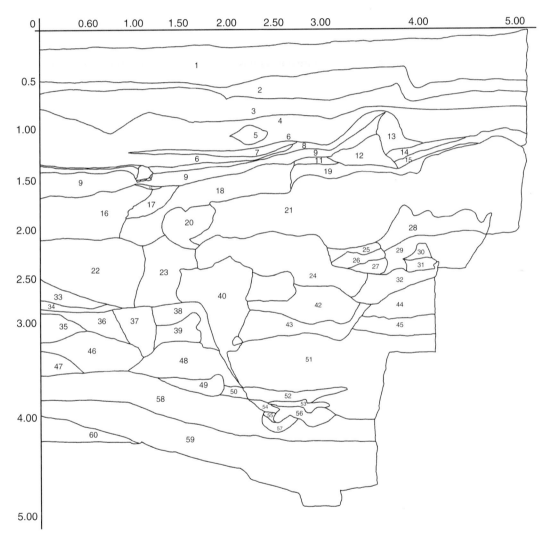

Figure 4.21 Hypothetical stratigraphic sketch showing some typical mistakes that result in meaningless successions (based on a compilation of stratigraphies that have been produced in archaeological sites):

- The lateral geometry of layer 2 is unrealistic as its abrupt step-like appearance is probably the result of a pit or a similar cut.
- The floating lens of layer 5 is stratigraphically meaningless unless further justified. For example, it can be a burrow or a local variation, but in the latter case it would be better presented with a dashed line because hierarchically it is of a lower stratigraphic order than the layer that contains it.
- The configuration of layers 12, 13, 14, and 15 is strange. The one-side vertical contact of layer 13 is unnatural. Layer 13 can only be a burrow or something similar, as several layers terminate around it.
- Layers 22 through 57 have a patchwork appearance and cannot produce any stratigraphic order.

(a) the feature is a local variation of the layer and therefore should be labelled as a subdivision of that layer, (b) there is a penetration (e.g. burrow, underground channel) into the layer, in which case it postdates the layer that includes it, and (c) there is a contact at the level of the feature that is not apparent in the field and the feature constitutes a stratigraphic unit in between two layers (Figure I.2). Without clearly indicating these possibilities the lenticular feature would remain a mystery for any future researcher to decipher.

The same is true with all contacts. There is no such contact that does not end in another line or feature (Figure I.2). In most cases, even if we cannot really follow the contact laterally, knowledge of the depositional characteristics of the feature enables us to predict the continuation of the contact. This is the reason why stratigraphy – and the recording of it with plans, sections, or text – cannot be conducted without knowledge of depositional processes. As an example more familiar to archaeologists, let's think of the possibilities of connecting wall remnants for reconstructing a former building. In most cases with just a few remnants, we can safely reconstruct the plan of a house, especially if it is a standardized one, such as that of a Roman villa. This reconstruction is based on our ability to predict the few possibilities of the development of walls in space in such a way that a meaningful arrangement of bounded areas is produced. Similarly, our ability to define boundaries of layers is governed by natural and anthropogenic sedimentary processes. The more we know about them, the fewer stratigraphic mistakes we will make.

Nevertheless, there will be cases where the contact of a layer cannot be safely followed. In these cases, it will be preferable to provide a plausible possibility, for example by drawing a dashed line, particularly to avoid the possibility of extending the contact in such a way that it will produce an unrealistic succession. The latter should always be our concern. Otherwise, we can put question marks rather than leave it open to all possible explanations.

A type of contact that often appears in excavations, but is not evaluated, is that of stone lines (Figures I.2 and 4.22). Very commonly, thick sequences of archaeological deposits look quite massive but are horizontally transected by lines of stones. These lines appear as discontinuous, one-stone-thick horizons with usually flat lying stones of various sizes. This stone arrangement is normally the product of deflation (erosional) processes on the earth's surface that may include wind action, rainwash, or even foot traffic (other stone lines may be the product of bioturbation, (e.g. Johnson, 1990) or high-energy slope deposits). All of these processes tend to remove the fine fraction from the surface sediment and enrich it with coarser particles, thus making a stone-rich pavement-like surface (Figure 4.23).

Figure 4.22 Stone line separating two layers (arrow) at the level of the flat stone block at the left of the image, indicating exposure, deflation, and temporal hiatus. A floor sequence with crude stratified appearance is also shown above (FL); a detail of a floor is shown in Figure 3.45. Scale is 1 m. EBA, Palamari, Skyros Island, Greece.

Some stone lines, therefore, represent a depositional stasis and gentle erosion before a new layer is deposited above, covering the stone horizon.

4.11.1 Erosional Contacts and Unconformities

For minimizing erroneous stratigraphic interpretations, a hierarchical approach of contacts should be followed (cf., MoLAS, 1994). A stratigraphic section normally will contain diffuse contacts, sharp contacts, erosional contacts, or generally contacts that truncate the sequence below. The last is the most important because they delineate major changes in the stratigraphic succession; in geological nomenclature they are called unconformities. In a variety of archaeological terminologies some of the contacts are called cuts, or negative contexts, the latter implying that the first material was removed and then filled in again (e.g. pits, trenches). Not all unconformities are cuts, and indeed the most important ones are those

Figure 4.23 Wind and rain deflation has produced this stone 'pavement' at the site of Palamari (see previous image for comparing it with the archaeological equivalent shown in the excavated profiles of the site).

that are far more extensive. They are often horizontal, hence not readily visible without following them in cross-section. It is advisable then, that a sequence has first to be divided into unconformity-bounded packets of layers and then filled in with the other type of contacts. For instance, if two thick packets of burnt remains are joined through an erosional undulating contact, it will be meaningless to focus on the subtle changes of the different burnt layers inside each packet and miss the erosional contact between them. The latter implies loss of material and probably loss of important parts of the sequence and/or a phase of abandonment of the area. Unfortunately, the Harris Matrix approach (see Harris, 1989), which calls for equal recording of all interfaces, leads to such an amount of recorded detail without any hierarchy of importance that the interpretation of the stratigraphy is overshadowed by the details (Berggren, 2009)

The approach we describe here is generally different from what some archaeological field manuals would advise. Often, it is suggested that all contexts defined during excavation should be found on the section and drawn accordingly; although not explicitly stated, someone could imply that anything else observed is irrelevant (MoLAS, 1994; Burke and Smith, 2004).

4.11.2 The Importance of Baulks and Sections

Harris (1989) totally dismisses the idea of drawing sections as being unnecessary. However, plan stratigraphy and section stratigraphy are essentially two different views that are often not totally correlated. This situation arises because not only do all layers not necessarily cross the profile – they can end well before – but also, for obvious reasons, vertical successions can be seen much better in vertical sections, whereas lateral variations can be much better appreciated in plan views. Since the chronological order of stratigraphy is based on the vertical succession of layers and features, drawing sections is a fundamental tool for interpreting stratigraphy. This is not to dismiss the value of plan views, but it should be made clear that it is the vertical succession on a profile that will confirm the observation on the plan view.

Indeed, it is expected that several of the recorded layers or contexts on plan view are just lateral variations not clearly separated by contacts and interfaces. In some excavations, large areas are simultaneously excavated to reveal details of the spatial organization and utilization of the area. However, even if we accept in plan view that meticulous observation of the lateral changes could reveal the stratigraphic relationships of the various excavated units, it would be impossible to assess the full geometry of the major contacts that transect the whole area and thus record, for instance, a depositional break or a stone line. Unfortunately, there are types of interfaces that can be correctly interpreted only in a vertical profile (Figure 4.24).

Likewise, in order to make sense of deposit variability present in most architectural sites, it is necessary to simplify what we see in plan view. Often, this variability masks major depositional types because lateral variation is just the product of minor fluctuations in content and fabric. When excavating, it is important to record all these variations. However, in order to reconstruct the history of deposits and understand the overall stratigraphy of a sector or a site, we should lump together several of these stratigraphic units within hierarchically higher aggregates defined by boundaries that record major

Figure 4.24 A faint change in overall colour and a discontinuous stone line defines the contact of layers on the profile (arrow). The stone line is at the continuation of a masonry stone fallen from a wall remnant shown in the right side of the photo. Protogeometric, 'Lower Town' of Mycenae, Greece.

changes. This hierarchy is very difficult to discern in plan view because the hierarchy is revealed only incrementally in a profile and disappears immediately; there is no way of seeing together several interfaces at a time in order to evaluate them. Only in vertical profiles can interfaces be clearly seen and remain there for repeated visits, meditation and discussion, and re-sampling if needed. The latter provides the possibility of applying new innovative approaches in identifying activity areas and resampling, for example, may be possible only after the site is partially or fully excavated.

Along the same lines, several of the most important natural sedimentary structures can be very difficult to recognize unless they occur in vertical profiles, for example grading (particularly inverse grading), cross-stratification, slumping, sliding, and generally most of the mass wasting processes that are very often found in every type of site (see Chapter 2). Their mere definition is based on vertical configuration of fabric changes as described in Chapter 1. Anthropogenic activities like dumping can often produce a series of vertically expressed features that can be hardly seen in plan view. Large-scale features that are associated with cutting a previous fill cannot be differentiated from features related to the back-filling process, unless seen in carefully chosen vertical profiles (see Karkanas et al. (2012) for fills in chamber tombs and other earthworks).

Therefore, a good excavation strategy is to leave as many baulks as possible, and often leave even temporary ones for regularly confirming the excavated stratigraphy. This practice might be somewhat disadvantageous in the recording of spatial data but it fosters correct depiction of the stratigraphy and interpretation of the site's history. Modern developments in the technology of recording (e.g. theodolite, geographic information system (GIS), photogrammetry and three-dimensional modelling) and

computational varieties of the Harris Matrix make all this possible (e.g. Forte et al., 2015).

4.11.3 Inclined Layers

Special care should be taken when digging inclined layers. In geology, the inclination of a layer is measured by its strike and dip. The strike of a layer is defined as the compass direction of the line formed by the intersection of the hypothetical plane of the layer with a horizontal reference plane. The dip angle is the angle measured from the horizontal reference plane down to the plane of the layer. This measurement must be made in a vertical plane. However, the true dip is always measured in the vertical plane that trends perpendicular to the strike of the plane. A dip angle measured in a vertical plane trending in any other map direction will always yield an *apparent dip* value, which is less than that of the true dip. An apparent dip measured parallel to strike always will yield a dip angle of 0°.

For visualizing the above, we may use the example of a book with one side lifted so it is inclined to the table. The lifted sides of the book record the true dip but the front side that still stands on the table appears horizontal. This is actually the strike of the plane of the book. Practically, it is important to realize in an excavation that there is only one direction that records the maximum inclination and there is another direction in which layers will appear horizontal and not inclined (Figure 4.25). There is one possibility of recording the true dip and this is by revealing an appreciable part of a layer and defining the maximum amount of the inclination and the compass direction of this inclination. An important consequence of inclined strata is that on the horizontal plane of the excavation, the younger layers are always found in the direction of the maximum inclination. Therefore, if we

Figure 4.25 Dipping layers appearing horizontal on the right section (see text for explanation). Grid is 50 cm. MP, Klissoura Cave I, Greece.

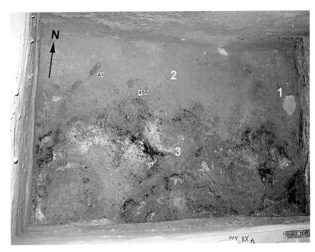

Figure 4.26 Plan view of dipping layers. Layers dip predominantly to the north/northeast, so that layer 1 appears in the direction of the dip at the east corner, layer 2 at the north part of the trench (the main dipping trend), and layer 3, the layer of the hearths, in the south part. In order to maintain the separation between the deposits of the different layers one must first dig in the area of layer 1, then in layer 2 and finally in layer 3. As one digs into layer 2 more hearths of layer 3 will be revealed. Layer 1 wedges out towards the west, something that can be seen in the northeast profile. Size of trench is 2.5 × 1 m. MP, Klissoura Cave I, Greece.

want to reveal sequentially older layers, we have to start digging from the youngest layer, which will be the one on the side of the trench towards the layers that are inclined (Figure 4.26). Figuring out the geometry of inclined layers is substantial for the visualization of the three-dimensional structure of a site and it improves our overall understanding of site stratigraphy

In the same way, the thickness of an inclined layer is not the difference of the maximum and minimum elevation measured for this layer but the difference measured on the vertical transecting the upper and lower contacts of the layer.

4.12 Concluding Remarks

This chapter deals with stratigraphy, the three-dimensional ordering of deposits in a site. As in Chapter 1, we have tried to show that without a thorough grasp of how deposits are assembled in the field it is difficult to extract the contextual data needed to interpret the history of the site correctly. Here, we did not elaborate on the specifics of how to 'do' stratigraphy because we believe that such practices can be found in a number of guides/readers that are devoted to the subject (see section 4.11). Rather, we tried to present some general theoretical principles to enable the reader to practice stratigraphy within a framework that encompasses a depositional perspective. Thus, we hope that the reader obtains a reasonably firm grasp of what stratigraphy is and why it is so important.

5

Non-architectural Sites

5.1 Introduction

To most non-practicing archaeologists, an archaeological site typically conjures up the notion of something like the Roman Forum, the Acropolis, Stonehenge, or the Pyramids in Egypt and the Maya world. These monumental sites exhibit impressive architecture and constructed spaces built by well-organized complex societies that lived during the last several thousand years. Generally, off the radar screen are sites made by (prehistoric) Palaeolithic hunters and gatherers tens to hundreds and thousands of years ago that – to oversimplify – left a scatter of stone implements, scraps of cut-up dead animals (some burnt), some rare botanical traces, and possibly some relatives. Non-archaeologists might also think that they built fires, but generally not in hollowed-out substrates or in constructed fireplaces lined with stones.

Yet oddly, these non-architectural hunter and gatherer sites constitute over 99% of human history. They represent bands of hunters and gatherers that moved about the landscape and lived in the open-air and in caves, and they only very rarely built small (a few square meters) features, which would hardly pass for architecture. Rare exceptions might include the Lower Palaeolithic (~300–400 ka) hearth structure at Qesem Cave in Israel (Shahack-Gross et al., 2014), the Aurignacian clay-lined hearts at Klissoura 1 Cave (Karkanas et al., 2004), or an Upper Palaeolithic (~40–45 ka) rock alignment at Üçağızlı Cave, Turkey, built of stones whose function remains unclear (Kuhn et al., 2009). Thus, in light of the material records of monumental sites mentioned above, the prehistoric non-architectural sites require a good deal of imagination to understand what was going on in the past: what did the people do; where, and how often; what and where did they eat and prepare food; is there any spatial arrangement of what they did, and how did any or all of this change over time?

In architectural sites (see Chapter 6), constructed structures and buildings provided shelter and represented loci for specific activities (e.g. rooms, courtyards and plazas, public buildings). Such sites are typically anchored to a particular place on the landscape through time. Furthermore, although they were periodically subjected to the effects of natural depositional and erosional events (e.g. stream flooding and erosion, tsunamis, landslides, sand storms), the overwhelming majority of sedimentary inputs and transformations/modifications is a result of human-related activities (e.g. initial adobe construction and rebuilding phases, dumping and cleaning of material, the generation of waste products) and not natural ones.

In contrast, non-architectural hunter and gatherer sites tend to be occupied at a given time over a shorter duration – or at least it appears that way because of our inability to recognize short time intervals so long ago. Moreover, if and when we find these sites, as many may be buried or eroded, they appear as dots over the landscape, which especially is the case for caves. These sites are generally smaller, less obtrusive, and less visible than architectural ones. Furthermore, non-cave, so-called open-air, sites are articulated within a larger scale geological depositional landscape and context. Thus, a fundamental distinction between the two is the fact that the matrix or 'sedimentary glue' that encloses the archaeological record in prehistoric open-air sites is a natural, geological one, with deposits of many types, including alluvial, lacustrine, aeolian (sand and dust), slope deposits of various types with inclusions of local soils, and even marine-related deposition (lagoons, tidal flats, beaches). All types can also be subjected to post-depositional processes that alter chemically and physically the original deposits. The most common process is that of soil formation (pedogenesis), which occurs during periods of non-deposition, that is, stasis of a site.

Below, we consider some of the principal aspects of non-architectural sites themselves, treating caves first and open-air sites after. We have purposely avoided comprehensive discussions of the geological dynamics of cave and open-air environments as these are treated elsewhere in considerable detail in archaeological and earth-science works (Butzer, 1976, 1982; White, 1988; Brown, 1997; Hill and Fortí, 1997; Reineck and Singh, 1980; Laville et al.,

Reconstructing Archaeological Sites: Understanding the Geoarchaeological Matrix, First Edition. Panagiotis (Takis) Karkanas and Paul Goldberg.
© 2019 John Wiley & Sons Ltd. Published 2019 by John Wiley & Sons Ltd.

1980; Woodward and Goldberg, 2001; Perry and Taylor, 2007; Reading, 2009, Gillieson, 2009; Goldberg and Macphail, 2006; Ford and Williams, 2007; White and Culver, 2012; Frumkin, 2013).

5.2 Open-air vs Cave Sites

Essentially, there are two types of settings in which hunter-gatherer, non-architectural sites are found (as opposed to constructed, enclosed spaces for most Holocene sites): open-air sites and caves (Straus, 1979). Not surprisingly, each possesses its own and very different set of site-formational processes in the largest sense, and depositional and post-depositional processes are radically different in each type. We first examine aspects of cave sites but for reasons of space we do not include rockshelters separately (see Mentzer, 2017a) as in several cases they are characterized by comparable formation processes.

5.2.1 Caves

Caves are confined spaces, and in addition to the types of depositional processes, the distribution of past activities is different from those at open-air sites. Caves are somewhat distinctive (geo)archaeological environments, and in many ways they behave as closed systems: what gets in there generally remains, although there are many post-depositional processes that can transform or remove these inputs. As such, caves serve as traps for both geological contributions, originating both within and from outside the cave system, and those derived from human activities (e.g. lithics, bones, organic matter) (Goldberg and Sherwood, 2006; Butzer, 1982, 2008). Thus, they offer the opportunity to yield thick sequences through long periods. They also have the potential to preserve detailed records through time and space of human activities (e.g. fires, middening, trampling) that do not appear – or at least survive – in open-air sites (see section 5.2.2).

There are many types of deposits, depositional and post-depositional processes in caves, and most are specific to this environment, although some also occur at open-air sites. They include geogenic, anthropogenic, and biogenic ones, and in Table 5.1 we summarize a number of them. There is much literature on these subjects, as mentioned above, and thus, rather than go through them systematically, we chose to highlight certain aspects that are not well published.

Facies in Caves
Cave sites can vary enormously in size from small niches/cavities that are set into the bedrock by a few meters or tens of meters (e.g. Roc de Marsal, Sandgathe et al., 2011a;

Grotte XVI, Karkanas et al., 2002) to caves occupying >100-m long phreatic tubes (e.g. Wonderwerk Cave, Chazan et al., 2008; Dadong Cave, Karkanas et al., 2008). Yet even in the smaller sites there is considerable variation in the types of sediments one encounters from place to place (microfacies in the sense of Courty (2001); see also section 4.10). These variations reflect the diverse types of sediment inputs in the confined spaces of caves. As such, they highlight a situation that is different from open-air sites where lateral and vertical lithological changes (including possible soil development, which is absent in caves) tend to be tied to certain types of sedimentary environments and predictable sediment types (e.g. floodplain clays and silts, gravel, etc.). Some of these facies are outlined below.

1) Entrance Facies
The deposits near the entrance are close to the dripline, including a few meters just inside and outside the cave. This zone is the locus of a number of different depositional and post-depositional processes from different sources. It is a zone subjected to exteriorly derived sediment of various geological sources and, depending on the situation, is a dynamic zone that includes deposition of windblown sand (as massive accumulations or puffs of individual grains) and windblown dust. In addition, it is a locus of deposition of large blocks of roof collapse, representing retreat of the brow and opening of the cave; large blocks commonly dot the surface and slopes in front of the present-day entrance.

In addition, deposits derived from sediments and soils along the surface and slopes above the cave entrance are transported by gravity and colluviation to accumulate beneath the brow (see section 2.3). This accumulation commonly results in the formation of a berm of fine material (mostly clay and silt) and stony material that slopes into the cave and away from the entrance (see Figures 2.22, 5.1, and 5.2). At the same time, if rainfall is sufficiently strong, cascading and falling water will not only transport soils, sediments, and organic matter derived above and along the slopes, but can also lead to the formation of channels and depressions beneath the falling water cascade (Figure 5.3a).

Associated with water falling/dripping from the brow of limestone terrenes is secondary cementation of the underlying sediment (Figure 5.3b). Carbonate-charged waters evaporate in this aerated environment producing so-called 'cave breccias' (O'Connor et al., 2017). Interestingly, other types of diagenetic alterations, such as phosphatization, are uncommon in this zone but they are more prevalent within the humid

Figure 5.1 Accumulation of material in the form of a berm at the entrance (E) of Dust Cave, Alabama, USA (Palaeoindian-Archaic). Blocks of bedrock (B) from the collapse of the cliff are mixed with finer grained material (red clay and reworked loess) washed down the hill above the cave.

Figure 5.2 Generalized sketch of depositional environments at Kebara Cave. See text for discussion.

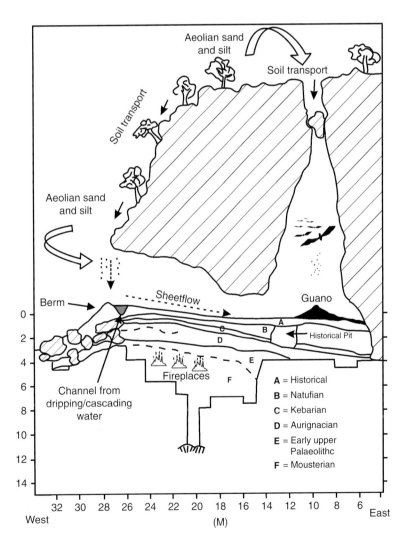

Figure 5.3 Entrance facies in caves. (a) Channel formed beneath the mouth of Hayonim Cave, Israel eroding the previous (MP) phosphatized (PH) sequence and later filled with dumped calcareous ash-rich silts and clays (CF). Width of profile is 2 m. (b) A conspicuous white area cemented with calcite column (CL) gradually produced beneath the dripline of PP5-6 rockshelter, South Africa (MSA). Note that sediments to the right are dipping towards the outside of the rockshelter in contrast to deposits to the left that are dipping towards the back.

environs of the cave interior where biological activity of bats is more common. Nevertheless, documenting the stratigraphic occurrence of calcite-cemented deposits can prove useful for detailing the retreat of the brow. At Pech de l'Azé IV (France), for example, successively younger cemented deposits are found increasingly toward the back of the cave, as the dripline collapsed (Turq et al., 2011; Goldberg et al., 2017a). At the same site, remains of intact hearths, as well as combusted products that were modified/redistributed by syn-depositional trampling and hearth maintenance activities (rake-out) (see also section 3.5) are found concentrated at the base of the sequence in an area that was just below and slightly inside the existing brow of the cave at that time. As discussed below, burning activities such as these tend to be located further back within the enclosed areas of the cave space.

The array of processes is generally different in the interior of caves, and includes depositional and post-depositional ones, especially anthropogenic deposition, and diagenesis (see Table 5.1).

2) Waterlain Facies

Waterlain facies of all sizes (cobbles, gravel, sand, silts, and clays) can be derived from within the cave system as part of much older phreatic deposits. Underground rivers flowing along cave passages can deposit clastic sediments that strongly resemble fluvial deposits in the open air, exhibiting similar textural characteristics (see also section 2.4.3). Boulders and cobbles can accumulate as poorly sorted, compact mixtures (diamictons) resulting from rapid

deposition; in very narrow channels, the deposits are well sorted and become finer upward. They also exhibit numerous small erosional channels filled with sediment (cut-and-fill structures), and gravels may show imbrication (Ford and Williams, 2007). Sandy deposits are moderately to well sorted and exhibit both fining and coarsening upward sequences; interbedding with silts and clays is common (Figure 5.4). Clays and silts – the most common type of clastic sediment in caves – generally exhibit finely laminated bedding.

At times of human/hominid occupation, erosional vestiges of much older phreatic deposits may have been present, as is the case at Pech de l'Azé II (Figure 5.5) (Laville et al., 1980; Texier, 2009). As such, they indicate a previous partial evacuation of the cave chamber, a common occurrence in caves (see also O'Connor et al., 2017). Furthermore, at the time of hominin occupation such remnants may continue to be eroded, transported, and redeposited as colluvium or solifluction deposits within the cave (see section 2.3); coeval with this reworking, any new anthropogenic accumulations may be incorporated and reworked into these new deposits. Moreover, alluvial sediments can also enter the cave from external sources, indicating that the cave entrance was at an elevation similar to that of the stream system at the time, even though at present the river may be well below the present entrance. The site of Zhoukoudian (China) is such an example, where Layer 10 and below appear to be waterlain deposits of the Zhoukoudian River (Goldberg et al., 2001).

Table 5.1 Important depositional processes in the cave environment (adapted from Goldberg and Sherwood, 2006; Goldberg, 2001; Butzer, 1982).

Geogenic	Anthropogenic	Biogenic
Gravity – roof/wall spall • Coarse blocks (éboulis) • Liberation of (quartz and calcareous) sand by granular disintegration of bedrock • Detachment of plants and organic matter growing on roofs and walls	Tracking in from exterior of: ○ soils/sediments ○ organic matter ○ fibres • Construction of hearths (ashes, charcoal and organic matter, sediments adhering to combustibles) • Organic matter (grasses, reeds, etc.) for bedding	Excrements and ejecta • Guano: ○ bird ○ bat • Owl pellets • Hyena and other carnivores: ○ gnawing and comminution of bone • Bear • Bird gastroliths
Phreatic deposition (typically relict deposits and/or reworked into archaeological ones) • Gravel • Sand • Silt • Clay	Preparation of matting/bedding • Plants • Phytoliths • Seaweed	Plant inputs • Debris from nesting birds • Debris from denning bears • Organic debris (grasses, twigs, seed coats etc.) derived from roof, cracks, growing below drips • Roots (trees) • Trees and bushes near entrance • Nesting rodents
Dripping water • Accumulation/infiltration of suspended clay and silt on substrate • Vertical percolation into underlying substrate and lateral migration of water from walls	Fire/pyrotechnology – hearths • Combustible: ○ wood ○ dung • Combustion by-products: ○ ash ○ charcoal ○ fat-derived char ○ clay aggregates around roots of grass kindling	Mineral inputs • Silt/clay from decomposing bird nests • Silt/clay from wasps • Bone • Shell (snails)
Sloping surfaces – typically from entrance inwards • Sheetwash – thinly bedded sediment • Mudflow, debris flow • Dry grain flow	Middening • Cleaning of fireplaces and lateral displacement of combustion products (hearth rake-out) • Dumping of: ○ charcoal ○ ash ○ dung ○ lithics ○ bones (burnt and unburnt) ○ plant organic waste from food preparation ○ edible and ornamental shells ○ human faeces	
Volcanic tephra • Massive accumulations or beds visible in the field • Crypto tephra, visible when isolated from the matrix	Constructions • Stone walls • Adobe (burnt) • Plasters • Lined fireplaces • Pits • Burials • Animal penning	

(Continued)

Table 5.1 (Continued)

Geogenic	Anthropogenic	Biogenic

Aeolian
● Sand:
  ○ distinct layers, or
  ○ incorporated into matrix
● Silt
	Accumulation of secondary materials related to onsite processing of:
	● worked shell and bone
	● lithics (debitage)
	● ochre
	● mobile art	

Streams
● In caves abutting against a fluvial setting
● Slackwater deposits:
  ○ sand
  ○ silt
  ○ clay		

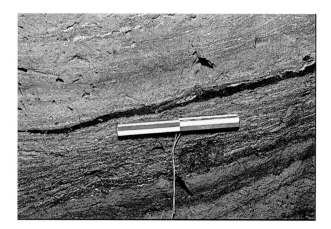

Figure 5.4 Finely interbedded waterlain silts and clays with organic-rich bands at the base of Kebara Cave, Israel (MP). The sequence is the result of phreatic activity before the occupation of the cave. Scale is 20 cm.

3) Interior, Gravity-induced Facies

A typical interior facies is exemplified by the accumulation of gravity-fall deposits derived from the roof and walls, and generally referred to as *éboulis* (= scree or rock rubble) (Laville et al., 1980). Eboulis is mainly dislodged from the bedrock by freeze and thaw, and, to a lesser extent, salt weathering, with local impetus supplied by earthquakes (Woodward and Goldberg, 2001). Eboulis are blocks of bedrock that vary in size from fine gravel up to large slabs several meters across (Figures 2.29 and 2.31) whose shape is a function of the type of the bedrock, including its composition, bedding (massive vs. coarse or fine bedded), and

Figure 5.5 Bedded to laminated fluvial sands (WL) from the base of the sequence at Pech de l'Azé II, France (MP). They are overlain with a sharp, unconformable contact by platy slabs of roof fall (E). Courtesy of Shannon McPherron, OSA team.

Figure 5.6 Large block fallen from the roof of the shelter (arrow) has deformed the sedimentary substrate at Laugerie Haute, France (Late UP; Magdalenian).

Figure 5.7 Remnant of cemented clast-supported scree deposits (SC) along the wall (W) of MSA Cave PP13b, South Africa. Scree is composed of angular roof-collapse material.

jointing. In any case, clasts of bedrock fall to the surface and if there are additions of finer material (e.g. colluvial, aeolian, or anthropogenic) they become incorporated into that material. Larger blocks of roof fall can be seen to deform the substrate on which they have fallen (e.g. Laugerie Haute, France; Figure 5.6). In any case, the size of the original blocks can be altered by fracturing on impact, as well as by subsequent processes, such as freeze-thaw, which will fragment pieces lying on the surface and leave usually angular platy debris (Gillieson, 2009). Post-depositional movement and churning by cryoturbation, for example, will tend to round the rock fragments and, in some cases, pulverize it (Laville et al., 1980).

Finer-sized material can also be produced from weathering of the bedrock. In the Dordogne region in southwest France, for example, the limestone is rich in quartz (and marine invertebrate clasts) (Laville et al., 1980) and upon disaggregation (e.g. dissolution, freeze-thaw), these grains fall to the cave floor below (Dibble et al., 2009; Texier, 2009). Consequently, these grains constitute the principal fine fraction for these deposits. Aeolian silt (loess) can also be deposited directly into caves and be embodied as massive loess deposits (e.g. Krajcarz et al., 2016) or as individual grains that are incorporated into the existing matrix or included as reworked remnants of soils that contain embedded aeolian silt, as at Tabun Cave (Israel), for example (Goldberg, 1978).

Once on the ground, rock fall accumulations exhibit different forms depending on their location in the cave and rates of sedimentation of finer grained components (both geogenic and anthropogenic). Accumulations of roof fall along walls produce clast-supported scree deposits that thin away from the wall and have a bedded structure (Figure 5.7) (see section 2.3); there is a relative paucity of finer material, although at a later time clay and silt may infiltrate from cracks/joints (filtrates; Ford and Williams, 2007) or they may percolate laterally through this material, resulting in clay coatings around clasts. Through time, these stony deposits will prograde away from the wall and into the open cave space, as at Liang Bua (Flores, Indonesia, Westaway et al., 2009a; Morley et al., 2017). Sediment gravity processes, such as slumps, debris flows and mudflows, or water flow processes such as sheetwash, channelized flows (rills) and hyperconcentrated flows, are responsible for the final emplacement of the clastic deposits along the cave interior (e.g., Kebara Cave, Goldberg et al., 2007; Dadong Cave, Karkanas et al., 2008; PP5-6 Cave, Karkanas et al., 2015a; Dust Cave, Goldberg and Sherwood, 1994; Sherwood, 2001; Obi-Rakhmat, Mallol et al., 2009). Each of these produce sedimentary bodies with characteristic geometries, fabrics, and sedimentary structures (see details in Chapter 2; Figures 2.13c, 2.14b, 2.19, 2.20, 2.35, 2.37. 2.38, 2.40, and 2.41).

Natural processes play a large part in modifying hearths and other deposits in caves. At the cave mouth and beyond, rainsplash and cascading water from the brow can obliterate the hearth structure, and erode and redistribute materials on a granular scale. Furthermore, accumulated water and materials can be carried downslope away from the cave as mudflow (Dykes, 2007; Karkanas et al., 2015a) and sheetflow deposits, or be eroded in the form of small gullies (Figure 5.8a). At Kebara Cave (Israel), for example, much of the Upper Palaeolithic deposits are finely laminated reworked hearth materials derived from Middle Palaeolithic deposits. These were redeposited as sheetwash flowing toward the back of the cave where the previous deposits there had slumped/subsided and produced a marked change in relief (Figure 5.8b; Goldberg et al., 2007). Similarly, at Dust Cave (Alabama, USA) hearths close to the entrance and dripline were transformed to massive diffuse clayey deposits and the pre-existing hearths and ashes there were transported by sheetflow toward the interior of the cave, resulting in accumulations of finely laminated ash and silty clay (Sherwood et al., 2004; Sherwood, 2001). At the same time, smaller artefacts (small flakes) can be moved and redeposited downslope. Furthermore, drips of water within the cave can remove any fine material between clasts of lithics (typically) or bones, leaving a lag-like concentration with little or no surrounding matrix (Figure 5.9). In sum, these processes make it difficult for us to recognize the limits of original cultural features such as combustion features or whether a fireplace was even there in the first place.

4) Anthropogenic Facies

Interestingly, in many cave sites occupation may be sporadic through time, but geogenic sedimentation is so slow that the net effect is that anthropogenic deposition of bones (if they are preserved) and lithics may equal or surpass natural sedimentation. We have discussed in Chapter 3 many of the types of anthropogenic contributions and components to deposits, so here we wish to highlight larger-scale aspects related to more general themes such as process, and human and natural activities within and around sites. As discussed below, although geogenic sedimentation is taken to be the norm for Palaeolithic/prehistoric sites, there are instances where this is not the case.

Fires

At many sites where anthropogenic deposits are preserved, occupation appears to be concentrated within the inner space of the cave, back from the entrance. The principal component of anthropogenic sedimentation is typically lithic debris, bones, and shells (where preserved), and these elements are the focus of most excavations. Another aspect of archaeological occupation and the deposits, however, is combustion features, which are discussed in detail in Chapter 3. Here, we will consider other aspects, such as their occurrence in space and time and their use.

It is worth noting at the outset that evidence of Palaeolithic hearths and combustion products is probably under-reported, both in terms of their original recognition in the field and, more so, in their making their way to publication. Many sites were excavated decades ago and when published did not meet the level of recording and documentation that we have today. Moreover, descriptions and information are published in journals and reports, many of which are difficult to find (the so-called 'grey literature' of cultural resource management archaeology).

Nevertheless, hearths and remains of them first appear mostly towards the very end of the Lower Palaeolithic (e.g. Qesem Cave, Karkanas et al., 2007; Shahack-Gross et al., 2014). However, they are found more typically within the Middle Palaeolithic in Europe (Rigaud et al., 1995; Karkanas, 2001; Karkanas et al., 2002, 2004; Goldberg et al., 2012; Schiegl et al., 2003; Miller, 2015; Dibble et al., 2009; Courty et al., 2012; Allué et al., 2012; Vallverdú-Poch and Courty, 2012), the Near East (Meignen et al., 2007, 2009; Shahack-Gross et al., 2014; Movius, 1966; Karkanas et al., 2007, Kuhn et al., 2009; Goldberg, 2003; Mentzer, 2011; Baykara et al., 2015), and South Africa (Karkanas and Goldberg, 2010a; Goldberg et al., 2009a; Miller et al., 2013; Karkanas et al., 2015a). These well-structured features

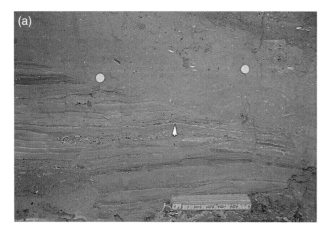
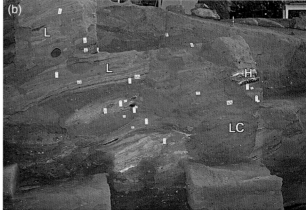

Figure 5.8 Waterlain deposits in caves. (a) Finely laminated and graded (arrow) sands, silts, and clays at the back of Wonderwerk cave (ESA, South Africa) truncated by massive sands most likely representing a hyperconcentrated flow. (b) Extensive sequences of finely laminated sheetwash deposits (L) and channelized flows (LC). The remains of intact hearths (H) that have been eroded by the channel are visible. See Figure 5.11 for detail. The excavated blocks are 1 m across. Transition from MP to UP, Kebara Cave, Israel.

Figure 5.9 Lag-like concentration of lithics due to washing out of the enclosing fine sediment by drip water. MP, Hayonim Cave, Israel.

(organic matter-ash couplets; see section 3.2) exhibit a wide variety of shapes and types. In the Middle Palaeolithic they may range from lenticular shapes that are centimetres across to thinner, more tabular, laterally extensive ones (e.g. Kebara, Roc de Marsal, Castanet; Meignen et al., 2007; Aldeias et al., 2012; White et al., 2017) (Figures 2.8 and 5.10); for the most part, hearths are not excavated into the substrate except in Upper Palaeolithic and later times, but exceptions can be found at Kebara and Abric Romaní (Vallverdú et al., 2012), for example. Constructed hearths fundamentally appear after the Middle Palaeolithic (Karkanas et al., 2004), although a rare, isolated construction can be found in the Lower Palaeolithic site of Qesem (Shahack-Gross et al., 2014). At the Upper Palaeolithic site of Abri Pataud, Movius (1966) documented numerous hearths delimited by either stones or rounded river cobbles, many of which were found in pits or basins.

In addition, despite a limited sample size, Upper Palaeolithic hearths appear to be far less numerous, and the reasons for this require systematic investigation. Moreover, they appear to differ from older ones not only

Figure 5.10 Varieties of hearths. (a) Bedded, tabular hearths dotted with yellowish phosphate nodules. MP, Kebara Cave. Scale markings are 10 cm. (b) Large hearth in the MP of Kebara Cave, most likely representing repeated firing at the same area. Long lines on the scale are 10 cm. (c) Fireplace on bedrock MP, Abric Romani, Spain. (d) Isolated, well-defined hearth on partially cleaned bedrock containing sand-sized splinters of burned bone. UP, Castanet, France.

in terms of shape, but also as being more spatially segregated and individualized. For example, at Abri Pataud, they occur as well-defined distinct features that commonly do not intersect or if they do, their borders are readily recognizable (Movius, 1966). At Kebara, Upper Palaeolithic ashy hearths are orders of magnitude less abundant than in the underlying Middle Palaeolithic ones (Meignen et al., 2007; Goldberg and Bar-Yosef, 1998; Goldberg et al., 2007); they are also more individualized, occurring within predominantly geogenic, sheetwash deposit (Figure 5.11). Exceptions occur at the Aurignacian site of Castanet (France), for example, where hearths are well defined, and are both isolated and nested (overlapping). Moreover, they are composed extensively of sand-sized splinters of burned bone and occasional pieces of fat-derived char, possibly suggesting their use of bone as fuel (Villa et al., 2004) (Figures 3.24, 3.26a, 5.10d, and 5.12) rather than the more characteristic ashy ones. Furthermore, at the Aurignacian of Klissoura Cave I (Greece), extensive sequences of burnt remains are recognized both as single constructed (Figure 3.3) and non-constructed hearths, as well as thick accumulations of raked out and trampled hearth remains (Karkanas, 2010b).

At Pech de l'Azé IV and Roc de Marsal, on the other hand – both Middle Palaeolithic sites in France – hearths are shingled, and their borders can be blurred and blend into each other (Goldberg et al., 2012; Aldeias et al., 2012). These effects appear to be a result of trampling, low rates of deposition, and hearth maintenance activities (e.g. rake-out). At Kebara Cave Middle Palaeolithic hearths are extremely abundant and exhibit mostly clear boundaries that are related to rapid rates of deposition, but there are also fuzzy ones that have been trampled or disturbed by rake-out (Meignen et al., 2007) (see section 3.5). Striking instances of laterally extensive lenticular and tabular combustion features are rather widespread during the MSA in South Africa (e.g. Sibudu, Klasies River Mouth, Diepkloof, PP5-6- (Miller et al., 2013; Goldberg et al., 2009a; Singer and Wymer, 1982; Butzer, 1978; Karkanas et al., 2015a)), although these are thought to be made by modern humans.

At PP5-6 Cave, both single hearth and thick hearth ash palimpsests are observed. Single and mostly intact hearth structures with complex internal microstratigraphy are numerous and they suggest the existence of several combustion events that existed for relatively short periods of time (Figure 5.13). The appearance of isolated but numerous hearth complexes has been interpreted to be the

Figure 5.11 Isolated complex of hearths (H) inside laminated, waterlain sediment (L). See Figure 5.8b for a general view. UP, Kebara Cave, Israel.

Figure 5.12 Thin section scan of large amount of bone showing various degrees of burning, presumably used as fuel. UP, Castanet Cave, France.

Figure 5.13 Sequence of individual hearth remnants interbedded with yellowish, fine fallen roof material. MSA, PP5-6, South Africa. From Karkanas et al. (2015a).

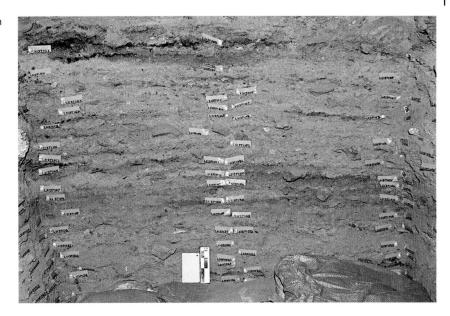

result of the high and relatively constant rate of geogenic input, particularly roofspall. In addition, low rates of anthropogenic activity (particularly trampling) prevented their destruction. This part of the sequence originated when the cave was close to the sea and enhanced salt-induced roof spalling prevailed. As sea level dropped and the coastline retreated, the geogenic input shifted to predominately-aeolian sediments. During this period, thick ash palimpsests were produced as the result of a high, but punctuated rate of aeolian input with intervening rapidly accumulating combustion features that were affected by extensive and relatively constant trampling (Figure 5.14; Karkanas et al., 2015a).

The above descriptions highlight a critical issue that remains to be elucidated, although it is currently being investigated by many researchers: what were the

function(s) of these hearths? At present, using a number of analytical techniques (Mentzer, 2014, 2017b), we are able to make a number of general observations on hearths/combustion features and their by-products, including:

- size and shape
- internal morphology/organization
- construction technique, if any
- composition, including
  - type of combustible used (wood, lignite, grass, bone)
  - freshness of the combustible (e.g. fresh vs. decayed wood)
- temperature(s) achieved during combustion (using Fourier transform infra-red spectrometry (FTIR), thermoluminescence (TL), organic petrology, etc. (see also section 3.2))
  - distribution of temperatures within structures as inferred from measurements on individual objects (bones, minerals)
- combustion conditions, for example oxygen availability, extinction by water or lack of fuel/oxygen
- post-combustion conditions, for example trampling, hearth cleaning and maintenance, reworking by wind, water, animals.

Yet, questions like 'What did they use this fire for?' (e.g. cooking, and if so, what did they cook, light, processing/treating materials, e.g. lithics, spears, hides, keeping predators away, signalling) remain elusive (Mentzer, 2014, 2017b). The use of experiments and ethnoarchaeological studies have been helpful but not definitive, and moreover show how little evidence of fire is preserved in the archaeological record (Mallol et al., 2007, 2013; Mallol and Henry, 2017). Knowing how fire was actually used and controlled, while intrinsically interesting issues, also

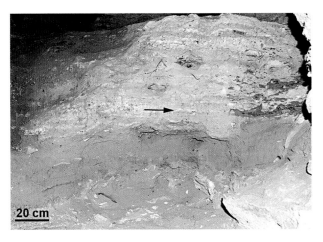

20 cm

Figure 5.14 Bedded ash palimpsest composed of raked-out and trampled ashes and occasionally remnants of *in situ* hearths (arrow). MSA, PP5-6, South Africa.

have ramifications for hominid evolution: was fire a requisite for hominin success (Wrangham, 2009), when and where did hominins learn to control fire, did Neanderthals in north and west Europe make it or just exploit it as some recent evidence might suggest (Sandgathe et al., 2011b,c; Sandgathe and Berna, 2017; Sandgathe, 2017; Roebroeks and Villa, 2011)?

Some understanding of the nature of fire – if not directly its use – can be gleaned from the spatial distribution of hearths within a site, the composition of hearths and that of the surrounding deposits, and knowledge of other aspects of the archaeology associated with hearths (e.g. lithics, phytoliths, etc.). As stated above, fires generally occur within the cave space itself and usually not near the entrance, although there are exceptions (see Aldeias et al., 2012). At the Middle Palaeolithic rockshelter site of Tor Foraj (Jordan), for example, oval hearths within excavated basins in the bedrock were exemplified by ash and charcoal, and burned and reddened sediments, rocks, chert artefacts (Henry et al., 2004). They follow a band that is ~0.8–1.5 m from the back wall and a set of hearths ~1.6–3.6 m from a wall situated inside the entrance alcove. The distribution of hearths is linked to that of artefacts and microbotanical remains, as well as concentrations of phosphorus. Moreover, different microbotanical remains point to different activities around different hearths:

- Phytoliths of date palms seed husks are localized around the central hearths, where 'date fruit and plant seeds were prepared and consumed' (Henry et al., 2004: 24).
- Monocot phytoliths are localized within the shelter and along the back wall. They occur in high densities between hearths, suggestive of the former presence of bedding (Rosen, 2003).
- A high concentration of dicot phytoliths occurs along a line paralleling the dripline. This may be a product of seasonal burning of natural shrubs or brush used as a windbreak there.

Thus at Tor Foraj, although the detailed aspects of hearths themselves were not studied in the same manner – as at the nearby Palaeolithic site of Kebara, for example (Meignen et al., 2007) – there is reasonable, but somewhat subjective, evidence that fires at the rear of the shelter associated with bedding could have been used for warmth. Although still not solving the issue of hearth function, phytolith studies from Kebara, Tabun, Hayonim in Israel, and Grotte XVI (France) reveal a preponderance of wood and bark phytoliths (Albert et al., 1999, 2000, 2003; Karkanas et al., 2002), at least pointing to the combustible used. At Hayonim specifically, wood/bark was found with a high abundance of dicotyledonous leaf plants, suggesting that '...the main fuel used for the fires

consisted of small branches of dicotyledonous plants associated with their leaves' (Albert et al., 2007).

Similarly, at the Middle Palaeolithic site of Abric Romaní in Cataluña (Spain) detailed observations of combustion features and their microfacies have provided some windows into the understanding of their use (Vallverdú et al., 2005, 2010, 2012). For example, in level N an inner zone displays a number of aligned hearths that are spaced at regular intervals (as at Tor Foraj, 1.3 m) and which contain lenses of 'carbonaceous' material. 'The dimensions of carbonaceous fire-use lenses evidence long-term repeated and accumulative fire use, probably with limited amounts of wood consumption', and their alignment and composition were interpreted to represent areas of sleeping and resting (Vallverdú et al., 2010: 142). In level I, activities around hearths showed the use of pinewood (Vallverdú et al., 2005).

Nevertheless, despite a clear classification of hearths based on morphology (e.g. flat, pits on topographic depressions) – including burned surfaces covering 6–9 m^2 – a clear notion of their function remains allusive and needs to be analysed further.

Kebara Cave, like Abric Romaní, is an excellent case study to illustrate a number of aspects of hearths, including their modification by diagenesis. The cave overlooks the coastal plain of Israel and at present has an exposed surface of ~500 m^2. The cave was excavated over several decades but the most extensive and multidisciplinary study took place during the 1980s (Bar-Yosef et al., 1992). Excavations revealed hundreds – if not thousands – of hearths and hearth by-products were revealed in over 4 m of deposits, although the true extent of the deposits is unknown (Meignen et al., 2001, 2007). The hearths are particularly concentrated in the area from the entrance to the central part of the cave, although major subsidence at the end of the Mousterian led to destruction of many of the original hearths. Nevertheless, the superposition of these structures through this thickness of deposition is striking.

Although the hearths at Kebara have been extensively studied, they are so numerous that additional work is needed to investigate more of them in order to obtain clearer estimations of how they formed, the temperatures reached, how were they maintained during the combustion (e.g. stirring the coals), and, ultimately, their function. Notions of their inferred function, however, can be constrained by a number field and microscopic observations:

- There are a variety of types of hearths that vary from thick ashy accumulations to much thinner lenses comprising ash/organic matter couplets. The functions of these two hearth 'end-member types' are probably quite different (Figures 5.10a,b and 5.11).

- Hearths and combustion features were clearly maintained. This is visible in the rake-out deposits, and in the dumping of ashy material against the cave wall, and perhaps the shifting of hot coals or ashes for a number of possible reasons, for example covering of nuts or tubers for roasting (Figures 3.14a and 5.15) (Meignen et al., 2007).
- The sheer number of the hearths and their imbricated nature show continued, repeated, and intense use of fire over ~20,000 years (Valladas et al., 1987), for whatever reason. Their abundance contrasts with relatively low numbers at other Neanderthal sites, and this makes it difficult to make inferences such as sleeping areas at Tor Foraj, as so much of the space has been covered by overlapping combustion features and their mobilized products. So, other than the faunal and ash dumps against the wall, notions of toss zone vs. drop zone, for example, cannot be verified.
- As mentioned above, phytolith studies show that wood is the primary fuel, with charcoal of *Quercus ithaburenis* and *Q. calliprinos* (Baruch et al., 1992). Although function remains to be verified, Meignen et al. (2007) suggest their use for at least light and heat.
- Upper Palaeolithic hearths were not studied in detail at Kebara but, as mentioned above, there are far fewer of them – by orders of magnitude – in comparison to Middle Palaeolithic ones: a few are exposed along one of the extant profiles and appear as single burning events, generally isolated in time and space (Figure 5.11).

One of the major issues concerning the Kebara hearths is why they are so well conserved. To be sure there are a number of trampled layers or zones, or areas affected by bioturbation, and these are expressed by homogenization of the hearth components (Figures 5.15 and 5.16). However, glancing at them in the field, they appear for the most part to be intact and maintain their original structural integrity. These observations are striking, as one would posit that with an internal cave landscape littered by fireplaces many more would be modified/destroyed/eradicated by trampling: just getting from the entrance to a place within the cave one would likely have had to tread on pre-existing fireplaces. Yet this doesn't appear to be the case.

Two explanations come to mind. The first is that the fireplaces were likely consolidated shortly after they were abandoned. This scenario would allow the hearths to maintain some internal consistency. In this light, the calcitic ash produced by the fires is reactive – particularly in a damp cave environment – and would encourage consolidation if not some cementation of the hearths brought about by partial dissolution and recrystallization of the calcite. Secondly, the hearths must have been buried rapidly in order to 'lower' them from the trampling zone

(see above the case of PP5-6 cave for a similar explanation). The details of the choreography that would allow for numerous fires over a large space without the inhabitants treading on them along the way to light the next fire, or simply walking around the cave space, needs to be additionally refined by experiments, for example.

5) Syn- and Post-depositional Effects and (Physical and Chemical) Diagenesis in Caves

A major facet of (prehistoric) caves is the presence of *diagenesis*, which can significantly alter the composition, structure, and overall aspect of the original deposits. Physical changes within the realm of syn- and post-depositional and diagenetic effects include slumping, faulting, subsidence, and mixing of deposits by the actions of plants and animals – bioturbation. Physical deformation of the deposits can also be brought about by dewatering of deposits and associated decalcification, as at Die Kelders Cave, South Africa. There, calcareous grains (shells and limestone) within sandy deposits were dissolved, leading to a reduction in volume of ~30%. This process resulted in the 'wrapping' of the deposits over the irregular bedrock walls and substrate (Figure 5.17) (Tankard, 1976; Tankard and Schweitzer, 1976; Goldberg, 2000a). Similarly, many caves show the effects of slumping as expressed by dipping strata

Figure 5.15 At base are horizontally bedded, homogenized raked-out ashes (R) with some intact hearth remains (H) overlain by dumped ashes against the cave wall that slope toward the wall in the direction of the back of the cave (D). For a detail of the dumped ashes see Figure. 3.14a. MP, Kebara Cave, Israel.

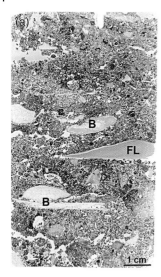

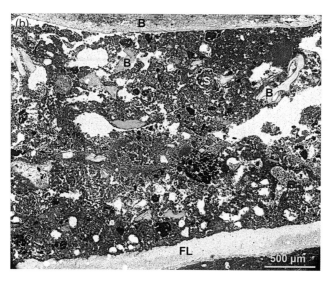

Figure 5.16 Raked-out hearths. (a) Thin section scan of dark hearth remains interbedded with flint (FL) and burnt bone (B). Moderate bioturbation is probably responsible for the aggregation of material and horizontality of the coarse bone and flint. (b) Detail of thin section scan showing a mixture of different materials shown in (a), including soil lumps (S). MP, Kebara Cave, Israel.

Figure 5.17 Large-scale sagging and slumping of occupational deposits due to decalcification. Detail is shown in Figure. 3.9. MSA, Die Kelders, South Africa.

(unconformably) overlain by later horizontal ones (Figure 5.18). Good examples of slumping can be found in a number of sites in Indonesia (e.g. Leang Burung 2, Glover, 1979) and Liang Bua (Morley et al., 2017; Westaway et al., 2009a,b), and at Kebara (Goldberg et al., 2007). A very striking example occurs at Tabun Cave (Israel), where the lowermost deposits are standing on end due to subsidence into a subterranean depression (Figure 5.18b) (Jelinek et al., 1973, Goldberg, 1973).

On the other hand, in more temperate and formerly colder climates, deformation of deposits can be brought about by processes such as cryoturbation and frost heave,

both associated with water freezing and thawing (see details in section 2.7.2). Macroscopically, cryoturbation is expressed as undulating, wavy, and contorted deposits; frost heave results in the production of patterned ground (e.g. stone circles). Small (~40–50 cm in diameter) heaps of subvertically oriented platy stones, for example, occur at La Ferrassie (France) where they were originally interpreted to represent structures produced by Neanderthals (Peyrony, 1934) but are now clearly seen to be patterned ground phenomena (Goldberg et al., 2016). In thin section, reorganization of the deposits by water/ice is visible in microscopic features such as cappings (fine, silty material

Figure 5.18 Scales of subsistence and slumping. (a) Small-scale deformation and slumping (arrow) of burnt layers due to loading by overlying sediment. MP, Denisova Cave, Russia. Courtesy of Mike Morley. (b) Large-scale slumping (inclined arrow) due to subsidence in underlying subterranean depression. Note that the dipping sediment gradually becomes vertical at depth but upwards becomes more horizontal and subhorizontally bedded sediment (horizontal arrows). LP, Tabun Cave, Israel.

resting on top of sand-sized and larger clasts) and banded fabrics (sorting of matrix into sandy and muddy laminae as a result of freezing and thawing), imparting a platy structure to the deposits (Figures 2.32 and 2.85) (van Vliet-Lanoë, 1985, 1998). In some cases, the platy fabric in the field can be misconstrued as bedding but in fact it is a post-depositional effect, as was the case at Pech de l'Azé II (France) (Goldberg, 1979a). Some spectacular field views of such deformations occur in Grotte XVI (Karkanas et al., 2002), La Ferrassie (France, Aldeias et al., 2014b) and Theopetra Cave (Greece, Karkanas, 2001) (Figure I.1b, 2.83b, 2.84, and 2.86).

Physical modifications of deposits – both in caves and in the open air – also can be biogenic, produced by plants or animals: bioturbation (Frederick et al., 2002, Stephens et al., 2005, Leigh, 2001). Plant bioturbation is largely by roots, which follow voids and cracks in the bedrock and exploit nutrients with percolating groundwater (Howarth and Hoch, 2012; Gillieson, 2009). In addition to dislodging bedrock fragments from the roof and walls (root wedging) (Springer, 2005), within the deposits themselves they also displace the sediment grains and enhance the porosity, producing a loose, granular deposit. In karstic settings, secondary calcification takes place along roots, resulting in the formation of calcitic rhizoliths (Figure 2.76). These features can occur as individual root

casts several centimetres across as at Contrebandiers Cave (Morocco) (Aldeias et al., 2014a) or as an anastomosing fine root mat at Cave PP13B (Mossel Bay, South Africa) (Karkanas and Goldberg, 2010a); both types of features are readily identifiable in thin section (Figure 2.75b) (e.g. Stolt and Lindbo, 2010; Durand et al., 2010). In either case, their formation indicates a stabilization of the deposits and a period of plant growth and wetter episodes.

Animal-related modifications are produced by numerous organisms that range from small invertebrates (e.g. small insect fauna) up to larger vertebrates (Kooistra and Pulleman, 2010). The former includes termites, ants, wasps, crickets, cicadas, mites, and snails, and their activities can produce millimetre or smaller up to centimetre-sized disturbances that are visible in thin sections as faecal pellets/excrements or so-called passage features (Figure 2.74), which indicate the path of a burrowing animal (Kooistra and Pulleman, 2010). In addition, earthworms are major burrowers in soils (Darwin, 1881; Canti and Piearce, 2003) and in open-air sites, although in caves their effects are confined to the entrance; earthworm activity is visible both in the field and in thin section as small (~1–8 mm) globular aggregates (Figures 2.71a and 2.72) that are usually clumped or coalesced. The net effect of most of these organisms is the homogenization of the fine fraction, making it difficult to recognize – or completely erasing – the original depositional structures (see also section 2.6). Anthropogenic features such as ashes and charcoal, which might record evidence of fire, can commonly be transformed to finely comminuted charcoal/organic matter and ash rhombs that are thoroughly blended within the matrix (Goldberg and Guy, 1996).

Larger fauna can be more dynamic in modifying deposits and the artefacts they contain. The most notable burrowers in caves – as well as in the open air – include small rodents (e.g. gerbils, moles, lemmings), which produce roughly circular burrows (krotovinas: Figures 2.71 and 5.19) that can be up to ~10 cm across depending on the animal (Pietsch, 2013). At Kebara and Hayonim caves, numerous krotovinas are visible within repeatedly burrowed hearths (Goldberg and Bar-Yosef, 1998).

It should be remembered that with all types of biological activity, even those involving smaller insect activity, an isolated burrow or biological action should be readily recognizable, leaving visible the burrow and the undisturbed surrounding matrix. On the other hand, deposits are more commonly subjected to repeated burrowing events, which can homogenize entire decimetre- to millimetre-thick profiles, and in the process displace and mix considerable numbers of artefacts and datable materials (e.g. charcoal) both in caves and in open-air settings (Goldberg et al., 2007; Johnson, 1990, 2002; Leigh, 2001; McBrearty, 1990; Stephens et al., 2005; Moeyersons, 1978). Although such burrowing can be recognized in the field (Figure 2.71b) and in thin section (rounded aggregated grains exhibiting of a wide variety of compositions, some showing original bedding), the loss of stratigraphic integrity is clear.

Larger vertebrates (e.g. porcupines and badgers) are common in caves and in the open air (Więckowski et al.,

Figure 5.19 Characteristic example of krotovinas (K), some of them superimposed, making large tunnel-like disturbances later reworked and filled with bedded fine waterlain sediment. LP to MP, Hayonim Cave, Israel.

2013). Porcupines can produce burrows over ~50 cm across and can extensively homogenize parts or entire deposits but also transport archaeological materials and destroy bones by gnawing (Więckowski et al., 2013). In addition, the surface of previously deposited sediments can be dug into and displaced by larger vertebrates, which commonly occupy caves, such as hyenas and bears. The activity of bears, for example, can result in the mixing of activity-specific objects over several square meters and in the destruction of burials, for example (see Camarós et al., 2017). Furthermore, the substrates of modern brown bear beds – clustering between 75 and 150 cm long by 50–125 cm wide and ~5–30 cm deep – comprise a variety of organic materials garnered from the forest (e.g. shrubs, mosses, grasses, rotten wood, etc.) (Mysterud, 1983). Numerous Palaeolithic sites are associated with bear activity (e.g. Gargett et al., 1989; Münzel et al., 2011), and at Yarimburgaz Cave (Turkey) hominid-modified bones are intermixed with stone artefacts and gnawed bones along with remains of bears and other animals (Stiner et al., 1996). After detailed study of the bone assemblages they concluded (p. 314) that:

> The most important force behind the apparent stratigraphic associations of these disparate materials, and the biological processes responsible for them, was slow or uneven rates of sediment accumulation over time. The bone and tool assemblages in Yarimburgaz Cave represent palimpsests of many relatively short-term depositional events, most of which were causally independent. Some of the material may have been locally redistributed later by the digging and scraping that is typical of bed preparation by bears.

The death of bears was unrelated to hominid use of the cave. Thus, the key issue for human interactions such as at Yarimburgaz is to evaluate microstratigraphically the relationships among biology, objects, and sedimentation, a notion that has been repeatedly been put forward by Gargett (Gargett et al., 1989; Gargett, 1999). Not surprisingly, every case needs to be considered separately, but with an increase in detailed studies such as those cited above, the movement/transport of bones and sediments by large animals such as bears should be clearer.

Similar issues arise for hyenas. As depositional agents, modern hyenas in Israel, for example, have been documented to transport to and deposit bones in their dens, such as dogs, sheep, goat, ibex, camels, equids, pig, cattle, porcupine, hare, birds and fox, and including humans (Horwitz and Smith, 1988). Similar instances are found within Pleistocene caves where in addition to bones, coprolites are not uncommon (e.g. Horwitz and Goldberg, 1989; Macphail and Goldberg, 2012; Boaz et al., 2000; Sanz et al., 2015). At Zhoukoudian, Locality 1, for example, coprolites are abundant, and they indicate that repeated occupation at the site by hyenas was interspersed with that of hominids. Moreover, Binford and Ho (1985: 425) suggest that 'What is known of the fauna from the Zhoukoudian deposits, including the hominid remains, is consistent with what we know of hyena-accumulated assemblages.' The lack of documented hearths and overwhelming evidence for geogenic sedimentation in Locality 1 (Weiner et al., 1998; Goldberg et al., 2001) – at least for Layers 4 and below – offer additional support for non-hominid deposition of the fauna.

At the hyena den of Bois Roche (Charente, France) infrequent stone tools occur throughout a sequence that consists of bedded silts and sands, sinter aggregates, and coprolite pebbles, all overlying a massive clast-supported layer with equant blocks of limestone and abundant bones of megafauna (Marra et al., 2004; Goldberg, 2001; Villa and Soressi, 2000). Again, the issue of evaluating the relative roles of hyenas and hominids as depositional agents and modifiers arose, and the micromorphological and faunal analyses indicated that the artefacts and bones accrued by independent processes: the bones through the action of hyenas and the artefacts by geogenic processes, such as gravity and slopewash (Villa and Soressi, 2000).

Chemical Diagenesis

No less subtle are the effects of chemical diagenesis. It can add, remove, or transform materials originally deposited, be they chemical elements, minerals, or organic compounds. Such diagenetic changes can have significant consequences for how we document and interpret the human and geological (*sensu lato*) history of archaeological sites.

Net additions in caves are best typified by the precipitation of carbonate minerals (usually calcite or aragonite) in a variety of forms characteristic of caves (White and Culver, 2012). Accumulations of carbonate bodies, such as stalactites, stalagmites, 'travertine', and flowstones (tabular and sometimes massive bodies of dense calcite) furnish the potential for documenting cave history, including chronology and palaeoenvironments using oxygen and carbon isotopes (Bar-Matthews et al., 2010; Bar-Matthews and Ayalon, 2011). Regardless of the remarkable quality of their data, they must be understood in their proper context. The stratigraphy and dating of many early hominid sites in South Africa have been controversial issues for decades despite refined, recent high-precision dates obtained from flowstones using uranium lead dating (e.g. Pickering and Kramers, 2010). However, if stratigraphic context and depositional/erosional history is fully documented, it is evident that the flowstones 'in association' with the StW 573

Australopithecus skeleton (Sterkfontein cave system) are clearly younger than the breccia which encases the specimen (Bruxelles et al., 2014: 46).

A more subtle accumulation of post-depositional carbonates is so-called cave breccia (Figure 2.99b), in which extant deposits (whatever their composition) become cemented by carbonate-rich water dripping from the roof/brow, or by flowing along the walls or floor of the cave (O'Connor et al., 2017). The cementation, as well as the passage of water associated with it, can annihilate original traces of bedding or the structure of features. Moreover, it can serve to 'fasten' sediments to the cave walls, and being indurated, they are difficult to erode, or at least completely. Thus, they commonly occur as vestiges stuck to the walls, well above the height of deposits that are visible when a site is first excavated (Figure 5.20). These 'hanging breccias' often contain artefactual material that is older than the lowest deposits in an excavated stratigraphic sequence at the site. They therefore testify to a previous level of infilling, cementation, and emptying out of the cave, eventually followed by deposition of the current observable fill that reaches the current cave floor. They provide evidence of occupations that are essentially no longer visible, and they furnish insights into understanding the dynamics of the infilling and erosional processes that may be linked to base level changes or more regional phenomena. Cave breccias are well known, but not well or systematically documented. O'Connor et al. (2017) provide examples of such instances from cave sites in Papua New Guinea and Timor Leste. Similar breccias are also known from cave sites in France, such as Pech de l'Azé I (Texier, 2009), for example.

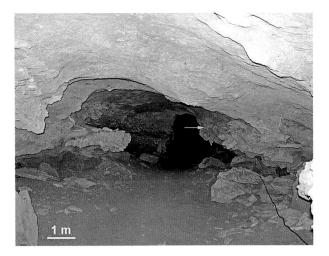

Figure 5.20 Hanging breccia on the wall (arrow) of Pech de l'Azé I Cave (MP, France), a remnant of a previous fill that has been eroded along with most of the sediment in the cave. The reasons for the massive erosion are not clear.

One of the most important types of chemical diagenesis in prehistoric caves involves the formation and transformation of different phosphate minerals (including bone), as well as the dissolution and precipitation of other minerals (Karkanas et al., 2000). Such transformations and activities have been well documented in prehistoric caves throughout the world (see details in section 2.7.2) (e.g. Mallol et al., 2010; Schiegl et al., 1996; Morley et al., 2017; van Vliet-Lanoe, 1986; Goldberg et al., 2015; Weiner et al., 1998, 2007; Karkanas, 2001; Karkanas et al., 2002; Quattropani et al., 1999). Here we discuss some of the implications for understanding cave histories and human activities.

Calcareous materials within deposits – for example, clasts of limestone, speleothems/flowstone, or calcitic ashes – at or slightly below the cave floor can be attacked by dripping or percolating acidic waters, which are produced by the decay of organic matter. An initial result is the partial or total dissolution of the carbonate, which may result in a substantial reduction in volume (Mallol et al., 2010; Weiner et al., 2007). In addition, phosphate – derived from bat (and bird) guano, plants, ashes, and bones – can react with the carbonate, resulting in the formation of the phosphate mineral dahllite (Ca_5 $(PO_4, CO_3)3(OH, F)$ (Weiner et al., 2002; Karkanas et al., 2000), which is similar to the composition of bones. Over time, a number of different phosphate minerals can form (e.g. montgomeryite, crandallite, taranakite, Karkanas et al., 2000; Karkanas, 2010a), as well as others produced by the breakdown of clays (e.g. opal-silica) (Weiner et al., 2002, 2007; Karkanas et al., 1999). The minerals that eventually are formed depend on local environmental conditions, such as pH (acidity), bedrock composition (calcite *vs.* dolomite), clay content and composition (Si, Al, Fe, K), and phosphorus concentration.

In addition to geochemical and mineral aspects, diagenetic changes can be significant for archaeology. Most obviously, they reflect the biological and hydrological conditions of the cave and how they might have changed. Thus, for example, without water, these reactions hardly take place, and periods of extensive diagenesis might reflect a change to wetter conditions, increased biological activity, and, most interestingly, occupation of the cave by humans who are major contributors to cave deposition (Karkanas et al., 2000; Goldberg et al., 2007; Straus, 1979; Butzer, 1982). Accordingly, in cases with a high frequency of repeated occupation, the pace of anthropic sedimentation exceeds that of the biological (i.e. guano) one, especially since human activities – including fire making – will result in less occupation by bats, for example. On the other hand, breaks in occupation and site abandonment expose the surface to guano accumulation and at least incipient or even more advanced diagenesis. In fact, in thin sections from Roc

de Marsal (France) and Geißenklösterle (Germany) it is possible to observe millimetre-thin phosphatic crusts and phosphatized zones developed on such abandoned surfaces (Goldberg et al., 2009c, 2012; Miller, 2015); these 'phosphate surfaces' are not visible in the field and just barely under the microscope (Figure 5.21).

As mentioned above, dissolution can result in significant volume reduction, as, for example, at Kebara (Weiner et al., 2007), Die Kelders (Goldberg, 2000a) and Theopetra Cave (Karkanas et al., 1999). However, at Hayonim, this was not the case: despite extensive diagenesis in the central area of the cave, backscatter plots of artefacts show horizontal distributions rather than depressed/basin-shaped ones if volume loss had occurred. Thus, each site has its own rules of the game and must be examined on its own.

Perhaps more noteworthy, however, is that diagenesis can result in the total removal of bone from the deposits and archaeological record. At Kebara Cave, for example,

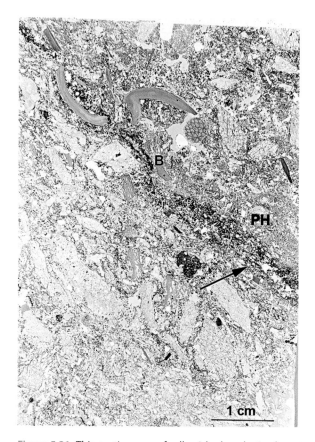

Figure 5.21 Thin section scan of yellowish phosphatized crust over burnt remains, which most likely represent a single burning event. They contain fine charcoal and burnt bone (B) above a thin oxidized substrate (arrow). Some bone, charcoal and oxidized patches below are probably remains of a previous hearth. The phosphate crust (PH) that formed by guano accumulation and decay implies exposure, and weathering and dissolution of calcareous ashes on the surface. MSA, PP135, South Africa.

the central area is lacking in bone. However, it is locally abundant, particularly in the northern part of the site, near the bedrock adjacent to the ash dump described above (Weiner et al., 1993). At first, it was not clear if this distribution represented Neanderthal behaviour patterns or some other explanation. However, detailed on-site FTIR analysis of the deposits revealed three areas: (a) sediments with calcitic ash from hearths, (b) sediments containing carbonate apatite (dahllite) + bone, and (c) those that contain other phosphates and no bone (e.g. montgomeryite) and siliceous aggregates (biologically produced amorphous matrix material that is rich in Si, Al, Fe, and K; Weiner et al., 1998; Schiegl et al., 1994, 1996). The results indicate that the phosphatic zone without bone is a result of bone dissolution, and not Neanderthal behaviour. Interestingly, in thin section charcoal in the phosphatic deposits can be seen to be broken and comminuted by the growth of phosphate minerals, thus pointing to modification and destruction of another part of the biological record at Kebara.

Finally, the different phosphate mineral assemblages have different elemental compositions of uranium, thorium, and potassium, and these elements constitute the major sources of radiation that influences dosimetry values used in dating by thermoluminescence (TL), optically stimulated luminescence (OSL), and electron spin resonance (ESR) (Karkanas et al., 2000; Weiner et al., 2007). Thus, diagenesis must be taken into account in attempting to gain accurate results using these techniques.

5.2.2 Open-air Sites

The situation and dynamics at open-air sites is considerably different from those at caves. Open-air sites are parts of open, dynamic, and higher energy systems: whereas sediments of various types can be deposited, they are equally liable to be eroded, taking archaeological material along with them. Thus, the preservation potential for archaeological materials, sites, and contexts is greatly reduced in open-air settings. Another different aspect of open sites compared to caves is that with the exceptions of very large and open caves (e.g. Niah Cave, Huah Fteah), the effects of constrained space in caves is not an issue (Straus, 1979). Nevertheless, we have to continuously remind ourselves that many of these non-architectural sites are produced by hunters and gatherers who are generally transient on the landscape, so they may not leave much to preserve in the first place, prior to be being covered or eroded.

Not surprisingly, the array of geological environments in the open air is vast (see Chapter 2 for a summary of sedimentary processes), and numerous processes and styles of deposition and erosion exist. It is not possible to provide here even a cursory discussion of open-air environments,

although major aspects of the principal types are summarized in Table 5.2 (see also Butzer, 1976, 1982; Goldberg and Macphail, 2006; Reineck and Singh, 1980; Brown, 1997; Reading, 2009). Note, for example, that the 'fluvial environment' consists of several subenvironments, such as the flattish floodplain, the raised levee and point bar next to the channel, an abandoned channel (oxbow lake), etc. Similarly, a sandy aeolian environment (Roskin et al., 2013) could be composed of massive sand sheets, dunes of various types (e.g. climbing dunes where sand is plastered against a bedrock rise) and also infill caves and rockshelters (Blombos and Mossel Bay, Jacobs et al., 2006; Karkanas et al., 2015a), and wetter spots localized within interdunal depressions. The latter are also loci of erosional windows into the underlying deposits (Fuchs et al., 2008).

Regardless of the specific environment, some of the major overarching issues of open-air sites concern whether a certain geological environment is attractive to human occupation in the first place, and whether it is more or less favourable to the preservation of human remains during and after occupation. For example, sites along steeper (say ≥30%) slopes are not particularly favourable for settlement due to the inherent instability of the slope, which can be enhanced by bioturbation or shrink-swell of soils. In addition, upland and mid-slope positions are relatively easily eroded by processes such as sheetwash, soil creep, solifluction, or rapid mass wasting such as slumps, landslides, and sediment gravity flows (Bertran et al., 2015; Field and Banning, 1998; Goldberg and Macphail, 2006). If erosion is only minimal, such processes can displace existing artefacts lying on the surface (e.g. Lenoble et al., 2008), or artefacts may be left on surfaces as lags (Rick, 1976). In most cases, the true extent of how widely artefacts have been moved in the past remains to be clarified and to be documented for each site.

Floodplains of meandering systems are complex, and the visibility and age of sites – both those at the surface as well as buried ones, for example – can be tied to rates of sedimentation and the sinuosity and rate of migration of features such as meanders (for details see, for example, Gladfelter, 2001; Brown, 1997; Huckleberry, 2001). Thus, sites can be destroyed by channel movement (e.g. meander migration, avulsion) or inundated by overbank flooding (Figure 5.22). At the same time, they can be subjected to groundwater rise, which not only induces saturated substrates and unsuitable living conditions, but also enhances pedogenic and diagenetic alteration of the deposits and archaeological materials. In a study of the rapidly migrating meandering channel of the Red River (USA), Guccione et al. (1988) documented a modern meander belt phase that is likely to be ≤300 years old. They demonstrated the removal of any earlier prehistoric sites by erosion; rates of overbank sedimentation

Table 5.2 Common sedimentary environments and deposits in open-air sites (inspired by and adapted from Butzer, 1982).

Environmental setting	Sediment types
Geogenic	
Fluvial, includes:	Boulders
• floodplain	Gravel
• margins and older, abandoned surfaces of alluvial fans	Sand
• delta margin	Silt
	Clay
Shore (marine, lacustrine), includes:	Gravel
• beach	Sand
• (foredune)	Organic-rich muck
• barrier island	Marl
• lagoons and marshes	
• lateral interfingering with:	
○ alluvium	
○ aeolian sand	
Springs	Chemical precipitates:
	○ calcium carbonate; iron and manganese deposits
	○ commonly around vegetation
	○ tabular to mound-shaped
	Organic-rich sediments
Aeolian	Sand:
• Sand	• dunes, sheets, veneers, some plastered on or against slopes
	• coastal (lakes and marine)
	• desert interiors
	• deflation from glacial deposits
• Silt	Loess
	Tephra/cryptotephra
• Volcanic	Boulder-sized bombs, down to silt-sized tephra (see above)
Slopes (and valleys), includes:	Mixed sizes and generally poorly sorted and bedded; clast and mud-supported deposit
• debris flows	Sheetflow: laminar flow with distinct bedding
• mudflows	
• granular flows	
• rockfalls and landslides	
• sheetflow	
Cold climates:	
• solifluction	
• cryoturbation	
• creep	
Anthropogenic	
Similar to those in caves but are more complex for non-hunter-gatherer sites, such as tells, mounds, urban areas, etc.	

are high, reaching an accumulation of 2–3 m and potentially burying sites of protohistoric age. On the other hand, older prehistoric sites occurring outside this meander belt are preserved because stream avulsion resulted in the abandonment of channels, thus eliminating the possibility of channel erosion.

Similarly, regional patterns of landscape change can make sites of certain ages practically invisible by burial. In the Central and Eastern Great Plains (USA), for example, Bettis and Mandel (2002) and Mandel (2008) have carried out extensive geomorphological/pedostratigraphic surveys of valleys with alluvial stratigraphy over large areas. Patterns of Holocene deposition and erosion are consistent over the area, despite ranges in climate from semi-arid to subhumid. The early Holocene (~11,000–8000 yr BP) is marked by slow alluviation in large and small valleys, whereas from ~8000 to 5000 BP, there was '…net erosion and sediment movement in small valleys, sediment storage in large valleys, and episodic aggradation on alluvial fans. During the late Holocene (post-5000 yr BP), alluvial fans stabilized, small valleys became zones of net sediment storage, and aggradation slowed in large valleys.' (Bettis and Mandel, 2002: 141). Whereas this information is interesting in itself as a record of environmental change, there are implications for the archaeological record: the lack of Archaic sites is conditioned by erosion and removal of sediments from the small valleys in early and middle Holocene times; thus if sites were there originally, they are not there now. On the other hand, Archaic sites are abundant in alluvial fans, which occur in large valleys and are buried during periods of greater rates of aggradation. Thus site visibility – including their absence – is conditioned by climatic and hydrological factors operating on a regional scale.

Similar scenarios exist for sandy aeolian environments where sand is mobile and energies are lower. Wind-blown sand can bury sites quickly so that wholesale erosion is less likely and there is a potentially greater probability that artefacts and features will remain intact if burial is rapid. Such was the case of the excavation of Late Palaeolithic hearths and sites in the sandy terrain of northern Sinai (Egypt), where migrating past and modern dunes served to cover and protect them (Figure 5.23) (Goldberg, 1977). On the other hand, wind erosion (deflation) can result in the abrasion of artefacts and shells, and the removal of the sandy matrix of a site, thus resulting in a lag or pavement of cultural debris, and the conflating of several previously individual archaeological horizons/strata (Butzer and Hansen, 1968; Olszewski et al., 2005;

Figure 5.23 Hearth remains *in situ* reworked by wind activity; grey ashes are still in association with charcoal. The remains are covered by aeolian sand. Note that some black speckles of charcoal are also dispersed in the overlying sand. This hearth is adjacent to the one shown in Figure 3.2. Epipalaeolithic, Moshabi XVI, Sinai, Egypt.

Figure 5.22 Fluvial deposits in sites. (a) Yellow-brown overbank deposits consisting of horizontal planar beds of silts and sands interfingering with light-coloured colluvial marls that thicken to the right, in the direction of their source. Intercalated within this alluvial/colluvial sequence are occupations and intact/disturbed hearths. The middle profile in the zig-zag is 2 m across. UP, Nahal Zin, Negev, Israel. (b) Massive bedded wadi silts (essentially reworked loess) burying a several meters long reddish occupational layer composed of burnt remains. UP, Qadesh Barnea, Sinai, Egypt.

Wendorf and Schild, 1980). On San Miguel Island (California, USA), for example, Rick (2002) showed for a Middle and Late Holocene archaeological site that aeolian activity – tied to human overgrazing – moved light fish bones, concentrated shells and heavy mammal bones, and significantly abraded, etched, and polished shells, artefacts, and bones. Such modifications were not confined to surface horizons and disturbances reached depths of up to 20 cm.

On the other hand, sites, objects, and features that are situated in areas of windblown silt (loess) can be preserved largely intact (Dodonov, 1995; Hoffecker, 1987; Lisá et al., 2013; Händel et al., 2014; Simon et al., 2014). These loess sequences in Central Europe are complex and several meters thick, and they are punctuated by numerous polygenetic soils (multiple periods of soil formation; Schaetzl and Anderson, 2005) and by episodes of reworking (Figures 2.62 and 5.24). They serve as remarkable palaeoenvironmental records (Terhorst et al., 2014) coupled with a highly intact archaeological record.

In sum, the 'rules of the game' of open-air-sites are much less confined than those of caves. Sediments can accumulate quickly, bury archaeological materials, displace them, or completely remove them from their original locus of accumulation (Goldberg, 1987), leaving little if any traces of their previous existence. Alternatively, non-architectural open-air sites can be remarkably preserved under suitable conditions: low-energy burial by aeolian dust, slackwater deposition associated with flooding, or back swamp areas of floodplains.

Figure 5.24 Massive beige loess sequence interbedded with light brown to brown palaeosols (arrows). Saint Pierre les Elbeuf, Normandy, France.

5.3 Other Stratigraphic Themes

5.3.1 Burials

A topic that is often overlooked and in need of close stratigraphic scrutiny is that of human remains and their contexts, principally burials. The issue of intentional Neanderthal burials, for example, is one that is currently debated (Gargett, 1989, 1999; Sandgathe et al., 2011b; Dibble et al., 2015; Rendu et al., 2014, 2016; Maureille et al., 2015), and the consequences for Neanderthal symbolic behaviour and cognitive ability are significant. The issue is particularly interesting in light of the mobility of hunters and gatherers on the landscape and the fact that so few complete specimens of Neanderthals have been found, whether they are burials or not. Moreover, in light of the antiquity of the remains and their associated deposits, as well as the small size of Neanderthal groups, geological processes tend to play an important part in examining and evaluating the issue of Neanderthal burials.

Although for each instance or site different arguments are amassed (e.g. Sandgathe et al., 2011b; Rendu et al., 2014), in general certain criteria or lines of reasoning should be followed to demonstrate more or less whether there is enough evidence to conclude that a body was purposefully put in the ground. Gargett (1989), for example, summarizes a number of criteria that can be used to infer burial: (1) position of the body and the extent to which it could not be that way due to natural causes, (2) depression dug into the substrate that is not the result of natural phenomena, (3) the occurrence of grave offerings, (4) protection of the corpse, if it can be demonstrated, and (5) magical or ritual inclusions. In sum, 'it is not enough to say that humans could have produced a given deposit; it must be shown that nature could not.' (Gargett, 1989: 161). It is interesting to point out that until now most of these criteria were focused on the objects and their associations, and less on the stratigraphic/microcontextual issues, other than excavation of a pit as a dug grave and whether the body might have been covered by cultural deposits and not natural ones.

Thus, in the case of Kebara Cave, the KMH2 skeleton was found in 1983 in a pit filled with yellow-brown sediment that differed from the surrounding charcoal-rich deposits associated with one of many hearths in the site (Bar-Yosef et al., 1992). The pit itself was not natural nor was the filling because no geological processes were observed at the site to produce a closed depression, and the fill was clearly unlike that of the surrounding deposits; colluviation, as suggested by (Gargett, 1999) does not fit with the deposits in this part of the cave (Goldberg et al., 2007).

On the other hand, geological arguments can be made to cast doubt on the burial of the 2–3-year-old child at Roc de Marsal (France). Found and excavated in the 1960s (Bordes and Lafille, 1962; Lafille, 1961), this relatively complete specimen has been described in depth and often cited as an intentional Neanderthal burial (for details about the site see Sandgathe et al., 2011b; Aldeias et al., 2012; Couchoud, 2003; Goldberg et al., 2012; Guérin et al., 2017; Turq et al., 2008). In the original excavation Bordes and Lafille (1962) noted that the skeleton was found within a ~70 × 90 cm pit feature in the bedrock floor that was inclined toward the west. The cranium was associated with a yellow-orange layer composed of decomposed sandy limestone whereas the remainder was linked to black deposits. The skeleton was facing down within the sediments, with legs bent upward and the left arm extending into the bedrock cavity; resting on bedrock was only the lower part of the body (Goldberg et al., 2017b; Sandgathe et al., 2011b).

Recent geoarchaeological investigations focused on the nature of the pit and the sedimentary context of the skeleton. The pit itself appears to be a natural cavity resulting from karstic activity of the site (Couchoud, 2003) and many similar types of hollows can be observed in the limestone today, within the cave and in the surroundings. The deposits themselves offer additional insights, as the original description by Bordes and Lafille (1962) indicated that the infant was covered by at least two different types of sediment: the lower parts were within black sediments, whereas the cranium protruded into the overlying yellow-orange deposit (geological level 4 of the recent excavations (Goldberg et al., 2012). Both sediments suggest that the skeletal remains were not completely covered at one time, and thus were not related to rapid burial, although it might be hypothesized that a period of erosion took place between the accumulations of both deposits; the fact that the depression is closed would tend to eliminate this explanation.

Moreover, the infilling of a nearby karstic channel (~2 m away) contains deposits which are similar to those of the main excavation (Aldeias et al., 2012; Goldberg et al., 2012, 2017b) and appear to match those described by Bordes and Lafille, particularly the uppermost yellow-orange decomposed limestone deposits. These occur throughout the site and appear to represent a marker horizon of cold conditions, as shown in thin section by cryoturbation and silt cappings on grains, and in the field by large blocks of roof collapse that locally covered it (Couchoud, 2003). Thus, from a sedimentological point of view there is no unique stratum that covers the skeleton, nor are the deposits distinct from others observed elsewhere in the site; both of these observations argue against a purposeful burial.

Both examples serve to illustrate that close microstratigraphic examination of the deposits can provide some insights and constraints as to whether a skeleton is evaluated as being part of a natural system of accumulation – including the Roc de Marsal child itself (possibly left to die or simply thrown into the depression) – and was covered by natural or cultural deposits or placed within a depression dug into the substrate as at Kebara.

As a cautionary note, it is worth underscoring that in the case of Roc de Marsal, for example, it was indeed difficult to reconcile fully our observations with the relatively sparse records (particularly photographs) of the original excavations, where complete documentation of context was lacking. This is a common theme in the modern re-examination of the Neanderthal burial issue as shown, for example, by the recent debate over the site of La Chapelle-aux-Saints (Dibble et al., 2015; Rendu et al., 2014, 2016).

Similar types of issues have plagued recent efforts to place the Neanderthal specimens of La Ferrassie 1 and 2 within their stratigraphic and 'anthropological' contexts (Aldeias et al., 2014b; Goldberg et al., 2016) as initially highlighted by Gargett (1989). In a forensic approach, A. Turq (formerly of the Museum National de Préhistoire, Les Eyzies) steered us to examine the sediment attached to the foot of the LF2 specimen. Mesoscopic and microscopic examination led us (1) to correlate it with one or two of the stratigraphic units that we and Peyrony (1934) observed and (2) to conclude that it is not within the same layer as LF1. Unfortunately, the scant – and sometimes contradictory – descriptions of Peyrony (1934) did not permit us to evaluate whether there was indeed a natural or anthropic depression associated with LF1. His 1909 description is as follows:

> 'Mais il n'avait certainement pas été enterré dans une fosse. (...)Peu à peu les débris usagés des habitants de l'abri (...) s'étendirent sur lui comme dans le reste de l'abri et furent vraisemblablement étalés et foulés par le va-et-vient des habitants de la grotte. Ainsi se constitua au-dessus de lui une stratification parfaitement régulière de dépôts archéologiques, tout comme dans le reste de l'abri, et se succédant d'âge en âge, suivant l'évolution humaine en ce lieu.'
> (Capitan and Peyrony, 1909: 409)

This is freely translated as:

> 'But it certainly was not buried in a pit. (…) Little by little occupation debris of the inhabitants of the shelter … extended over him as in the rest of the shelter and were probably spread out and trampled by the comings and goings of the inhabitants of the cave. Thus a perfectly regular stratification of archaeological deposits was formed above him, as in the rest of the shelter, which continued over time, following the human evolution in this locality.'

which fits well with the most recent excavations at the site (Guérin et al., 2015). In sum, as we have stated numerous times, the observation and documentation of context at all scales (regional, site, depositional) is critical to understanding and interpreting site history.

5.3.2 Palimpsests

A subtle aspect of sediments of all sorts is trying to come to grips with gaps in the sedimentary record: are there any, how much time do they represent, how can we observe them and document the number and amount of missing time? These issues are not simply related to archaeology; they are also well known among geologists (Ager, 1993; Bailey, 2017).

Just looking in the field is not a sufficient guide to 'seeing' breaks in the stratigraphic record, let alone ones that occur at the microstratigraphic/microscopic scale (e.g. Goldberg and Berna, 2010; Goldberg et al., 2012; Karkanas and Goldberg, 2008). A trunk of cycad buried in ~30 cm of Carboniferous channel and deltaic deposits (in what is now Nova Scotia) clearly demonstrates that these deposits accumulated on the order of years to decades while the trunk was still standing and had not decayed; at this locality calculated accumulation rates based on this tree sequence would be about 50 m/kyr whereas '… net rates of accumulation of the thick Coal Measure sequences in which the buried trees occur are nearer 5 cm/kyr' (Bailey, 2017: 2). In contrast, with shorter time intervals relating to human prehistory in the Pleistocene, sedimentary accumulations within Palaeolithic deposits can be rife with small, nearly invisible temporal gaps (Goldberg et al., 2012) (Figure 5.25). Even in much more recent shell middens, gaps can be recorded (Aldeias and Bicho, 2016).

The above preamble has implications concerning an array of fundamental interrelated issues, including:

- the notion of so-called living floors and occupation surfaces (see also section 3.4.1)
- the coherence of archaeological 'assemblages': repeated occupations on a stable, non-aggrading surface can lead to 'mixed assemblages' that do not reflect individual events

 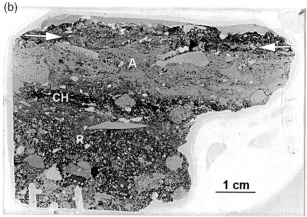

Figure 5.25 A seemingly simple combustion feature from the base of the Mousterian layers of the cave of Roc de Marsal, Dordogne, France. (a) Field view showing reddened base and overlying whitish grey ashes. Note the small size of the feature with the scale at front. (b) Labelled thin section scan showing a much more complex history, with the presence of two hearths. The lower is shown by a reddened former substrate (R) overlain by traces and patches of white ash (not visible in this view, see Goldberg et al. (2012) for details) that is overlain with a sharp contact by the next hearth with a basal charcoal layer (CH) and centimetre-thick band of whitish ashes (A), which terminates with a sharp contact by a millimetre-thick layer (between arrows) of trampled and weathered bone and phosphatized ashes. The history of just the last, uppermost layer comprises deposition of bones and ashes (possibly by dumping) on the eroded surface of the upper hearth, trampling of the surface resulting in *in situ* bone fracturing, and ultimately non-deposition and exposure of this surface with phosphatic weathering mitigated by bat guano. In all then, there are two erosional breaks and one period of stasis yielding weathering of bone and phosphatization of the ashes.

- the balance between the 'rates' and 'intensity of occupation', and those of deposition before, during, and after occupation
- ultimately the issue of time and frequencies between depositional and occupational events.

Many of these issues are incorporated under the phrase 'palimpsests', which has been and continues to be a much-discussed topic in archaeology; the ultimate goal – whether subliminally or explicitly expressed – is to try to isolate human events in time with as much temporal resolution as possible (Figures 5.14 and 5.26). As cited and discussed by Bailey (2007: 203):

> In common usage, a palimpsest usually refers to a superimposition of successive activities, the material traces of which are partially destroyed or reworked because of the process of superimposition, or 'the traces of multiple, overlapping activities over variable periods of time and the variable erasing of earlier traces' (Lucas, 2005: 37).

Bailey in particular has discussed many of the aspects of palimpsests (Bailey, 2007; Bailey and Galanidou, 2009), and has delineated a number of different types (Table 5.3).

The literature discussions on palimpsests have been and continue to be active (e.g. Straus, 1979; Henry, 2012; Malinsky-Buller et al., 2011; Mallol and Hernández, 2016; Ferring, 1986; Bordes, 1975; Dibble et al., 1997), and are too abundant to be summarized here. Instead, we take a somewhat different approach that falls within the theme of this book. Typically, the emphasis in investigating palimpsests has been on the use of objects – and occasionally features (e.g. hearths) – as a means to concentrate on short-term events. Typical examples include artefact refits,

Table 5.3 Types of palimpsests as recognized by Bailey (2007).

Palimpsest type	Description from Bailey
True palimpsests	'sequence of depositional episodes in which successive layers of activity are superimposed on preceding ones in such a way as to remove all or most of the evidence of the preceding activity.' (p. 204)
Cumulative palimpsests	'…[when] successive episodes of deposition, or layers of activity, remain superimposed one upon the other without loss of evidence, but are so re-worked and mixed together that it is difficult or impossible to separate them out into their original constituents.' (p. 204)
Spatial palimpsests	'…spatial palimpsests [are] a variant of the cumulative palimpsest but distinct from it and defined as a mixture of episodes that are spatially segregated but whose temporal relationships have become blurred and difficult to disentangle. As with true palimpsests, the boundary between cumulative and spatial palimpsests is not a sharp one.' (p. 207)
Temporal palimpsests	'…an assemblage of materials and objects that form part of the same deposit but are of different ages and "life" spans….[It differs from] the cumulative palimpsest, [in which] the association of objects of different ages is really an aggregation due to the effect of mixing together what were originally distinct episodes of activity or deposition. The temporal palimpsest comprises what, from the point of view of a cumulative palimpsest, might be viewed as a single episode, a so-called "closed find" such as a shipwreck, a burial chamber or the room of a house, where all the materials are found together because they are constituents of the same episode of activity or deposition.' (p. 207)

Figure 5.26 Palimpsest composed of bedded burnt remains from MP features at Roc de Marsal, France. (a) Thick sequence of bedded lithics within superposed combustion features that rest on a residual, rubified clay on the bedrock. (b) Oblique view of white ashy combustion feature that overlies and underlies blackish organic-rich remains from combustion events. Note the richness and density of the lithic materials in both photos.

the projection of artefacts to examine tightness of fit, and the study of hearths. All represent short-term almost mini Pompeii events (Goldberg and Berna, 2010; Berna and Goldberg, 2008; Miller et al., 2013; Karkanas and Goldberg, 2008; Machado and Pérez, 2016); magnetic studies of hearths have also been used to establish – or at least refine – synchronicity of combustion features (e.g. Sternberg and Lass, 2007; Carrancho et al., 2016). Newer approaches (see, for example, papers in *Quaternary International*, volume 417, 2016; Mallol and Hernández, 2016) still focus on objects, but employ different parameters, such as raw material units (Machado et al., 2013) or the zooarchaeological and taphonomic study of the fauna as chronological indicators (e.g. Machado and Pérez, 2016).

Interestingly, little attention has been paid to the deposits themselves and what they can reveal about the temporal framework of archaeological materials and their ability to record short-term events that can be correlated across space within a site (Sañudo et al., 2016). In other words, a shift is taking place from concentrating on the objects to their encasing matrix (e.g. Polo-Díaz et al., 2016). Micromorphology and archaeo-stratigraphic analysis (Bargalló et al., 2016) have been key tools in looking at finer intervals of depositional time and past human activities (see Figure 5.26).

Nevertheless, even with all or many of these multipronged approaches, some thorny issues always remain since the coupling of object and its microstratigraphic context is often impossible to figure out. For example, an artefact can be (a) within a depositional unit, (b) on a depositional unit, or (c) span several depositional units. Furthermore, objects are commonly thicker than many of the individual deposits or microfacies units (cf. Goldberg et al., 2009a), so to which stratigraphic unit or units do they belong? In other words, it is extremely difficult to locate precisely in the field the exact position of say, the lower contact of a flake.

Similarly, there are problems related to determining microstratigraphic resolution in the field to distinguish individual depositional units and their contacts, which can be spread over large areas, ranging from tens to hundreds of square metres. Moreover, erosion and redeposition can easily displace deposits and objects, and deflation and winnowing of finer matrix materials can lead to lags of artefact accumulations that have nothing to do with each other. Also, trampling can push material down and treading can kick material up into an overlying layer. Finally, there are cases where objects or materials are at the contact between two depositional units. As was shown in Goldberg et al. (2009c), pieces of charcoal at the Middle Palaeolithic–Upper Palaeolithic boundary can represent either a lag leftover of Middle Palaeolithic charcoal exposed at the surface, or in fact be the first Upper Palaeolithic charcoal to arrive on the scene.

In a way, the above was envisaged decades ago, and in this light it is interesting to read afresh the words of Laurence Straus (1979: 334) extensively citing François Bordes, who started out as a geologist:

> Bordes rightly points out that under any circumstances caution must be exercised in arguing that all the objects found together on a 'surface' were deposited strictly contemporaneously in the context of related manufacturing, use or dumping activities. Pleistocene prehistorians cannot yet adequately control for 'micro'-chronological differences (which may involve scores, hundreds or even thousands of years represented within the thickness of a single archaeological level), nor can they ever fully control for changes in the rate of sedimentation at any type of site. Indeed, prehistorians must always be aware of the possibility that their cultural deposits, no matter how thin and compact, might be palimpsests; the results of repeated occupations of the same spot either during a short period of time, or during a period of very slow sedimentation. As Bordes points out, these problems obtain [sic] in open-air sites, where even the absolute contemporaneity of two structures exposed on the same 'surface' at Corbiac, for example, cannot be readily guaranteed.

So what is the bottom line? At this point, it is clear that palimpsests – whatever their form – are something we will have to live with. Pompeii's are rare geoarchaeological occurrences at the large scale, and at the small scale 'mini-Pompeiis' – as revealed by microstratigraphic analysis and micromorphology – are both very subtle and very difficult to recognize (see Figures 5.21, 5.25, and 5.26). Likewise, even if very discrete and individualized depositional units and events can be observed – as at Sibudu (Goldberg et al., 2009a) – it is another thing to operationalize them: (a) they maybe observable at a microscopic scale but not at the field scale and (b) lithics and bones, for example, are likely thicker than any of the depositional units. Finally, if we think of sites where rates of sedimentation are optimistically cited as being 'rapid' (e.g. mm/year), to which layer or depositional event do you ascribe a 2–3 cm thick fragment of bone, pottery, or chert?

On the other hand, palimpsests '…provide the key to how we should go about investigating the longer-term, larger-scale dimension of the human condition…' (Bailey, 2007: 220). In that sense, a palimpsest as a feature/layer is part of the stratigraphy of a site and therefore has to be interpreted in relation to the other layers, which are also the product of accumulation and erosion. Therefore, the sum of the characteristics of the palimpsest in relation to the overlying and underlying layers can reveal a great deal about human behaviour. The mere existence of palimpsests and their variation is perhaps what we should focus on. Small-scale analysis using microstratigraphy and micromorphology can reveal glimpses of isolated events in a palimpsest and perhaps give a general idea of the type of activities taking place; coarse scale stratigraphic relationships and variations of palimpsests can reveal longer-term behavioural changes and depositional histories (Karkanas et al., 2015a).

In the end, we should probably forget that there is such a thing as a readily recognizable and operational 'mini-Pompeii': we are always dealing with mixtures of objects to some extent. On the other hand, we should strive to collect and observe the data as carefully as possible in the field, and to do this consistently at most sites, especially those that play a particularly important part in human history. Turning again to Straus:

> Excavation, while aiming at the uncovering of materials deposited on surfaces, must be conducted so as to avoid stratigraphic confusion as much as possible, through the use of balks, the technique of working metre squares back from a section, alternate squares, etc. (Straus, 1979: 334).

5.4 Concluding Remarks

To sum up, this chapter was about what hunters and gatherers do at a site and how they leave their marks. It also tries to link these activities within the overall framework of natural geological agents, which can bury (i.e. preserve), erode, or transform them, either physically or chemically. The emphasis has been on caves, which being closed systems – although geochemically active ones – tend to provide more faithful records of specific activities or behaviours that the ancient residents practiced. In contrast, hunters and gatherers operating in the broader, less spatially constrained expanses of open-air sites likely carried out similar types of activities as we observe in caves, but due to the dynamics of the geological environments (see Chapter 2), such actions are likely to be more poorly preserved, completely buried, or absent altogether, as we have seen in the examples of meandering floodplains.

We infer what people were doing at sites by carefully and systematically observing and documenting all the aspects of the deposits, as we have discussed previously: their composition, texture, internal organization (sedimentary structures) and fabric, and lateral and vertical extents. At Kebara, we saw that heaps of dumped ashes grade into a midden of bones tossed against the back wall. Such house cleanings were contemporaneous with abundant pyrotechnical activity as epitomized by thousands of stacked hearths spread over an area of more than $150\,m^2$. Integrating these essentially geological data with information from the fauna (partially destroyed by diagenesis), botany (charcoal and phytoliths), and lithics provides a holistic view of site formation processes and history, as well as Neanderthal lifeways during the Late Pleistocene (Speth, 2006).

6

Architectural Sites

6.1 Introduction

Architectural sites are characterized by the permanent demarcation of areas and the erection of permanent man-made structures that include buildings as well as non-building structures such as tombs, tumuli, and mounds. They are mostly associated with Holocene sedentary societies and therefore post-Palaeolithic cultural periods.

Permanent habitation involves characteristic uses of space, such as indoor, roofed enclosures vs outdoor, unroofed or sometimes partially roofed spaces. In well-preserved sites with still-standing walls, it is often easy to separate indoor from outdoor phases. However, even in such sites, the uses of rooms or whole buildings have changed during the course of their history, and even temporary structures have ended up as open spaces. In other cases, the architecture of the walls is quite complicated and even when successive phases are detected, it is difficult to assign the deposits to a certain building phase, primarily because it is not easy to identify the nature of the sediments. In such sites, often the deposits all look similar, particularly after a couple of hours under the sun in the middle latitudes, where the profiles immediately become dusty and where they acquire an overall grey brownish tint. It is therefore the backbone of any correct interpretation of the archaeological record in an urban site to be able to differentiate natural from anthropogenic sediments, indoor from outdoor ones, floor sequences from construction fills (levelling and foundation/packing fills) and occupational fills (Table 6.1).

The interpretations in the examples in the following sections are based on the characteristics of the depositional sequences and not on architectural features such as walls and other erected or permanent structures or the archaeological content of the deposits. Although the last is of chief importance, they can often provide an elusive picture of the actual formation processes. As we have stressed several times in this book, the structure, fabric, and all other sedimentary characteristics of the deposit are the decisive criteria for interpreting the way it was laid down on the earth's surface and not its content.

6.2 Roofed Facies

In general, a roofed facies comprises sequences of floor, construction fills, and occupational deposits that often terminate with post-abandonment phases (Table 6.1). The latter may include midden deposits and ultimately sediments from the roof (with total roof collapse) and walls; just before the abandonment, there may be changes in the use or secondary (dis)use of the space.

Occupational deposits are the result of slow accumulation of debris from everyday activities and in many ways resemble constructed floor surfaces, which may include a lower packing/levelling layer. Often, the natural substrate has been used as a floor as is, or perhaps after some cleaning and levelling. All these scenarios represent actual floors in the sense that they acted as occupational surfaces. Therefore, in a hierarchical approach it is of principal importance first to be able to differentiate between *in situ* occupational deposits – including constructed floors – from packing/levelling fills. Although probably an assumption that has to be proven in separately each case, any massive sequence inside rooms that does not show any subhorizontal arrangement of elements and where their coarser clasts are chaotically distributed must represent a construction fill. Occasionally, inclusions of overturned floor or plaster fragments are evidence of dumping processes and hence are good indications of construction fills (Figure 3.70). An upper horizontal surface also implies levelling, at least of the upper part of the fill (Figures 2.2b and 3.69). In contrast, floor sequences and their associated occupational deposits always show some kind of horizontality, although this has to be confirmed very carefully, probably on a meticulously cleaned profile. The horizontality can be expressed at all scales of observation, but often this might be

Reconstructing Archaeological Sites: Understanding the Geoarchaeological Matrix, First Edition. Panagiotis (Takis) Karkanas and Paul Goldberg.
© 2019 John Wiley & Sons Ltd. Published 2019 by John Wiley & Sons Ltd.

Table 6.1 Types of deposits and facies in urban sites.

Roofed (indoor) facies	Destruction and abandonment facies inside houses	Unroofed (outdoor) facies

• Floor sequences
  – Constructed floors
  – Beaten floors
  – Occupational debris
    ○ Special activities (e.g., dung from indoor penning, pyrotechnological vestiges etc.)
• Construction fills
  – Packing fills
    ○ Destruction layers*
  – Foundation fills
	• Debris from secondary uses (penning, storage)
	• Middens
	• Roof and wall collapse
	  – Destruction layers*
	• Natural sediments (see unroofed facies)
		• Middens
		• Gardens (anthrosol)
		• Street deposits
		• Dung (pens)
		• Natural sediments
		  – Puddles, ponds
		  – Sheetwash
		  – Landslides, debris flow, water flows

*'Destruction layers' are typically produced during destruction/abandonment phases but are often levelled and used as construction fills.

crudely displayed in the field as a vague horizontal linearity of the coarser elements or a discontinuous banded appearance (Figures 3.45, 3.46, and 6.1).

The Middle Bronze Age site of Palamari in the island of Skyros, Greece offers a good example for illuminating the above (Karkanas, 2015). Palamari is a fortified settlement with complex architecture including impressive fortification walls, bastions, buildings, streets, and open areas. The substrate of the site consists of aeolian and pedogenetically altered sands, and in general the archaeological sediments are dominated by an aeolian sand component. This situation makes the differentiation of the various sedimentary facies even more difficult because the sand imparts an overall homogeneous appearance to the matrix of the deposits. At the same time, in such sandy terrains artefacts and generally large features are more noticeable in the field as well as

remnants of clay-rich materials, which have a contrast in colour and appearance (like mudbricks fragments, for example).

Inside rooms, alternating sequences of massive and banded to bedded deposits were identified. This difference was not observed during excavation but only after thorough cleaning of the profiles (Figure 4.22). The sedimentary content of both facies is almost the same and the grain-size distribution is noticeably similar (Table 6.2 and Figure 6.2); thus, only the arrangement of the particles and their geometric distribution are responsible for the subtle differences shown in the field. Banded and bedded facies consist of thin, centimetre- to millimetre-thick layers or lenses, each one consisting of different grain-size distributions, fabric, texture, and mineralogical composition. Commonly, there is a coarse sand to granular-sized layer rich in green schist particles that is

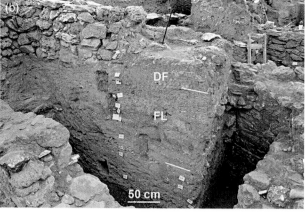

Figure 6.1 General appearance of floor sequences in constructed spaces. (a) Distinct banded appearance of a series of reddish brown clay floors (FL). (b) Crude banding of mostly worn-out clay floors (FL). Their clay content and reddish colour show up clearly in the section. The grey layer above that constitutes the foundation of the overlying wall is rich in burnt remains and is interpreted as a construction fill composed of demolished and destruction material (DF). EBA, Palamari, Skyros Island, Greece.

Table 6.2 Granulometry of Palamari sediments (see also Figure 6.2).

Sample	63–125	125–250	250–500	500–1000	1000–2000	2000–4000	>4000	>63	>2000
				Grain size ranges (µm)					
Constructed floors									
PL35	8.13	10.47	15.25	8.93	7.47	8.44	3.88	62.56	12.32
PL36	8.98	9.47	12.24	7.92	7.87	7.99	11.13	65.6	19.12
PL39	8.74	10.81	12.57	7.48	6.67	7.61	17.96	71.85	25.57
PL40	9.42	10.45	16.03	9.63	7.92	7.13	5.67	66.25	12.8
PL24	7.46	8.88	13.63	6.80	5.01	6.06	11.57	59.4	17.64
Construction fills									
PL34	8.76	11.44	18.46	9.34	6.47	6.19	2.79	63.45	8.98
PL41	8.55	12.19	19.40	7.94	5.23	5.91	4.12	63.35	10.03
Gravity flows									
PL42	5.66	8.21	15.58	7.69	6.50	7.72	10.06	61.41	17.78
PL43	7.37	9.97	15.91	7.81	5.98	6.94	7.14	61.12	14.09
PL44	6.53	9.02	15.43	7.28	5.99	6.57	11.53	62.35	18.09

PL35, PL36, PL39 and PL40 are constructed floors; PL24, PL34 and PL41 are construction fills; PL42, PL43, and PL44 are gravity flows. No systematic differences in granulometry are observed for the three different sedimentary facies.

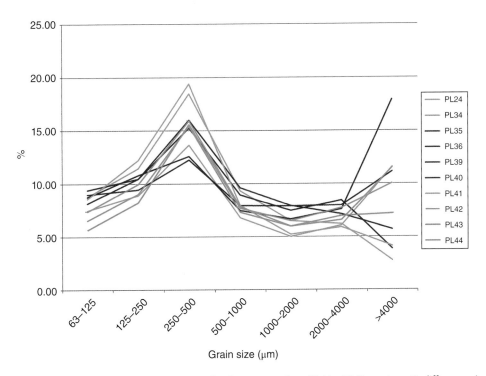

Figure 6.2 Grain-size distribution curves of sediment samples of Table 6.2. No systematic differences in granulometry are observed for the three different sedimentary facies: red = constructed floors; blue = construction fills; green = gravity flows.

capped by a brown aeolian sand-rich layer or reddish to brown clay (red sandy clay layers regularly intervene in the sequence; Figure 6.3). Their structure is dense with few pores, and a microscopic 'mosaic' structure is observed in the coarser layers (Figure 3.49a). In general, all microscopic and macroscopic features are evidence of constructed floors as detailed in section 3.4. Not surprisingly, the bulk grain analysis (Table 6.2) did not pick up these differences because several microlayers were lumped together during sampling for granulometry. The

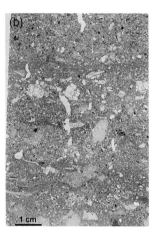

Figure 6.3 Scans of resin impregnated slab (a) and corresponding thin section (b) showing a series of prepared floors (arrows) composed of greenschist and coastal sand admixture (G) overlain by sand coating (S) in some cases rich in red clay. For a field view of this type of floor see Figures 3.45. 3.46, and 4.22; for detail of their microstructure, see Figure 3.49a. EBA, Palamari, Skyros Island, Greece.

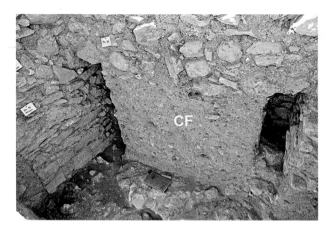

Figure 6.4 Massive construction fill (CF) with chaotically dispersed large stone in an unsorted matrix. Notebook for scale. EBA, Palamari, Skyros Island, Greece.

material of the floors is similar to the construction fills in terms of content and grain size, as most of them probably represent destroyed and reworked floor deposits and/or materials from outdoor deposits that in fact represent redistributed material from the decay of houses. Indeed, grain sizes of fine-grained outdoor facies are also similar to those of the fill and floor.

Nevertheless, both construction fills and outdoor facies often contain larger amounts of very coarse clasts. These grain sizes are not typically discerned in regular granoulometric analysis, which would require a vast amount of sediment for a representative sample (see section 7.5.3). They are estimated visually, but such estimations often do not provide conclusive evidence as, again, it is more their overall distribution rather than their frequency that defines formation process in most cases. In Palamari, massive fills are characterized by centimetre- to decimetre-sized clasts with a chaotic distribution and orientation (Figures 4.3 and 6.4). Similar clasts do appear occasionally in floor sequences, but overall they follow the horizontal pattern of the floor surfaces. Furthermore, microscopically massive fills consist of an agglomeration of sediment aggregates with variable porosity and no particular organization. Their upper contacts are

horizontal and associated with a new building phase (Figures 4.9 and 6.1b). They are therefore interpreted as packing and levelled fills.

We should also stress the fact that the above differences were not always visible in the field even after thorough cleaning. This is because floor sequences were often laterally disturbed or homogenized during their use and thus their characteristic horizontality and banding had been destroyed. So here the study and understanding of the better-preserved sequences helped us to establish a pattern that was applied to the other, more decayed parts of the site. More generally, it is evident that both extensive sampling and a strategy to formulate a model of site-formation processes acting in each site is a prerequisite to understanding all deposits of an urban site (see section 7.4 on sampling strategy).

An additional interesting feature observed in Palamari is the occurrence in the profiles of linear arrangements of coarse clasts (Figure 4.22), which define surfaces that are one clast thick and extend for several metres. The stone surfaces are often impossible to identify during excavation because the stones are in some cases quite separated from each other, and the stone line can be followed as such only in the profile. Stone surfaces always separate sequences of bedded floors and massive fills, and may represent deflation surfaces on abandoned occupational sequences. Stone lines are often very helpful in identifying small hiatuses or discontinuities in urban sequences (see also section 2.7.1).

Another example is the Bronze Age site of Mitrou in Greece, which illustrates the dissimilarity between different types of constructed floors and occupational debris (Karkanas and van der Moortel, 2014). Impressively thick sequences of overlapping worn-out floors and occupational deposits appear to the naked eye as having a characteristic finely layered appearance

(Figures 3.43 and 3.47). Moreover, there is no clear association of a single floor level with a specific building phase but rather thick sequences of floor build-up in each architectural phase. Purposely constructed floors are planar bodies with sharp upper and lower contacts, and made of a dense compacted matrix and constituents not found regularly in the site bedrock or its catchment. Ash-rich or lime-rich varieties of these floors were often recorded. Above and in between constructed floors are layers and lenses of grey to dark grey, fine, loose, and aggregated debris that is very porous and with a chaotic appearance at the microscopic scale. Their main content is ash and fine food debris, such as burnt and unburnt bone, and charred plant remains; they are thus interpreted as occupational debris produced by everyday activities. Fragments (rip-up clasts) from the underlying floors are at times incorporated into this overlying debris.

These facies have been interpreted as representing a non-compacted, last-moment levelling layer prior to the application of a new floor. They can also represent remains of demolition debris mixed with occupational debris that was mostly cleared away, that is, sweeping facies. In most cases, occupational debris when continuously trampled appears more compact, but fissures and occasionally elongated fragments are oriented horizontally (see section 3.5.3). As such, they resemble in many ways the 'beaten floors' reported by Macphail et al. (2004) that comprise non-constructed floors made of trampled occupational debris most likely formed 'by gradual deposition of debris on a sequence of accruing surfaces' (Milek and French, 2007: 342). In the case of Mitrou, such occupational deposits additionally show a layered appearance resulting from incremental and not continuous build-up (Figure 6.5) (Karkanas and van de Moortel, 2014). Upper contacts of these microscopic sublayers exhibit organic staining or show a more compacted and cemented surface, which implies that they were surfaces that had been exposed for some time. Their sediment also shows evidence of ductile deformation that formed during application of the material in a wet state. Therefore, this type of occupational debris is interpreted as having been both maintained and made into surfaces.

6.3 Diachronic Spatial Organization

A common pattern that emerges is that house sequences not only represent archives of the history of individual buildings but also reflect the social organization, practices, and beliefs of entire neighbourhoods and urban areas that, in turn, help define cultural periods.

Figure 6.5 Scans of resin impregnated slabs with floor sequences from the EBA of Mitrou, Greece. (a) Bedded sequence of floors with little standardized construction technique. Some of the floors are composed of a mixture of reddish clay and serpentinite, and some others include occupational debris. (b) Example of occupational debris made into surfaces (D). They overlie a constructed clay floor (C) – with evidence of burning on its surface (darkened surface) – that in turn rests on top of loose occupational debris (OD). Note the fine horizontal stratification in the upper floor (D) and the whitish ash lens (arrow) that microscopically shows evidence of stretching and deformation, such as sweeping over the substrate (see Karkanas and van de Moortel, 2014).

At Mitrou, Early and Middle Bronze Age phases are characterized by continuously and meticulously maintained multiple plastered floors. As described above, the associated debris was recycled inside houses and made into constructed surfaces. At the beginning of the Late Bronze Age more standardized construction techniques of floors is observed, which encompassed thin floor sequences and clearing away of occupational debris. This radical change is connected to the rise of a prepalatial elite just before the debut of the Mycenaean palatial society, and it can be interpreted in terms of more controlled and systematic building behaviour (Karkanas and van der Moortel, 2014).

At the Neolithic site of Çatal Höyük, domestic and ritual activities are differentiated on the basis of the thickness of plaster floors and their associated occupational deposits (Matthews et al., 1997). Thick plaster floors show evidence of heavy trampling and wear (horizontal cracks post-depositionally impregnated with salts). The overlying occupational deposits are relatively thick, at 2–3 cm, and are characterized by large amounts of food remains including bone, charred husks, and grass phytoliths, as well as fragments of heated oven plaster and ash. In one of the buildings, this domestic sequence is followed by a later sequence of thin white and orange plaster floors often characterized by a thin finishing coat of white plaster, which was kept clean. These fine plaster layers cap a large human grave that was dug into the previous thick plaster floor sequence. The sequence of thin plaster floors is also associated with moulded plastered sculptures, and a cow jaw and horn found collapsed on the last floor. These features suggested that this building changed use from a domestic residence to an ancestral shrine.

In addition, horizontal and vertical changes in microstratigraphic sequences as described above have been identified in several buildings in Çatal Höyük (Matthews et al., 1996). In particular, sequences of plaster floors characterized by white plaster finishing coats were always almost clean and not associated with occupational deposits on their surface. Sequences of this type are mostly observed in large or elaborate buildings. Thick plaster layers with few or no finishing coats are always followed by relatively thick occupational deposits characteristic of domestic activities as described above. Nevertheless, it appears that in many buildings domestic activities (food storage, preparation, and cooking activities) were often contemporary but always spatially separated from ritual features (burial platforms, sculptures, and wall paintings). The dirtier plaster and occupational sequences were separated from ritual, cleaner areas by steps, ridges, or platforms (Matthews, 2005).

6.4 Unroofed Facies

Unroofed facies comprise a major part of the depositional sequence of a site that includes not only the areas in between buildings but also the abandonment facies, which regularly occur after the collapse of the roof.

6.4.1 How to Recognize an Unroofed Area

Differentiating roofed vs. unroofed areas is a commonly recurring theme in architectural sites. Unroofed faces are dominated by natural processes, which in urban sites mostly affect previously deposited anthropogenic deposits. A fundamental exercise in defining natural processes in archaeological sites is to recognize the landscape restrictions in every area of a site under investigation. We define here landscape restrictions as the

geomorphology of the open space, including shape, inclination, and geometry of the substrate (conditioned each time by the previous deposit), and the sources of sediment that either surround the human structures or occur upslope from them. All these parameters and sources constrain the possible nature of the depositional processes that can act in a particular microenvironment. Therefore, high-energy water flows and debris flow can occur in settlements built on a slope or close to a river. In horizontal areas, the most characteristic natural sedimentary processes away from slopes are low-energy rain-derived sheet wash, ponding, and small-scale freefall and debris flows associated with decayed mud walls. Small-scale landslides and medium-scale debris flows occur on temporal sloping surfaces formed by land management during the life history of a settlement (e.g. peripheral areas of tells). It is almost inevitable that open spaces in or close to sloping areas of a settlement will contain debris flow or water flow deposits.

Natural deposits are characterized by (a) repeated or rhythmic layering with lateral continuity, (b) bedding or laminae defined by different sizes of particles, (c) sorting of particles, (d) occasionally rounding of large particles, (e) grain-size grading along or across layers, (f) orientation of elongated particles along the slope surface, and (g) tiling up of large clasts (Figures 2.10a, 2.12a, 2.13, 2.14, 2.24, 2.26, 2.45, 2.46, 2.51, 2.53, and 6.6). These features can be observed in the field but in some cases they can be identified only under the microscope. Indeed, often in archaeological sites outdoor facies are recognized by the existence of microscopic sedimentary crusts normally exhibiting graded bedding (see section 2.4.1), and small pockets of well-sorted sand-sized sediment are probably the result of water flow. In addition, detailed field observation may reveal ripple structures or crude cross bedding (Figure 2.14). Layers of coarse clasts showing down slope or normal and reverse grading are probably the result of dry grain flows or debris flows (Figures 2.13c and 6.7). All these features, however, have to be placed in the surrounding context and with the configuration of the landscape (see also section 2.2.5 for a more detailed discussion).

Additionally, microscopic features accompanying the appearance of water percolation features, such as pendants, bridges and dusty clay coatings, are likely to be associated with outdoor facies. Clay coatings, however, can occasionally be found indoors as a result of sweeping and trampling, and dripping water. In the Greek Classical site of Molivoti, compact, massive clay-rich layers inside houses show typical dense porphyric-related distributions of the coarse clasts (mosaic structure) attributed to prepared clay floors. In contrast, street deposits are coarser and also characterized by intergrain microaggregated (enaulic) microstructure that is accompanied by

Figure 6.6 Examples of natural processes in sites. (a) Debris and mud flow carrying coarse and fine sediment downslope, including sherds (arrows). Note the chaotic orientation of stone in the upper part of the debris flow and the alignment of coarse particles along the flow in the lower part. White tag is 10×7 cm. EBA, Palamari, Skyros Island, Greece. (b) Small gravelly channelized flow below the wall showing some crude imbrication and a few sherds (S). Hellenistic-Roman, Ancient Messene, Greece.

Figure 6.7 Debris flow with characteristic reverse grading (tip of arrow) from fine grained at the base to coarse grained at the top. Note also the crude sorting at the base consisting mainly of fine gravel. Hellenistic-Roman, Ancient Messene, Greece.

Figure 6.8 Photomicrograph of trampled street deposits: complex organic-rich, dusty clay coatings and intergrain aggregates. Classical, Molyvoti, Thrace, Greece.

the extensive appearance of silty features that produce pendants beneath clasts or bridges between clasts (Figures 3.74 and 6.8). Microaggregates are mostly organic rich and presumably represent street refuse.

6.4.2 Destruction and Abandonment of Buildings

During abandonment, buildings can be used as middens, left to decay slowly, be completely or partially demolished, or filled and levelled for erecting a new construction. Several combinations of the above can take place during the history of an abandoned building. Collapse phases can be gradual or abrupt due to a conflagration or a natural disaster (e.g. earthquake, subsidence, earth slide). After the collapse, a building structure continues to decay and be lowered by natural processes; however, very

often people dismantle and rob it for its stone, wood, and tile components that accelerate its destruction.

Construction (packing) fills have already been described above, and their sedimentary characteristics are detailed in section 3.5.2. Middens, by definition, comprise unusually large amounts of occupational debris that is almost ubiquitously rich in ashes and generally burnt materials, as well as other perishable materials that lead to decay, differential compaction, and sagging (described in detail in section 3.5.2; Figures 2.81a and 6.9). Nevertheless, middens can be used partially as packing fills for construction of a new building phase above it. Middens with a layered structure are most likely the result of *in situ* gradual accumulations of dumped occupational debris (Figure 6.10). Massive middens with a macroscopic aggregated structure and vertical

Figure 6.9 Sequence of thick midden deposits comprising mostly ashes and burnt remains. Finely bedded ashes (B) are characterized by large amounts of burnt fine plant material rich in phytoliths. The upper part of the sequence, which appears massive (M) in the field, in fact consists of relatively unsorted slopewash of ashy, dung-rich sediment that is microscopically bedded sediment. Neolithic, Promachonas/Topolniča, Greece.

orientations of large clasts indicate rapid and massive dumping, and probably represent packing fills.

To be cautious, there is always the possibility that the previously described massive sequences having the characteristics of packing/levelling fills may actually represent natural deposits. Massive sequences are normally deposited by gravity flow processes, such as mudflows and debris flows. These deposits must have found their way through an opening in the room because they do not override structures, unless arguments can be made for a very catastrophic event, something that will be immediately identified from walls that have fallen in or have been taken down. Habitually, they do not have a horizontal and planar appearance but rather display a sloping lobe or a channelized form. In addition, debris and mudflows often have characteristic sedimentary features such as clustering of large clasts with fine-grained tails or coarser sides (levees) (see also section 2.3).

In the Palamari Bronze Age fortification site, debris flows are associated with a series of collapses of the fortification walls upon which a new phase of walls was erected every time. These debris flows always occur on an inclined substrate and their large, decimetre-sized clasts float in an unsorted matrix that also follows this general trend. Rarely, crude imbrication of clasts is observed, as well as clustering of large clasts with fine-grained tails that slope upwards. Often, debris flows follow a first-stage accumulation of free-fall wall stones at the base (Figure 2.24). A detailed description of the structure and fabric of debris and mudflows is provided in section 2.3.4.

In the later phases of the history of tells, gravity flows and landsliding occur as a result of the inclination of the

Figure 6.10 Bedded midden deposits composed of burnt remains and construction material refuse. Most layers consist of heterogeneous aggregated materials inside a fine matrix. Each bed represents a distinctive mixture of occupational refuse perhaps reflecting different areas from where the material was derived. Scale is 50 cm. Neolithic, Imvrou Pidadi, Thessaly, Greece.

ground. The anthropogenic nature of tell deposits, with their high amounts of organic matter that are prone to decay and compaction, produces unfavourable ground conditions in the peripheral, sloping parts of a tell. At Çatal Höyük this is evident in frequent failures of constructions and various measures to repair and strengthen the buildings (Baranski et al., 2015).

As a result of the previous discussion, the problem of identifying the nature of massive fills can be further refined by differentiating between gradual collapse deposits due to natural processes and more rapidly produced levelled/packing deposits. In profiles normal to walls, natural processes should appear coarser with some sloping particles and objects. However, this will depend on the orientation of the decay processes, with some walls characterized by more pronounced features. As have been shown in ethnoarchaeological contexts, when a roof degrades and collapses, wall degradation also accelerates (McIntosh, 1977; Milek, 2012; Friesem et al., 2014a,b). Relatively few sites show clear evidence of roof collapse. Such evidence consists of roof tiles (baked or unbaked) with occasionally timber elements that are overlain by mudbricks or stone. Plant materials covering roofs, such as reeds, sedges, seaweed, and similar organic matter, are rarely preserved. In cases where the entire structure was burnt down, the fired floor or tiles can be recognized more easily and their original position determined; burnt clay structures preserve timber and generally plant imprints (Figure 3.53). It is possible to reconstruct the details of the conflagration event – particularly if there is a pattern in the way the structural elements are burnt – by noting the direction of the intensity of fire indicators, that is, the position of the reduced (blackish) and oxidized (reddish) parts relative

to the ground as well as any timber imprints: the blackish parts correspond to the buried surface and the reddish to the exposed surface. Forensic fire investigation is very rarely conducted in archaeological sites, although such an approach can not only shed light on the fire dynamics but also reveal if the fire was an accident or a result of arson (Harrison, 2013).

When a wall collapses individual mudbrick and other masonry elements such as mortar may still be recognizable and even if thoroughly decayed and melted within the rest of the collapse sediment, and they usually appear as distinct coloured clay and cemented masses. Slow mudbrick wall degradation produces talus cones at the foot of the wall, overlying whatever has remained of the collapsed roof (Friesem et al., 2014b). Gravitational slurry movement characterized by fine-scale slope processes produces crude layering and sorted lenses parallel to the walls and inclination along the talus slope (Figures 3.55 and 3.56). Therefore, overall slow decay processes can be differentiated from packing deposits by the sloping geometry of layers away from the walls and the existence of gravity flow-related sedimentary features close to the walls (Figures 1.3a and 6.11).

In the case of an upper floor, overturned fragments of clay floor are often the lowest element to be observed lying on the ground floor surface. In general, the characteristic appearance of an upper floor collapse is a chaotic assemblage of architectural features lying on a horizontal surface representing the original floor in between walls (Figure 6.12). Even if these features are welded in the form of an amorphous mass, its aggregated nature – comprising relatively large structural elements (mortar, plaster, clay structures, timber, and stone) – is still identifiable, especially when it has not been later reworked. The

Figure 6.11 Talus cone of decayed wall materials. Wall construction elements (outlined with dashed line) floating in a mass of fine unsorted material. Part of the fine fill has been excavated, but the overall inclined geometry away from the walls is still visible (arrows). Hellenistic, Hatziabdulah, Palaipaphos, Cyprus.

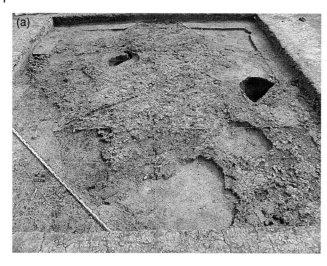
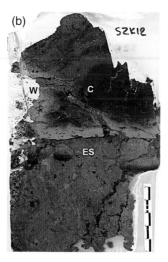

Figure 6.12 (a) Upper clay floor that collapsed *en masse* due to a conflagration. Note preservation of the general outline of the building by the shape of the floor. Grid is 1 m. (b) Resin-impregnated slab of a detail of the fallen burnt clay floor (C) on the earthen surface (ES). Note that the highest burning temperature is recorded at the base of the fallen block, suggesting that the fire started in the ground floor and heated the overlying, now collapsed floor. Curvature at the left end of the clay block is from wood impression (W). BA, Körös region, Hungary. Courtesy of Körös Archaeological project.

upper surface of this mass is uneven, with protruding angular to subrounded blocks of architectural elements.

Nevertheless, there are cases when these amalgamated masses are compacted and levelled for the construction of a new building phase and therefore it is often difficult to discern the sequence of events in the field. In addition, there is always the possibility that natural deposits were later levelled to produce a foundation/floor surface or cases where natural material and/or material from outside was brought in to build a foundation layer. No doubt all these possibilities are of archaeological significance, although they basically represent reworked deposits. The latter, originating inside the excavated room, have different implications in the interpretation of the archaeological record as compared to deposits brought in from somewhere else. In cases where natural deposits have originated inside the room and later levelled, normally the lower part of the construction fill must show features associated with *in situ* gravity flow processes. Often, it is almost unavoidable that only microscopic analysis can provide evidence of such processes and thus a correct interpretation of the nature of the fill.

Natural processes continue to act after the collapse of the architectural features. In stone constructions, walls can survive as partially standing objects for several thousands of years, particularly when seismic or other catastrophic events have not occurred. It is almost certain that robbing has occurred when some of the masonry stone is missing and only the foundation stones are found (Figure 6.13).

The types of sediments that cover abandoned settlements are able to yield details on the process of

Figure 6.13 Remnants of foundation wall at the 'Lower Town' of Mycenae (Hellenistic), Greece. In such cases where all masonry stone of the rest of the walls is missing, it is almost certain that robbing has occurred. The surrounding foundation walls in closed spaces should have trapped some stones if high-energy flows had occurred.

destruction of a site that can be used in deciphering details of its depositional history and stratigraphy. Such an example is provided by the abandonment of the Geometric-Archaic settlement recently found at the foot of the Mycenae citadel, Greece (Karkanas, in press). The post-Geometric Period is characterized by a type of sediment that appears to have its origin in the slopes of the small valley to the west of the acropolis. The first phase of deposition occurred when most of the buildings were still standing, at least to a certain height (Figure 6.14). The standing structures were blocking the free water flow, which in turn led to deposition in between the structures of only fine infiltrated silty and clay sediments, therefore the standing structures acted as sedimentary traps. Indeed, the fine-grained sediments of this initial slope deposit can be possibly used to delimit building structures even when their walls are missing due to later robbing (Figure 6.13). In addition, given the low energy of deposition of this phase it can be deduced that most of the associated coarser archaeological finds should not have been seriously reworked and can be considered more or less *in situ*.

After the filling of most of the standing structures the area became an even-sloping surface with only a few

Figure 6.14 Natural types of sediments associated with abandonment. The sediment filling the space between the remnants of the walls (arrow delimits the top of the filling) is mainly trapped, infiltrated, or overbank fine-grained sand, silt, and clay. The sediments above the remnant walls are coarser grained water gravity flows coming downslope generally from the right, over the walls and the fine fill. Hellenistic, 'Lower Town' of Mycenae, Greece.

Figure 6.15 Small gully (dotted line) formed after all spaces between abandoned structures have been filled in and an evened and smooth sloping surface has been created. Note gravel lag at the base of the gully (arrow). Hellenistic, 'Lower Town' of Mycenae, Greece.

obstacles (still higher standing walls), thus leading to higher rates of flowing water, formation of small gullies, severe erosion, and deposition of coarse deposits (Figure 6.15). The deposition of this coarse facies implies that most walls had collapsed or had been demolished before that time. It should also be stressed that the carrying capacity of these slope deposits was not sufficient to remove the large building blocks of the buildings (Figures 6.13 and 6.14). Thus, the fact that the majority of the building blocks are not found in the site strongly suggests that they were taken apart by people of the area before the accumulation of the second phase of slope deposits.

6.4.3 Courtyards, Gardens, and Other Open Spaces

Understanding how the open spaces of a house lot were used is of major social and economic significance (Hutson et al., 2007). Being associated with permanent house constructions, courtyards are found early in the Neolithic of Anatolia (Garfinkel, 1993) and after that they appear in many cultures, regions, and periods. Consequently, courtyard deposits will be the result of different anthropogenic processes characteristic of the practices people were engaged in for each particular period. Knowledge of anthropogenic processes as described in Chapter 3 is therefore a prerequisite for interpreting courtyard sequences, which are often the compound result of many of them.

Courtyards

Courtyards host different features, including hearths for cooking, ovens, trash or storage pits, middens, and privies. Some animals (e.g. chicken, hens, ducks) may range

freely in the courtyard, whereas others may be kept in more restricted areas but not in dedicated enclosures (e.g. sheep, goat, pigs). Thus, courtyards contain the remains of many different activities, including penning, food processing and preparation, and craft activities. Courtyards may have untreated surfaces or be paved or constructed with plaster floors (Hutson et al., 2007; Matthews, 1995). At the same time, courtyards are prone to natural sedimentary processes such as rainwash, ponding, and aeolian processes (deflation, dust, or aeolian sand accumulation), which often yield a stratigraphically complex sequence characterized by interfingered tabular to lenticular layers that overall show considerable extension and lateral repetition (Figure 6.16). This interfingering is the result of the different activities that take place in the same or in different areas on a daily or seasonal basis. In between, natural sedimentary processes (rain wash and wind) rework and deposit layers, which themselves are then reworked and trampled by foot and animal traffic; the result is alternating areas of compacted and looser sediment from trampling and trash accumulation, respectively. The latter should show higher organic contents, including food remains (bone, shell, ashes, and pottery). Therefore, courtyard sequences overall make for a single stratigraphic body but at the mesoscale are characterized by the agglomeration of several types of sedimentary features and layers of contrasting nature and irregular periodicity.

In the Neolithic site of Koutroulou Magoula, Thessaly, Greece, open spaces between houses are characterized by stratified deposits that include bedded dung microlayers (= trampled penning deposits), reworked wood ashes, and dung-rich midden-like layers, along with small, flat red clay constructions that in some cases are

associated with burning activities (Figure 6.16) (Koromila et al., in press). This irregular alternation of small-scale penning activities, midden layers, and domestic activities (hearth and constructed working areas) are characteristic of open spaces around houses as based on ethnographic observations.

At Abu Salabikh, a mid-third millennium BCE city in Iraq, courtyards include alternating cycles of constructed plaster or bitumen-lined pathways and overlying thick, loose occupational deposits with characteristic features of dumping and sweeping (Matthews, 1995; Matthews and Postage, 1994) (see also sections on dumping and sweeping in section 3.5).

Yet, such a rich stratigraphy is not always obvious, particularly in arid areas where trampling is less efficient in building up layered sequences. In the arid zone of the Negev Highlands, Israel, Early Iron Age courtyards of oval compounds consist of highly reworked grey sediment that originate only from degraded livestock dung. What characterized these dung deposits is the lack of large amounts of grass phytoliths that will enhance stratification. In contrast, these sediments are very rich in dung spherulites, which are better preserved in dry environments than in humid temperate regions (Shahack-Gross and Finkelstein, 2008). Furthermore, the degraded dung-derived organic matter of this site shows elevated $\delta15N$ values, which are characteristic of modern and archaeological livestock enclosures (Shahack-Gross et al., 2008a).

In general, courtyards are regularly dirtier than house interiors and therefore their general aspect will be grey, organic-rich sediment, mostly compacted due to trampling; for the most part they are quite laterally extensive and larger than a house room, for example. Note that the

Figure 6.16 Courtyard deposits at Neolithic Koutroulou Magoula, Greece. (a) General appearance of greyish, ash- and organic-rich deposits (below black arrow) in the open space of a building (W). A temporary clay construction is observed (white arrow). (b) Detail of the same open area in another section showing midden-like deposits. These are composed of a mixture of microlaminated dung remains – related to intermittent penning activity – and charred bedded refuse material (B) rich in dung that probably result from maintenance activities (see also text). Scale is 30 cm.

indications for recognizing an unroofed area as described above also hold for courtyards. It is true, though, that some courtyards are partially roofed and therefore the intensity of the natural processes affecting an unroofed courtyard will be much lower and encountered only occasionally.

Gardens

Gardens inside house compounds are often used for planting economically useful plants on a household scale. They often consist of mixed plant communities such as herbs, fruit trees, and staple crops (Hutson et al., 2007). Overall, gardens appear as an extensive, homogeneous, planar body of fine-grained sediment, with dark hues and a gradual lower contact with the original soil substrate. Garden soils have pedogenetic features characteristic of cultivation (Courty et al., 1989: 130–133; Macphail et al., 1990), such as dusty clay coatings, internal slaking, and loose fine matrix. They are regularly enriched in organic refuse, sometimes directly related to household activities (e.g. kitchen gardens). Hence, they regularly have dark colours and are rich in food refuse (fragments of decomposed bone), manure, fuel residues, and turf (in northern humid environments), and they are often colonized by large numbers of snails (Golding et al., 2011; Macphail et al., 1990). In addition, they are rather devoid of large archaeological objects, particularly those with sharp edges such as large pottery fragments, flint, and metal. Any archaeological object that eventually finds its way into the garden will gradually become rounded by frequent reworking during garden maintenance activities.

Phytoliths and macrobotanical remains are characterized by *in situ* plant growth and decay features. As ash and burnt remains are very nutrient-rich materials for plant growth, gardens are particularly enriched in microcharcoal and general pyrogenic carbon. Bioturbation is very intensive, and thus anomalously high biological activity is often characteristic of a cultivated soil (Courty et al., 1989: 133). As a consequence, no bedding or layering is observed, but instead a massive grey soil that grades down to the natural substrate (Figure 6.17). Bioturbation and human-induced turbation result in finely comminuted organic particles and microcharcoal. High concentration of soluble phosphorus is also related to gardens, but other activity areas such as byres, courtyards, and middens have also high phosphorus concentrations (Hutson et al., 2007; Wilson et al., 2008; Davidson et al., 2010).

The distinction between a garden and courtyard when the latter is homogeneous and massive in appearance is not always easy. Courtyard deposits can also be affected by pedogenic processes when the area is abandoned so pedogenesis will blur some of its layering and also

Figure 6.17 Garden soil (G) outside a house wall, underlying decayed mudbrick (M). Dark, organic-rich soil with generally massive appearance and occasionally containing stones and other trash. A red pottery inclusion is shown (arrow). Scale on the wall is 10 cm. Abandoned village of Kranionas, Greece.

produce pedogenetic features in the sediment, albeit somewhat different from those of cultivated soils. Nevertheless, only micromorphology can decipher the nature of such homogeneous-looking sediment.

6.4.4 Street Deposits

Streets, like courtyards, are encountered very early in the history of urban sites, some of them from the pre-Pottery Neolithic in Anatolia (Garfinkel et al., 2012). Street sequences in architectural sites are not difficult to recognize, as often they are clearly demarcated by buildings or small structures on both sides; they have a continuous, monotonous elongated stratigraphy for tens of meters if not more. In later periods, for example during Roman times, well-paved streets are common, but their construction technique is beyond the scope of this book.

In the Pottery Neolithic site of Sha'ar Hagolan, Israel, dated to ca. 6 Ka BCE, Garfinkel et al. (2012) identified plastered streets up to 3 m wide and several tens of centimetres thick. The deposits consist of layers of fine rounded fluvial gravel laid in a packed, clay-rich matrix. Streets comprising several such plastered layers of the same composition indicate frequent resurfacing.

Relatively similar sequences are found in Bronze Age sites in Greece. At the Bronze Age site of Mitrou, central Greece, thick (up to 80 cm) sequences of pebbled streets were identified and dated to the beginning of the Late Bronze Age (Karkanas and van de Moortel, 2014). These are actually composed of alternating couplets of gravel and mud layers (Figure 6.18). The gravel layers are clast or matrix supported. Overlying these gravelly substrates are successive layers of plastered mud, suggesting frequent maintenance. In one case, a mud plaster was capped by a thin coating of

Figure 6.18 Street deposits. (a) Thick sequence of pebbled streets consisting of alternating gravel and mud layers. This is a typical method of street construction where the fine material is used to level the surface and the gravel to promote drainage. Scale is 30 cm. (b) Detail of the same street in another section showing different types of gravel in between mud layers. Dark gravel is predominately serpentinite pebbles and lighter ones are limestone. Such sorting according to rock type of gravel is not observed in natural environments. Scale is 30 cm. LBA, Mitrou, Greece.

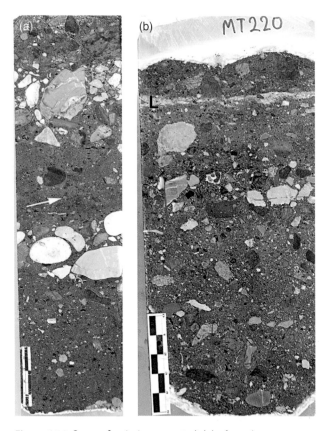

Figure 6.19 Scans of resin-impregnated slabs from the street deposits in Figure 6.16. (a) Clear separation of mud and gravel beds, the latter consisting of white limestone or greenish serpentinite. Note that the mud layers show evidence of replastering and maintenance (arrow). (b) Diffuse bands of gravel-rich beds containing a mixture of types of rocks. The white thin lime plaster near the top (L) might suggest occasional ceremonial use of the street (see also text).

lime wash (Figure 6.19). Several couplets of gravel and mud were identified as the result of regular reconstruction of the street. This well-constructed street deposit, although resembling constructed house floor sequences, rather implies a ceremonial use, as the road passes in front of an elite building. However, other Middle and Early Bronze street deposits in the site are made much more carelessly, consisting of irregularly strewn gravel littered with trash.

Not all street sequences are made in similar ways. In the Classical site of Molyvoti, Greece (ca. fifth century BCE), street sequences consist of homogeneous mud deposits without any evidence of construction. The streets are characterized by large amounts of trash in the form of burnt and unburnt shell and bone fragments, as well as rounded to subrounded pottery fragments. There are cases where whole shell pieces were unearthed that presumably reflect dumped trash deposits. Careful field observations corroborated by micromorphology showed that the street sequences actually consisted of at least three couplets of loose and compacted sandy clay sediment. The loose substrate of each couplet contained large amounts of shell and similar trash, whereas the overlying compacted and trampled layer was delineated by a diffuse pottery 'carpet' that produced characteristic 'stone lines'; the latter represented winnowing, or washing-out of fines that were exposed on traffic surfaces (Figure 6.20). Such stone lines can be used to estimate the phases of street use in otherwise undifferentiated sequences.

At Abu Salabikh, streets are mainly composed of heavily trampled deposits (Matthews, 1995; Matthews and Postage, 2004). The alternation of wet and dry

Figure 6.20 Street deposits. (a) Field photo showing diffuse banding of lighter- and darker-coloured beds (white and black lines), the latter representing trash-rich phases gradually accumulating at the street surface. Lighter mud layers probably represent phases of construction/repair of the street. The upper arrow shows a linear arrangement of pottery, whereas the lower shows a cluster of pottery. Such arrangements occur with walking surfaces, particularly when they are associated with contacts between light and dark bands. (b) Scan of resin-impregnated slab showing linear bands of angular and subrounded sherds suggesting trampling and reworking at the street surface. White linear features are shell refuse. Hellenistic, Molivoti, Thrace, Greece.

conditions created randomly distributed and densely compacted sediments that microscopically show a por-phyric-related distribution. In the field, the deposit appeared homogenized to depths of up to 10 cm.

The porphyric (embedded) related distribution contrasts with the 'intergrain aggregated' distribution that is observed in Molyoti site (Figure 6.8). The latter is considered characteristic of dumping activities (Matthews, 1995), which suggests that the rate of dumping at Molyoti was higher than the rate of trampling of the sediment (see section 3.5.2 for more details on dumping). Nevertheless, in addition to the rate of sediment accumulation and the intensity of trampling, the nature and humidity of the sediment certainly play a role. Therefore, clay-rich deposits will more readily develop homogeneous reworked porphyric (embedded) related distributions, something that may be wrongly attributed to constructed surfaces.

Moreover, at Abu Salabikh, a high percentage of plant voids (up to 20%) was observed, which implies that streets were dumping areas of fresh and decayed vegetal matter. Only occasionally was naturally deposited sediment encountered in the form of sorted and graded waterlain surface crusts (see section 2.4.1 for identification of such features). Erosional horizons were not identified. The lack of water flow and ponding features may appear surprising but this lack was also observed in Molyoti street deposits and has to be attributed to the intensity of trampling (Matthews and Postage, 1994).

6.5 House Pits, Pueblos and Kivas

Deposits in architectural features that are less complex than those in tells and urban areas occur worldwide (Macphail, 2016) and in Europe include grubenhäuser, which have been studied with micromorphology (Macphail, 2016; Macphail and Cruise, 1998; Rentzel, 1998). Here, we concentrate on house pits of the Pacific northwest, and kivas and pueblos from the American southwest. Although these structures generally show the same facies as described above (indoor, outdoor, floor sequences, and construction fills), they are also characterized by some unique arrangements of facies that serve as an example of the great variability of architectural sites.

6.5.1 House Pits

House pits have been intensively excavated in the Pacific northwest of the USA and Canada, and the site of Keatley Creek in southwestern British Columbia stands out for its overall spatial extent, the large number of house pits, and its role in the rise of residential corporate groups (Hayden and Spafford, 1993; Hayden, 2000). The village site has 115 semi-subterranean houses that exhibit a bimodal size distribution (7 m and 15 m) with maximum diameters up to 25 m (Figure 6.21) (Hayden and Spafford, 1993). Details of dating of the site and its role in the origins of complexity in this part of the world (Prentiss et al., 2005, 2007; Hayden, 1997, 2005) are not discussed

Figure 6.21 (a) Circular house pits from the site of Keatley Creek in southwest British Columbia, Canada (Late Prehistoric: Shuswap, Plateau, Kamloops Horizons). These crater-like features show the central depression of the house pit and the surrounding berm of material dumped during the excavation of the house pit. (b) House Pit 7 at Keatley Creek showing excavations in the centre of the structure and prominent rim. See Figures 6.22 and 6.23 for discussion of microstratigraphy/micromorphology.

here and we simply accept that '...the entire aggregated village existed between 1614 and 747 cal. BP' (Prentiss et al., 2007: 303). We focus on the microstratigraphy and infilling of the structures.

The infilling of the structure as observed by excavations is complex and results from a number of activities, including creation of floors, household activities, hearth construction, and burning down of the structure as a cleaning mechanism to reduce infestation of pests; subsequent construction reoccupation adds complexity to the sequence as many of the materials are recycled.

Excavations can reveal several successions of occupation surfaces/floors (with different lateral microfacies) and roof collapse. One house showed a plastered floor, which is virtually unique for these houses (Figure 6.22). In addition, some house pits exhibited smaller 'sub house pits' (Prentiss et al., 2003).

The pit houses were created by partly excavating a pit into the glacio-fluvial deposits of the Fraser River, ejecting the excavated material that accumulated at the base of a rimmed berm that surrounded the structure (see Figure 6.21b); through time, the rim grew with successive

- roof deposits are characteristically darker because of charcoal inclusions and contain poorly sorted rock clasts of granule to pebble size with abundant artefacts resulting from in-place lithic reduction on the roof, and also dumping of debris from the interior
- rim deposits are complex and represent interfingered layers of dumped materials that include
  - redeposited material from the house pit floor
  - loose deposits rich in ash, fire-cracked rock, burned and unburned wood, matting material from the roof, and lithic artefacts accumulated on the rim but derived from the roof
  - deposits near the surface that are richer in rock fragments and impoverished in faunal and botanical remains.

Whereas these deposits can be perceived in the field at a macroscopic level, micromorphological observations can partially refine these observations (Goldberg, 2000b). However, it should be pointed out that many features in these deposits have been blurred by a number of post- and syn-depositional processes that involve digging (e.g. dogs and other carnivores), collapse of the roof (intentional (burning) or unintended) that contains midden material previously cleaned from the interior, insect burrowing, particularly cicadas (Figure 3.11), and dissolution under acidic soil conditions (Macphail, 2016).

Nevertheless, a few examples reveal processes and events not evident in the field (Goldberg, 2000b). For example, a sample of rim deposit from House Pit 7 (Figure 6.22) showed a composition of poorly sorted rock fragments and some charcoal in a matrix of non-calcareous silty fine sand (Figure 6.23a). In addition, within the fine fraction are finely divided shreds of organic matter and charcoal; phytoliths are relatively abundant. The latter, and the fine, splintery nature of the charcoal in the fine fraction, as well as the coarse pieces of charcoal are suggestive of the remains of an ash deposit that has been decalcified under the current acidic soil regime. The fine-grained nature of the sediments overall suggests they are the fine sweepings of a hearth just adjacent to the rim. Moreover, some rolling features with the sediments and fine silty coatings around the coarser rock fragments suggest dumping.

Roof deposits are typified by a relative abundance of lithoclasts, as well as coarse and fine pieces of charcoal that have been mixed together by the action of cicada burrowing. Phytoliths (seemingly mostly grasses) are also quite abundant, and their presence may point to either sod or grass thatching on the roof, perhaps for insulation. With the burning of the structure, wooden beams and sod/thatching would have been combusted to leave a mixture of charcoal and phytoliths; on the other hand, the phytoliths could represent grass bedding on roofs or the remains of material used in the chinking of

Figure 6.22 Field view of House Pit 7, Keatley Creek, BC, Canada. Shown here is a sequence of deposits from inside the pit comprising roof, floor, and wall deposits. A variety of sediments are present with the following descriptions:

88-12, roof: Dark grey brown stony/gravelly silt overlying slightly finer and lighter brown fine gravelly silt floor deposit. At the contact is charcoal resting on a reddened substrate.

88-13, floor with disturbed hearth: Hard, compact pebbly silt (clay) with many charcoal flecks and reddened mottles. Reddening generally in middle. Jumbled texture/fabric and pieces of reddened earth, which are possibly dumped or slumped hearth material.

88-14, wall slump: Jumbled dark brownish black, homogeneous, angular stony silt with clay resting upon sterile silty sand; laterally some charcoal. Clay content suggests slump from wall is clayey.

additions of waste from the interior and collapse of the roof. Poles were inserted into the rim to form a conical shaped roof, which was covered with sediment/soil as well dumped midden material derived from occupation within the structure. A hole in the roof served as the entrance –with a ladder leading into the covered house – and as a smoke hole (Hayden and Spafford, 1993).

Re-excavation of the site by Prentiss et al. (2003) elaborated on details of the initial excavations such that:

- floors were typically composed of well-sorted grey-brown sandy silt, with broken and smeared charcoal, botanical and faunal remains, as well as small flakes the building. Remains of human activities are also encapsulated within the roof deposits. They include bone and yellowish and reddish brown coprolite fragments, possibly those of a dog, which could have been perched on the roof, gnawing and splintering bone.

The three units shown in the section of Figures 6.22 and 6.23b include the floor and substrate deposits and can be characterized as follows:

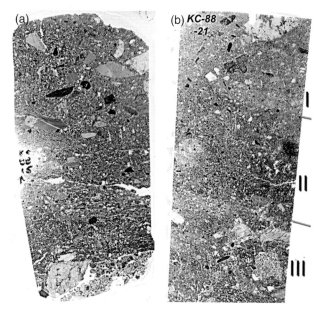

Figure 6.23 (a) Thin section scan of rim deposits from House Pit 7 at Keatley Creek, BC, Canada (Plateau/Kamloops Period). They are composed of various sized pieces of charcoal in a silty and sandy matrix. Coatings on grains show that they were dumped and rolled downslope. (b) Sample 88–21 from House Pit 105 at Keatley Creek, BC, Canada. It encompasses lighter and darker roof deposits (I and II) roof, floor with fish bone at the contact between II and III, and basal deposits of hard compact grey tan reworked till.

I–Loose surface material with fluffy aggregate structure that contains some larger pieces of charcoal and bone but little charcoal in the matrix. Bones are scattered throughout the unit.

II–Floor, which is darker and much richer in fine-grained charcoal than I. It is locally compacted by the action of cicada burrowing. Bone occurs with some ash (only in the upper part), which adheres to bone surfaces. The matrix contains fine splinters of organic matter as well as phytoliths. Sand-sized fragments of brownish amber coloured (isotropic) that are similar in colour to material found in recent brown bear turds suggest that some coprolite material occurs in the sample.

III–Sterile base, essentially dense glacio-fluvial sediment, burrowed by cicadas.

6.5.2 Plastered Floors from Structure 116

One building from Keatley Creek (Structure 116) yielded a number of plastered floors (Villeneuve et al., 2011) (Figure 6.24). Differences in archaeological content varied so, for example, the uppermost plaster was devoid of remains, whereas numerous fish bones and small pressure flakes were found both within and on the second plaster layer, possibly reflecting differences in function through time.

Briefly, thin section analysis of some of the plasters (Figure 6.25) shows similar, repeated sequences in which the plaster rests on a gravelly or sandy base. This is evident in the lower part of the thin section where the micritic plaster is placed on a centimetre-thick upwardly fining fine gravel and sand. The plaster layer is composed of dense micrite (microcrystalline calcite), with inclusions of quartz sand and silt, and is similar to lime plasters produced by the slaking of lime (see section 3.4.3).

Figure 6.24 Plastered floors from Structure 116, Keatley Creek, BC, Canada (Plateau Horizon).

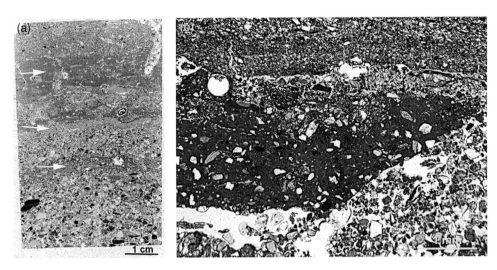

Figure 6.25 (a) Thin section scan of plaster floor sequence from Figure 6.22. Shown here is a basal fine gravel overlain by fine gravel and sand interbedded with lighter-coloured and generally compact lime plaster layers (arrows). (b) Detail of middle plaster floor showing dense calcareous matrix with inclusions of sand-sized material.

Interestingly, the nearest outcrop of limestone is ~20 km distant, so a great deal of effort went into making these plasters (no lime kiln has been found yet at the site to our knowledge). Interestingly, the lowest plaster is thinner, more discontinuous, and less well defined than the ones above it, and overall the plaster layers become thicker upward and the intervening gravel and sand layers become thinner in this direction. The bedding and grading with the gravels and sands is clearly due to water, but not enough exposures were available to evaluate whether this was natural or anthropogenic.

6.5.3 Pueblos and Kivas

Villages (pueblos) and structures in the American southwest have been excavated for decades, especially Pueblos.

> 'The principle element of Pueblo architecture is the room, used primarily for residence and storage. Rooms are organized into contiguous, sometimes stacked clusters or roomblocks that maintain shared walls, floors, and roofs. In northern New Mexico these architectural elements are built primarily with earth, stone, and wood. Walls are constructed using 'coursed adobe', in which local clay is mixed with water and applied in layers as the wall is built...' (Duwe et al., 2016: 21)

A great deal of the focus involving these structures is aimed at their closure and behavioural interpretations associated with them, particularly that of ritual (Adams and Fladd, 2014). Much of the evidence for these inferences is based on excavated objects associated with sequences of sedimentary bodies (Walker, 2002).

Interestingly, only recently have researchers turned an eye to examining the details of the deposits themselves to evaluate these socially targeted inferences (e.g. Adams and Fladd, 2014). For sure, people filled in these spaces for a reason, but examination of the sediments must put constraints and focus on the nature of the interpretations.

A shift to a more detailed approach in the study of Pueblo infilling was made by Adams and Fladd (2014), who investigated the composition and stratigraphy of Hopi Kiva (a subterranean ceremonial room) infillings, focusing particularly on the ashy deposits. They tied the ash to objects and associated them with purification. Although this study fosters the approach 'deposits as artefacts', microstratigraphic details were not presented, so it is difficult to envision what are the compositional differences between 'ash cones' vs. 'ash lenses', for example; in other words, are all ashes in the deposits 'the same' and were they derived from different types of fires?

In one of the few microstratigraphic studies of these types of deposits, van Keuren and Roos (2013) used micromorphology and geochemical analyses (organic carbon, carbonate, and phosphate contents) to investigate stratified deposits from a Kiva at Fourmile Ruin, Arizona, in order to illuminate demonstrate how ritual closure can be 'read' from the sediments.

The Kiva is a shallow depression associated with a plaza and large Pueblo context and was constructed and lived in in the last half of the fourteenth century (van Keuren and Roos, 2013). Five stratigraphic units were recognized, consisting of bedded silty clays, clays, and clay loams, with localized accumulations of cobbles and some wood charcoal. Additional features include calcareous plaster in Unit II and a pebbly mud in Unit IV that contains traces of bones and chipped stone artefacts. Unit V at the top is loose and bedded sands, and dates to the 1980s.

Field indications did not suggest any ritualized behaviour in the closure of the Kiva. On the other hand, the micromorphology of Unit I, for example, revealed a great deal of heterogeneity in composition, including young alluvial soils, sandstone slabs, ashes, and iron-rich clasts derived from the underlying bedrock, the latter coming from off-site. All of these materials formed a 40-cm thick accumulation at the base of the structure, and imbricated stones near the wall suggest that the roof was no longer there at the time that these deposits accumulated. An anthropogenic charcoal layer on top of the base Unit I was overlain by Unit II, consisting of decomposed architectural debris (plaster) that seems to represent '... purposeful and rapid disassembly of portions of the exposed kiva walls and their deposition in a continuous layer above the burned material' (van Keuren and Roos, 2013: 622). This is in turn covered in a rapid fashion (charcoal and ashes would likely have been washed or blown away if not) by Unit III, decomposed wall material. It lay exposed on the surface and as such represents a gap in cultural sedimentation. Deposition resumes with Unit IV, the final phase of closure. The rounded and sandy nature of the bone fragments and the lack of internal bedding suggest that the material that makes up Layer IV had been previously exposed on the surface somewhere else at the site and was then dumped in place rather quickly. Pedogenesis occurred on the top of Unit IV, as shown by bioturbation, organic matter accumulation, and the formation of secondary carbonates. Based on the microstratigraphic evidence from the field, thin sections, and geochemistry, van Keuren and Roos conclude that although the Kiva does not display evident aspects of ritual closure, the sedimentary, pedogenic, and chemical data point to '... a series of orchestrated ritual activities' that may have lasted over years or decades (van Keuren and Roos, 2013: 624). Interestingly, the microstratigraphic evidence here furnishes a contrast with the 'singular' construction event of the mound at Poverty Point, for example, described below, where no evidence of temporal breaks was observed.

6.6 Tombs

Tombs are confined structures enclosing internment space. Some tombs contain very little accumulated interior sediment, such as vault (*tholos*) and chamber tombs, or free-standing, aboveground structures (e.g. pyramids). Often though, graves are filled with sediment and show indications that they have been opened and used more than once. Underground bedrock-carved tombs may carry signs of maintenance. In addition, bone taphonomy is highly affected by the conditions of soil burial. Nevertheless and unfortunately, tomb deposits have been studied only very rarely.

In cases where the tombs were repeatedly used, stratified deposition can occur in association with both their continual opening and closing, and also with their maintenance. For example, maintenance practices were identified in the form of fine lime plaster floors in the interior of Mycenaean chamber tombs (Karkanas et al., 2012) and this investigation allowed for the separation of burial groups based on their relationship to floor surfaces. A thin distinct layer containing charred remains was identified at the base of a shaft tomb and presumably was the result of a ritual activity.

Corridors leading to chamber tombs are usually filled with sediment. These sediments show typical sedimentary structures attributed to grain flow and debris fall processes associated with the formation of small piles during shovelling of debris in the tomb (Wright et al., 2008; Karkanas et al., 2012) (see also dumping processes in section 3.5.2). In addition, large-scale truncations that separate packets of deposits with different degrees of compaction and types of fill mark reopening phases of the tombs (Figures 3.72 and 6.26). Understanding these processes enables the differentiation of features related to reopening and regular backfilling.

Rarely, chemical signatures in the soil, such as iron- and phosphorus-rich compounds (Žychowski, 2011) or elevated manganese concentrations (Keely et al., 1977), are produced during cadaver decomposition. Although not widely tested, it is possible that where remains have already decomposed and no visible remains survive, these soil silhouettes of trace chemical residues enable the detection of pre-existing burials (Keely et al., 1977). Such silhouettes can also lead to a probable differentiation of primary from secondary burials.

Several parameters regarding the preservation conditions of the bone include the type of substrate in which the tomb was dug, as well as the nature of the fill. Above all, the hydrological regime characterizing the substrate in addition to the one that has been created by the mere existence of the tomb is a controlling factor for bone dissolution. If the tomb is dug into impermeable clay-rich sediment water will preferentially stagnate in the tomb. Dry stone-lined tombs will preferentially lead water seepage to the ground surface of the tomb. The loose fill of tombs induces quick water percolation from the surface, and depending on the nature of the substrate will be stagnated (clay-rich substrates) or quickly drained (sand and gravelly substrates). Calcareous deposits help in preservation of bone whereas acid soils lead quickly to bone dissolution. However, dripping water in tombs within an open chamber may be acidic as the water that

Figure 6.26 Section of backfill of a corridor at the entrance of a chamber tomb showing several distinct packages of sediment (the lower part of each being marked with arrows) that indicate reopening of the tomb. LBA, Barnavos tomb, Nemea, Greece. From Karkanas et al., 2012.

6.7 Monumental Earthen Structures

Tumuli and mounds are monumental earthen structures used mostly for ritual purposes. Tumuli are mounds raised over one or several tombs and are also known as barrows or burial mounds. Local names are also in use, such as the kurgans of the Central Asia steppes, *cerritos* and *atteros* across South America, or the Greek 'tumbos' from where the word tomb is derived. Tumuli from the Western European Neolithic are also known as megalithic tombs and are one of the most impressive monumental

features in the landscape of this region. Different types of tombs and graves are included inside these monumental mounds, and their shape and form define major types of tumuli (e.g. long barrow, court tomb, passage tomb). The mounds of the eastern North and South America built by ancient Native American peoples also attain monumental proportions. The oldest mound dates to ca. 5500 BCE but mounds continued to be built throughout pre-Columbian times. The purpose of these mounds is not always clear but it seems that in most cases they are associated with burials, ceremonies, and other public events (Sherwood and Kidder, 2011).

The focus of most tumuli studies is on the internal structures of the monuments (graves and tombs), and surprisingly little has been done in understanding the building of the earthen constructions themselves (Canti, 2009). However, the tumulus is a fundamental component of the symbolic funerary architecture (Robin, 2010) and its stratigraphy provides critical information about its use and construction. It seems that this type of tumulus has a unique stratigraphic organization representing several built stages and different types of materials. In the study of the passage tombs in England and Ireland, Robin (2010) suggests that most of these tumuli represent a circular system composed of several concentric layers made of distinct materials such as turves, stones, and sand, which are delimited by circular structures, including kerbs, stone facings, and ditches.

Similarly, tumuli from Scythian, Thracian, Macedonian, and Etruscan cultures show that elaborated embankments are often raised above graves using complicated and well-conceived earthen structures. In a study of Thracian tumuli Evstatiev et al. (2006) found that big tumuli are multilayered and were built in stages. One or a few small mounds were first built above graves and then these were included under a larger mound. In some cases a base layer was laid down, over which the stone sepulchres were built. Occasionally, different types of sediment were used to produce alternating layers of sediment; however, frequently the fill appears undifferentiated. In many cases, the materials in the deposits were carefully selected so that they were not readily eroded away, as in the case of the Lofkënd tumulus in Albania. Here, clay-rich schist soil was added to the loose sandy soils of the substrate of the tomb for building a stable mound (Papadopoulos et al., 2008).

Laona, Palaipaphos (Cyprus), a mound of Hellenistic times (ca. fourth century BCE: Iacovou, 2017), exhibits an impressive complex and colourful stratigraphy (Figure 6.27). The mound is asymmetric, with the gentle side of the fill consisting of horizontal lenses of beige angular marl fragments having little interstitial fines alternating with red relatively massive clay-rich soils. Some intervening grey sandy lenses with well-rounded gravel are

passes quickly from the humus top soil through open fractures will be acidic irrespective of the nature of the geological substrate. Occasionally, however, dripping water produces complex calcareous thin encrustations that thus record the palaeochemistry of the groundwater (Jones and Fallu, 2015).

(a)

(b)

Figure 6.27 Mound stratigraphy resulting from dumping activities during the construction of the mound. (a) Sequence at the core of the mound showing horizontal couplets of beige crushed angular marl (M) and red relatively massive clay-rich sediment (R). (b) Sequence at the side of the mound consisting of dipping tabular beds of mixed sediment. Hellenistic Laona, Palaipaphos, Cyprus.

observed, presumably derived from old fluvial sediments located in the surrounding area. Each couplet of marl and red soil is compacted and levelled before a new couplet is laid on top of it (Figure 6.27a). The chaotic open-work appearance of the marls suggests *in situ* loads probably dumped by a wheelbarrow or a similar means of transport. On the other hand, parts of the mound are formed by sloping tabular beds that consist of the afore-mentioned couplets or mixtures of both sedimentary materials (Figures 3.71a and 6.27b). Finally, the mound was sealed with a carefully prepared and pugged thin layer of red calcareous soil. Later on the mound was truncated on the top and refilled with a cap of crushed marl deposits. The stratigraphy of the mound and the study of the formation processes is still under investigation, but it is obvious that the mound fill was carefully planned and constructed in order to produce a stable earthen structure.

From a geoarchaeological perspective, some of the American mounds are the best studied earthen features in terms of deciphering stratigraphic complexities, details of sedimentary building processes, and chronostratigraphy (Sherwood and Kidder, 2011; Ortmann and Kidder, 2013; Kidder et al., 2009; Cremeens, 2005; Cremeens et al., 1997). For studying these mounds, excavations were accompanied by detail field stratigraphic observations, coring, geophysical prospection, sedimentology, and soil and sediment micromorphology. In the case of Mound A at Poverty Point, one of the largest mounds in North America, it was found that the mound was built very rapidly by first constructing the 22-m high steep-sided conical section, then the 10-m high, flat-topped platform, and finally the earthen ramp that joins the cone and platform elements (Kidder et al., 2009; Ortmann and Kidder, 2013). The mound was built over a swampy swale that was first cleared and burnt, and then covered with a relatively thin white to grey fine silt layer.

Similar, unique, and probably ritualistic colour-specific preparation layers have been identified in several other mounds (e.g. Cotiga Mound, Cremeens, 2005). Following an initial construction episode, the mass of the mound was constructed using variously coloured silts heaped together in basket loads. Specifically, initially 'haystack'-like piles of earth were made that were joined by horizontally laid silty deposits, most likely by broadcasting sediment. Micromorphological analysis confirmed that the lower white-grey substrate as well as the basket loads do not show any evidence of weathering, thus implying rapid build-up of the mound. The basket loads consisted of various materials that were excavated separately and then mixed in the mound area before being emplaced. The interface between individual basket loads is sharp, indicating rapid application (see also Cremeens, 2005). Finally, a ramp was constructed using very fine weathered sediment. Occupational surfaces, construction hiatuses, and erosional deposits were not observed, thus indicating that the mound was built in a very short period of time and thereby required a substantial workforce in order to carry out the job so quickly. Pedogenesis has affected the upper part of the platform, destroying the basket load features. The presence of argillic (Bt) and eluvial (E) horizons suggest a mature stage of soil development attesting to the old age of the entire mound (ca. 1260 BCE).

Generally, tumuli and mounds are formed by dumping processes and therefore recognizing the various elements of this process is fundamental for understanding the building history of these features (see section 3.5.2). It is of great importance to differentiate features and sedimentary structures associated with individual fills from features that separate packages of fills delineated by erosional, weathering, or cut-and-fill interfaces. Some fill packages are not necessarily separated by construction hiatuses, as in the case of the 'bridging loads' between piles of basket loads in American mounds. Construction hiatuses are associated with weathering interfaces or interfaces that truncate the underlying stratigraphy during the digging of an older fill, such as in the case of reopening a tomb.

6.8   Concluding Remarks

Architectural sites are the domain of the infinite repertoire of human activities. As a consequence, urban sites are known for their complex stratigraphies and depositional histories. Nevertheless, natural processes play a significant role in the final emplacement of deposits. They are always acting in the background, altering human constructions, redistributing anthropogenic sediments, and occasionally burying parts or whole sites with sediment during exceptional natural disaster events. They are also particularly active during abandonment phases as the ultimate fate of a site is to become part of the natural landscape.

Natural processes and their deposits are part of the history of a site and as such they provide the background context of anthropogenic activities and the real context of a vast array of archaeological objects that have been affected by geogenic processes. In this chapter we have tried to review the main natural processes acting within the realm of these types of sites and presented some criteria for differentiating them from some anthropogenic activities that produce seemingly similar deposits.

An attempt has also been made to put some order in the immense variability of anthropogenic sediments in urban archaeological sites by grouping together deposits that appear to share some basic similarities in fabric. This similarity is actually the product of a few major types of anthropogenic activities that characterize the majority of the sites. In roofed or indoor faces the distinction between floor sequences, construction fills (levelling and foundation/packing fills), and occupational fills is based on differences in fabric, microstructure, and overall geometry of the beds.

Unroofed or outdoor facies are characterized by the occurrence of sediment deposited by natural processes. Nonetheless, the type or types of associated anthropogenic activities is the decisive criterion for correlating a deposit with a particular unroofed environment such as a yard, a garden, or a street, for example. Abandonment facies of a building is also included in the outdoor facies and their recognition and interpretation is of major importance in interpreting context in occupational sequences. These facies are most often characterized by the occurrence of gravity processes and occasionally dumping activities, including midden formations. A unique combination of abandonment and dumbing facies and ritual closure of houses characterizes some architectural sites in the American southwest.

In earthen structures such as tombs and mounds, deposits show typical dumping features that in many ways are similar to natural sediment gravity processes. Nevertheless, the geometry of the sedimentary bodies produced by human actions is not linked to the associated geomorphic system, and the range of sedimentary structures is rather monotonous and does not show the variability expected in natural slope systems. However, the information gained by studying the sedimentary history of earthen structures gives important insights into ritual practices and histories.

In sum, although the details of the depositional history of architectural sites are fundamentally difficult to discern and interpret, the geoarchaeological approach we advocate here provides a pathway to reconstruct their stratigraphy and formation processes.

7

Some Approaches to Field Sediment Study

7.1   Introduction

In order to carry out the goals that we have proposed in previous chapters, it is clear that samples must be collected from the sites under consideration, since as we have repeatedly stressed, one cannot make robust inferences about site-formation processes from field observations alone. Below we present some guidelines on sampling procedures that include what to collect, how much, where, etc. These recommendations are based on our experiences working with large, complex urban sites with many big excavated areas and baulks. What we suggest is not meant to be a universal recommendation, since we realize that investigators working with smaller hunters and gatherer sites, for example, will need to adjust some of these ideas. Nevertheless, as we discuss below, the principal tenet of intensively sampling facies is applicable to all sites.

Sampling is a significant part of an excavation and most handbooks on geoarchaeology and archaeological soils have a dedicated section on this topic (Courty et al., 1989; Holliday, 2004; French, 2003; Goldberg and Macphail, 2006). Protocols for sampling archaeological materials such as soil, seeds, charcoal, radiocarbon dating samples, and phytoliths, for example, are widely used in most excavations (Balme and Paterson, 2014; Kipfer, 2007). However, here we discuss the philosophy and general strategy of sampling a site as a whole; we do not focus on specific archaeological features or details of sampling a particular type of material (e.g. seeds). The sampling approach that we advocate here is an integral part of both the excavation procedures and the task of interpreting the stratigraphy.

Simply stated, the primary focus is to understand how the site has been built and how the stratigraphy has been produced. Consequently, we present in more detail those methods – such as sediment and soil micromorphology – that are particularly useful in interpreting site-formation processes. Special attention is given to sampling baulks and sections because they can often be used as sources of samples for innovative microarchaeological approaches,

even after the excavation is completed (Weiner, 2010). Bulk sampling approaches are more relevant to a specific content of the strata and the detailed study of a specific material. We also discuss the issue of representative sampling, something that is neglected in most sampling programs, that is, how many, how much, and what kind of samples. No matter which techniques will have been used in the analyses, at the end all analytical information is combined to gain a better understanding of site-formation processes.

7.2   Drawing

At the outset, sampling first requires a rough understanding of the stratigraphy of the area to be sampled. Regardless of the stratigraphy that is accessible and exposed during excavation, additional detailed description of the sediments is strongly recommended along with a sketch/drawing of the sampling area. In this way the stratigraphy, composition, and fabric of the sediments are independently recorded and they can be later compared with notes made available by the excavators. In addition, drawing the stratigraphy forces one to think over the stratigraphic variability observed in the field. Decisions about the geometries and configurations of the lines that depict the boundaries of depositional bodies are guided by the realization of the possible processes that made these arrangements. In this fashion, the sampling plan may be later revised in the field following this preparatory work.

Before sampling, the profile and the general area to be sampled should be cleaned because during the cleaning process one can feel the texture of the sediment and at the same time think about the major and subtle details of the stratigraphy and its variations. Cleaning is a technique by itself. Scratch features and lines should be smoothed with suitable equipment (e.g. a sharp knife) but not smeared. Unfortunately, there are different types of tools available in each area and the type of soil determines the tool and the method of cleaning to be done.

Reconstructing Archaeological Sites: Understanding the Geoarchaeological Matrix, First Edition. Panagiotis (Takis) Karkanas and Paul Goldberg.
© 2019 John Wiley & Sons Ltd. Published 2019 by John Wiley & Sons Ltd.

The goal is to reveal the true fabric of the deposit and therefore the surface must remain rough enough so that individual grains are still visible.

In general, sharp and fast horizontal and vertical movements with a trowel are generally effective. Perceived strata boundaries *must never be engraved or scratched into the profile because this practice not only introduces bias but it forces the observer to see the profile in a certain, possibly incorrect and unchanging way.* After finishing, removal of dust is recommended but not with a brush since this smears the dust on the surface and blurs the features. Hand air blowers, bellows, BBQ fans or similar equipment work well. In very hard clayey deposits, digging tools produce compressed and polished surfaces, so gently picking of the profile surface with a knife – a common practice in pedology – will freshen it without doing much damage. As mentioned in the Introduction, the arrangement of the lines of the interfaces should make sense, and if not, the stratigraphy is wrong. Therefore, it is recommended that after finishing the drawing, one should try to describe and tabulate the history of deposition based on the drawn lines. It is necessary to record the position of a given sample with accuracy, both on drawn profiles and on photographs (see section 7.3). This practice preserves the context of the sample, thus providing the ability to relate it to other materials and features of the excavation.

7.3 Photography

Photography is a necessity, and obtaining a good photo requires some effort. In our experience, the most important condition for good photography is lighting. First of all, we would not recommend excavating and sampling in direct sunlight since the conditions of excavation can significantly influence the accuracy of the results. Similarly, light is as important as the excavating tools and the skills of the excavator. Consequently, photography always has to be conducted under shady/shaded or cloudy conditions. In any case, an additional series of photos has to be taken during early morning, evening, or twilight when the site is not under direct sunlight. Several photos at different distances from the profile should be taken, always with a scale for reference. One must put the sample in the general context of the area and then progressively zoom in.

In order to acquire high resolution and sharp images, digital SLR cameras and lenses of relatively good sharpness are recommended. Today, most digital lenses have adequate sharpness for taking photos of excavation areas. Sharpness generally decreases with increasing aperture number (f-stops), where the f-number is the ratio of focal length to effective aperture diameter. Increasing the number of the aperture stop increases the depth of field. Consequently, when photographing relatively flat sections where depth of focus is not needed it is better use low aperture numbers. As a rule of thumb, f-stop numbers between 6 and 9 are very good for photographing profiles. Under most lighting conditions such aperture values can be used without the need of a tripod, particularly with new cameras, which have image stabilization technology that minimizes blur caused by camera shake.

Nevertheless, utilization of flash is also recommended for taking photos for stratigraphic analysis even under very good light conditions (obviously, always in shade). Flash smooths out roughness details and therefore enhances the overall contrast of strata that are difficult to

Figure 7.1 Profile photographed under different lighting conditions: (a) under bright sunlight with a rough-and-ready shade in which only a top dark layer is observed capping the sediment below and (b) during dusk with a flash. Under these lighting conditions a lower layer (dotted line) is clearly seen that rises up and caps the wall remnant to the left. It is apparent that the latter once filled the space up to the base of the dark layer (arrow), and was eroded and covered later by the deposition of the middle layer. Post-Hellenistic, 'Lower Town' of Mycenae, Greece.

discern otherwise. It is often a surprise to take photos with a flash in the evening and see how clear the stratigraphy is depicted (Figure 7.1). However, this does not work very well in some colourful profiles.

7.4 Sampling Strategy

There are two general types of sampling. The first type is regular sampling for extracting a variety of materials that cannot be collected by hand during excavation (e.g. seeds, charcoal, microfauna, debitage). The second type is what we call here strategic sampling, which is employed in order to understand site-formation processes and the stratigraphy of a site using a variety of techniques such as micromorphology, chemistry, soil analysis, dating methods, and occasionally microremains, which are also collected regularly during excavation. Here we discuss only strategic sampling, as regular sampling is dictated by the methodology and state of the art of each specific field.

As a first rule, the amount and intensity of sampling has to follow the pace of excavation from the very beginning. At the outset nothing is clear because the variety of facies is not apparent, and the nature of sediments is not yet fully understood. In addition, the top tens of centimetres are highly bioturbated and perhaps reworked by modern activities (e.g. ploughing, foot traffic) so deposits cannot reveal much information. As a result, sampling has to be at a minimum at the beginning, although it is expected that impatience will be at the maximum.

As the excavation proceeds and the first general picture of the excavation begins to appear, the need for maximizing the intensity of sampling becomes apparent. This is the right time to analyse the stratigraphy through field observations and information gathered by the study of the samples that have been already analysed in the first stages and to discover how many facies make up the site. Facies represent types of sediment that are the product of a unique set of activities and behaviours, and thus they should be somewhat different in every cultural period. Therefore, the maximum variability is expected when each cultural period has been excavated enough to reveal a representative portion of it. This is not to deny that there will be spatial differences since the type of activities will vary laterally. However, we believe that some facies are typical of each cultural phase and they are normally encountered in most areas of a site.

Setting the facies as the unit of sampling and not stratum is a crucial aspect of the philosophy of sampling we advocate here. In this way, representation of sampling will not refer to a certain layer, as this will lead to taking a large number of samples in order to understand how each layer has been formed. Certainly, there will be layers that will be the focus of special sampling: in certain cases it is important to identify details about the nature of certain boundaries of major interest, or to document variations within a particular feature. We want to stress here that at this stage the focus should be the variability of the deposits, that is, their facies, and not the hundreds of beds, layers, contexts, or features. The number and types of facies will determine the number of samples and the type of sampling. Microscopic examination of the main facies types eventually will lead to the creation of subfacies, and in this way a refinement of the original sampling strategy will follow. As has been nicely summarized by Courty (2001: Figure 8.3), a long-term iterative field/laboratory strategy over several seasons provides the best representation of materials for interpretation.

Sampling should not concentrate only on the unknown deposits. It should be equally focused on the best preserved and least disturbed types of deposits, as it is of crucial importance to construct an internal reference collection. Therefore, if it is clear that a deposit is a packing fill, or an intact hearth based on their fabric and content in the field, they should be sampled in order to be used later as a reference for other areas where the sediments do not show obvious field characteristics of a packing fill or a nice fireplace.

In all sites, the variety of soils and sediments surrounding the site in large part dictates the fabric of all deposits of a site. Therefore, sampling the substrate of a site is of major importance, as is sampling the major soil phases of the site catchment. Indeed, understanding the characteristics and properties of the nearby soils and the bedrock of the area helps clarify the appearance of the various facies in the site.

7.5 Representative Sampling

7.5.1 Sampling Methods

Methods of sampling along strata in a section are different from those conceptualized during the excavation when strata are just gradually being revealed. In the latter case, sampling can even encounter the entire layer or feature. However, there are methods for representative sampling of sections and baulks. The most often used is channel sampling, which is very accurate and universally applicable. It involves cutting a continuous shallow channel, usually 2–3 cm in diameter, in the direction of the greatest variation of the layer through its entire thickness (Figure 7.2) (Kuzwart and Böhmer, 1984: 263). Channel sampling was effectively employed in sampling charcoal from excavated profiles of the Palaeolithic site of Theopetra Cave (Ntinou and Kyparissi-Apostolika, 2016.).

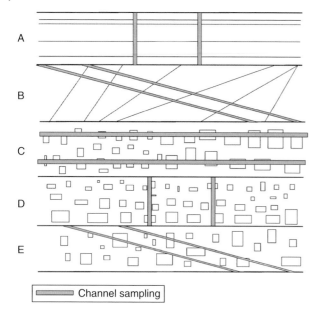

Figure 7.2 Sketch of channel sampling in various stratigraphic configurations. Case A represents a horizontally stratified or laminated sediment and therefore a vertical channel sampling should include the maximum variation of the layer. In case B, the layer contains a lot of diagonally variations (e.g. cross-stratification) and therefore channel sampling should cross-cut them. Case C represents an example of horizontal lateral variation and therefore maximum variation is expected along its extension. In case D, where sediment shows vertical grading, channel sampling should follow the same trend. In cases with totally random heterogeneity, like E, channel sampling should traverse the layer to include the variations across the layer as much as possible. Note that features/inclusions inside layers are not to scale and will not necessarily be stones or artefacts – they can be soft aggregates or any matrix variation as well.

Often, however, the nature of the material and specific research questions require other types of sampling, such as grab sampling, core sampling, or continuous sampling across the stratigraphy. For example, identification of the first appearance of a specific type of material (e.g. cryptotephra of a volcanic eruption) demands continuous vertical sampling at intervals of 1 or 2 cm (Lane et al., 2014; Karkanas et al., 2015b).

7.5.2 Number of Samples

We often hear the question, 'How many samples we should take?' As we previously proposed, the number and type of facies will determine the number of samples (Courty, 2001). A facies represents lumping of variations in a group. Some variations are not critical as they represent small deviations in the fabric and content of the sediment that reflect idiosyncratic changes in activities and processes, such as distribution of burnt remains around a fireplace or intensity of trampling. It is important that sampling includes as many variations of a particular facies as possible

to be sure that the facies designation is correct or if a new facies/subfacies designation is required (Figure 7.3).

In this light, during the initial stages of sampling each stratum chosen as representative of a facies must be sampled in several different locations over its extent, and the samples should contain all different sedimentary features that characterize the content of this layer. If a layer is characterized by the presence of whitish aggregates, the full variety of these aggregates should be included in the samples. If the sampling is for micromorphology then critical areas that show the maximum variability of the layer should be included in the samples. The same is true for sampling for soil analysis, chemostratigraphy, or mineralogy.

As for the total number of samples per layer, context, or feature, the criterion is the heterogeneity, that is, the distribution of the material under investigation. Here, we present some empirical suggestions also based on sampling methods in mines and ores (e.g. Kuzwart and Böhmer, 1984). In archaeological sites, regular distribution of most objects/features is expected to occur very rarely, for example an ash layer of an intact hearth feature or the chemistry and mineralogy of a fine-grained, homogeneous construction material (clay and lime plaster). Two samples per feature may be considered enough in such cases. Nevertheless, in order to understand the variability and define the characteristic composition of this kind of feature (facies) more than one feature has to be sampled. A starting point would be to sample at least three features, which will provide a secure indication of the assumed homogeneity or variability of the facies.

In the case of *in situ* reworked hearth remains, occupational facies on constructed and living floors, and similar types of deposits, normally the materials of interest are irregularly distributed. This is the case for phytolith, magnetic susceptibility, and occasionally charcoal, seed, and fine construction material inclusions (lime lumps, mudbrick fragments), which are randomly distributed within the sedimentary matrix. Sample spacing in these types of deposits can be about every 1 m for extensive layers or two or three samples in total for a smaller feature. Two micromorphological samples will normally reveal a good deal of the variability of such a layer. In severely reworked occupational facies, and in coarse and highly unsorted sediment and heterogeneous archaeological materials, such as dumps and very coarse fills, sample spacing should decrease to 0.5 m. Three or more micromorphological samples are needed to understand the nature and variability of such occupational facies.

In terms of reproducibility, it is recommended that two or three features or layers of the same facies are a good starting point (see Figure 7.3). However, we believe that often more than two or three will be needed because one of them will probably show a variability that cannot be matched in the other sampled layers or features of the

Figure 7.3 Micromorphological sampling of an urban profile. The sampling strategy has taken into consideration the three major facies observed in the field: (1) the fills of the two robbed walls (samples 7 and 8), (2) the dominant reddish talus at the lower central and right side of the image (samples 9, 10, and 11), and (3) the grey, ash-rich sequence developed on both sides of the upper stone wall (samples 1, 2, 3, 4, 5, and 6). The number of samples in every major facies is dictated by the number of strata and their variations. Thus, the samples in the grey upper facies include all contacts (and the sediment above and below) but also some variations/features like the whitish lens of sample 4. The three samples of the reddish talus include features like the red lens of sample 10, the main representative part of the facies (sample 9), and its lateral downslope variation (sample 11). Only one sample was taken from each fill of the robbed walls, mainly to identify their content and source in comparison to each other and to the upper grey facies. Note that this is probably the minimum number of micromorphological samples needed in order to understand how the deposits of this profile were formed. For example, in order to fully understand all the natural and anthropogenic processes responsible for the formation of the talus additional sampling would perhaps be needed. However, once fully understood, all similar features in the site would require only occasional sampling in order to confirm the model or elucidate some variations. Late Roman to recent, ancient Corinth, Greece.

same facies as it was defined in the first place. Gradually increasing the sampling number is known as 'interactive sampling technique' and has provided very good results for a minimum number of samples (Lewis and McConchie, 1995: 49–51). In interactive sampling, analytical data obtained in the field or in the laboratory are used to modify the intensity and the general sampling program.

In some cases, areas of concentration of a specific material or composition (e.g. phosphorus, calcium, iron, manganese, magnetic susceptibility, and other chemical proxies) cut across layer and facies boundaries but generally remain within the boundaries of an occupational level. In these instances, a different sampling density method is recommended: sample spacing of one third or one quarter of the smaller dimension of the potential concentration area is considered to give very good results, at least for identifying the major concentration features. Further reduction of the sample spacing to one tenth of the concentration's smallest dimension will reveal smaller concentration areas (Goudge, 2007). For example, a 0.5 m sample spacing will accurately map all features smaller than 5 m in length (one tenth), whereas 1 m spacing (one fifth) will identify but not accurately map all concentrations. Thus, sample density will be defined by the nature of the component under study, as well as the size of the concentration area that is considered significant.

7.5.3 Size of Samples

We think that the issue of sample size is somehow ignored because often the nature of archaeological materials and exposed deposits (including witness sections) impose restrictions on the amount of the material that can be sampled. The appropriate sample weight that is considered to be representative may be smaller in homogeneous, well-sorted, and fine materials. However, when sampling coarse and heterogeneous sediments (which is often the case with archaeological deposits) large masses of sample are needed if it is to be a representative sample.

Empirical equations for estimating the mass of samples are based on the size of the largest particle in the sediment to be sampled. A first approximation can be the quantities suggested by Krumbein and Pettijohn (1938) in Table 7.1.

Various more elaborate equations as a function of the D_{max} particle size are recommended (for a review, see Bunte and Abt, 2001):

Table 7.1 Quantities of sediment samples as a function of size of the material as suggested by Krumbein and Pettijohn (1938).

Material	Size	Weight to be sampled
Clay*	<0.002 mm	125 g
Silt	0.05–0.002 mm	125 g
Sand	2–0.05 mm	125–500 g
Granules	2–4 mm	1 kg
Pebbles	4–64 mm	2–16 kg

*Note that here the silt/clay boundary is 0.002 mm as opposed to 0.004 mm in the Wentworth scale (cf. Table 2.3).

$$Ms = 1,388,000 D_{max}^3$$

where Ms is sample mass expressed in kilograms and D_{max} is the size of the largest particle in metres. This equation is recommended for particles sizes less than 32 mm.

As an example, the weight of a sample with the largest particle size of 10 mm will be 1.39 kg and for 20 mm, 11.1 kg, therefore sampling coarse deposits is rather impractical. Soil scientists recommend using only the <2 mm fraction for most purposes except grain-size analysis. The reason is that the fraction above 2 mm is mostly dominated by inactive lithic fractions and therefore can be safely removed before analysis by wet sieving.

In archaeology, this refers to materials that are much smaller than 2 mm (phytoliths, pollen, diatoms, volcanic shards, and chemical features that are related to the fine, mostly clay-sized fraction of the sediment).

7.5.4 Micromorphological Sampling

It should not be a surprise that we strongly believe that micromorphology – in addition to field observations – is the backbone of any interpretation of site-formation processes. Archaeological deposits, in most cases, are difficult to study and interpret by field observations alone: they have a complex macrostructure, are for the most part highly unsorted and fine-grained, have diverse organic and minerogenic compositions, and commonly lack obvious macroscopic sedimentary structures. They form mostly by sub-macroscopic processes and therefore their microscopic study is a physical continuation of macroscopic observations of their morphology (Courty et al., 1989; Karkanas and Goldberg, 2017a).

Based on our experience, the use of Kubiëna soil tin boxes is not suitable or adequate for most excavations. If we want to study the details of variations of deposits, including their interfaces in the sample, the box must be big enough to contain all these features. We are also of the opinion that it is possible to sample almost all types of sediment, including stony layers and very loose sands

Figure 7.4 Micromorphological sampling. (a) Pebbled street deposits. The rather large size of this sample (~35 cm high and in places 20 cm wide (scale is 30 cm)) is needed in order to avoid collapse and include the maximum variation of stony layers as well as at least two mud layers in between. LBA, Mitrou, Greece. (b) Occupational sequence including, from bottom to top, a whitish hard constructed surface, a sequence of very fragile, thinly laminated sediment, and, at the base, a relatively thick loose grey ashy sediment. The profile was exposed for some time and ready to collapse, so in order to secure safe removal a ready-made cardboard mould was wrapped around the sample into which watery plaster of Paris was poured. The sample was held in place for a few minutes to allow the plaster to dry. White tag is 10 cm high. MBA, Vésztő-Mágor, Körös Region, Hungary.

(Figure 7.4). Sampling these unwieldy deposits normally requires some innovation and experience (Goldberg and Macphail, 2003) but not fear.

Some discussion on the number of samples was presented above, and we wish to reiterate that the number of samples must be adequate to describe a facies and to find as many microfacies as possible that can account for the variability observed in the field. As a reminder, micromorphology is mostly a qualitative method and therefore identifying characteristic features of a process is by itself important and does not necessarily need confirmation by other samples. For example, a massive homogeneous silty sediment that in one sample shows microscopic waterlaid features can be safely interpreted as being deposited by water. Perhaps an additional sample will reveal other processes that have affected the layer, such as trampling, or aeolian reworking, but these additional observations cannot dismiss the validity of the identification of its being waterlaid.

After resin impregnation, the sample has to be cut into several parallel slabs (hence, large samples are better) for recording mesoscale variations not easily identified at the microscopic and macroscopic level (Figure 7.5). Flatbed scanning or static high-resolution image acquisition of whole thin sections are additional tools for observing features at the mesoscale (Figure 7.6) (Arpin et al., 2002; Carpentier and Vandermeulen, 2016).

A new very promising development is the utilization of industrial micro CT (computer tomography) scanners that can produce continuous sectioning of the whole micromorphological sample (Kilfeather and van der Meer, 2008; Tarplee et al., 2011; Huisman et al., 2014). The resolution in samples that are about 10–15 cm thick can be in the range of a few tens of microns, which very well reveals the void pattern, and the arrangement of fine bone, charcoal, several construction materials, and lithics based on their density differences.

Micromorphological analysis often puts all microremains in context, and therefore it is advisable that some additional sampling of these materials must be conducted from the profiles after the microstratigraphy and formation processes are understood (see below).

7.5.5 Microarchaeological Sampling

According to Weiner (2010: 1) the archaeological 'microscopic record is composed of the materials of which the macroscopic artefacts are made, as well as the sedimentary matrix in which the artefacts are buried'. A vast array of microscopic materials comprises the sedimentary matrix. In addition to mineral and rock components, there are biological materials such as bone, teeth, phytoliths, pollen, diatoms, eggshells, otoliths, mollusc shells,

Figure 7.5 Two slabs taken from the same resinated sample showing lateral variations of floor sequences at the scale of a few centimetres. The right sample is from the centre of the sample and the left from the outer part. The thin tabular layer of red clay (1) is seen only as a diffuse reddish feature in the left slab. Indeed, more than one red constructed surface is shown in the right sample. In contrast, the dark thin occupational layer (5) observed in the left sample does not appear in the right one. The nicely preserved intact thin ashy layers of the left sample (3) appear only as dark raked-out burnt remains in the right sample. The burnt remains above (3) continue up to the red clay worn-out surfaces. In the right sample it is clear that another floor (4) was constructed in between. EBA, Mitrou, Greece. From Karkanas and Van de Moortel (2014).

biological molecules and macromolecules, and pyrotechnological products, such as lime and ceramics.

Representative sampling of most of these materials falls under the same principles described above. In general, some of these materials are included in the 'soil' samples collected for general or specific purposes during excavation and therefore the material of interest can be extracted any time after that. However, the type of research questions related to a specific material often requires a unique strategy for sampling. Microsampling of very small variations that are located in critical areas of a feature or layer demands an almost surgical sampling procedure. It is often conducted in combination with micromorphological sampling in which material from the back of micromorphological

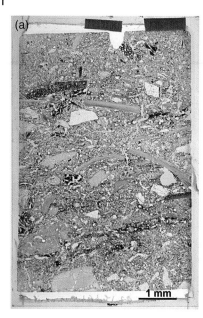 

Figure 7.6 (a) Static high-resolution image acquisition of an entire thin section that is enlarged in (b), a detail, which has been magnified about four times without losing its sharpness. The high-resolution image thus provides the possibility of better observing and studying the sample on a monitor. MSA, PP5-6, South Africa.

samples is collected for the analysis of phytoliths, dung spherulites, or the chemistry of sediment, for example (e.g. Matthews, 2010; Shillito et al., 2011a,b; Mentzer and Quade, 2013). In such cases, the specialist should personally sample the materials of interest, even though, as is often the case, it is impractical for a specialist to be at a site continuously, or sometimes even during a whole season.

Finally, we note that new innovative approaches in identifying activity areas may be possible only after the site is partially or fully excavated. Thus, a good strategic plan is to leave baulks or unexcavated areas for later sampling. Such an approach seems to be the only way to be certain that in the end the site will reveal the maximum of its archaeological potential (see section 4.11 for a discussion of this).

Concluding Remarks

In this volume, we have tried to continue in the spirit of Renfrew: '... since archaeology, or at least prehistoric archaeology, recovers almost all its basic data by excavation, *every archaeological problem starts as a problem in geoarchaeology*' [our emphasis] (Renfrew, 1976: 2). After all, the 'dirt' that is removed every day at an excavation is geology in its essence. This statement appeared just about the time PG was starting his career, and it was an aspiring assertion. In a look back over these four decades it is surprising that this statement has not been acknowledged in practice since then.

Nevertheless, in this book we have attempted to revive his advice by trying to instil in the reader insights into what they are digging, how the material got there, how it was modified during and after it accumulated, and how it is possible to 'read' and 'translate' these accumulations into an idiom that practicing archaeologists and geoarchaeologists can use to interpret site and human history.

The goal in this volume has been to sensitize the eyes of excavators and researchers in general to seeing, recognizing, and understanding that the deposits that enclose traditional archaeological records have a life history of their own. They must be treated just like artefacts, features, architectural elements, fauna, etc., and we are obliged to take them into account and incorporate their information into the entire archaeological experience that eventually results in interpretations, models, and excavation strategies (e.g., Schiffer, 1983; Adams and Fladd, 2014).

Even if we are not completely successful in our hopes, this book will with any luck help to raise the bar in archaeological practice and analysis. However, this text is only part of a larger issue. A decade ago, one of us published a somewhat confrontational paper, which, among other suggestions, tendered the idea that '... some familiarity with sedimentology, pedology, and microstratigraphy is essential for the archaeologist to appreciate what is being excavated and how it got there...In sum, a basic course in geoarchaeology should be a minimal requirement in archaeological studies to ensure that stratigraphic data are registered correctly' (Goldberg, 2008: 333). It was aimed at US curricula since archaeology in the Old World is commonly a part of the humanities curriculum (art history, philosophy, geography, history, classics, linguistics and languages, etc.). Whereas individual responses of any readership have so far been less than forthcoming since its publication, one reviewer of the original manuscript declared, 'I look forward to Goldberg's proposal to my Dean for a new faculty line in geoarchaeology' (reviewer's comments on submitted manuscript, 2005).

The reviewer's statement obviously sidesteps the issue, but it serves to highlight that there is considerable resistance for such a required course to take up a slot in an already jam-packed undergraduate course load. Similarly, in recent conversations with Classical archaeologists in Athens, they pointed out that their curricula have to be tailored to the ultimate employability of their students, and using a valuable course slot for geoarchaeology course is certainly not going to help them land a job.

This is a serious conundrum that is not readily soluble, and is beyond the immediate scope of this book. Nevertheless, we would be happy to think that our text will at least take a step in the right direction of resolving this challenge. In any case, something has to change, as the deposits are a vital intellectual resource that is being overlooked, if not neglected and thrown directly into the backdirt pile.

It has been said that football games are won on the practice field, and a similar statement can be made for archaeology: it is a field discipline first and foremost, and we all have to be sensitized to looking carefully at deposits before we get to the field and then apply this knowledge and associated skills. Understanding how archaeological data come out of the ground within their contextual setting is the first stage in carrying out proper archaeological practice. If the geological context is not understood properly at the time that the 'archaeology' – in all its aspects – is being retrieved, later analyses cannot resuscitate poorly collected/biased data and contexts.

This book is perhaps important for another reason. Since the study of sediments in historical sites and urban sites in general is rather underdeveloped, our knowledge

Reconstructing Archaeological Sites: Understanding the Geoarchaeological Matrix, First Edition. Panagiotis (Takis) Karkanas and Paul Goldberg.
© 2019 John Wiley & Sons Ltd. Published 2019 by John Wiley & Sons Ltd.

of the type of sedimentary fabrics and structures that anthropogenic activities produce in these sites is rather meagre. We attempt in this book to provide some fundamental aspects of depositional processes in sites with architectural remains. At the same time, we also develop the idea that major types of deposits have certain characteristics irrespective of their cultural period, as the human actions that they produce are generally similar. We admit, though, that we are far from reaching the point where we have fully mastered the details of these depositional processes. However, we are confident that we are moving in the right direction and that it is only a matter of time before we reach the same level of understanding as with natural depositional processes.

References

Adam, J.P. (1994). *Roman Building: Materials and Techniques*. Bloomington: Indiana University Press.

Adams, M. (2000). The optician's trick: An approach to recording excavation using an iconic formation processs recognition system. In: Roskams, S. (Ed.) *Interpreting stratigraphy. Site evaluation, recording precedures and stratigraphic analysis.* BAR International Series 910, pp. 91–101.

Adams, E.C. and Fladd, S.G. (2014). Composition and interpretation of stratified deposits in ancestral Hopi villages at Homol'ovi. *Archaeological and Anthropological Sciences*, 9, 1101–1114.

Ager, D.V. (1993). *The Nature of the Stratigraphic Record*, 3rd edn. Chichester: John Wiley & Sons.

Akeret, O. and Rentzel, P. (2001). Micromorphology and plant macrofossil analysis of cattle dung from the Neolithic lake shore settlement of Arbon Bleiche 3, *Geoarchaeology: An International Journal*, 16, 687–700.

Albert, R.M., Lavi, O., Estroff, L., Weiner, S., Tsatskin, A., Ronen, A. and Lev-yadun, S. (1999). Mode of occupation of Tabun Cave, Mt Carmel, Israel during the Mousterian Period: A study of the sediments and phytoliths. *Journal of Archaeological Science*, 26, 1249–1260.

Albert, R.M., Weiner, S., Bar-Yosef, O. and Meignen, L. (2000). Phytoliths in the Middle Palaeolithic deposits of Kebara Cave, Mt Carmel, Israel: Study of the plant materials used for fuel and other purposes. *Journal of Archaeological Science*, 27, 931–947.

Albert, R.M., Bar-Yosef, O., Meignen, L. and Weiner, S. (2003). Quantitative phytolith study of hearths from the Natufian and Middle Palaeolithic levels of Hayonim Cave (Galilee, Israel). *Journal of Archaeological Science*, 30, 461–480.

Albert, R.M., Bar-Yosef, O. and Weiner, S. (2007). Use of plant meterials in Kebara Cave: Phytoliths and mineralogical analyses. In: Bar-Yosef, O. and Meignen, L. (Eds.) *Kebara Cave Mt. Carmel, Israel: The Middle and Upper Paleolithic Archaeology, Part 2.* American School of Prehistoric Research Bulletin 49. Cambridge: Peabody Museum of Archaeology and Ethnology, Harvard Univeristy, pp. 147–164.

Albert, R.M., Shahack-Gross, R., Cabanes, D., Gilboa, A., Lev-Yadun, S., Portillo, M., Sharon, I., Boaretto, E. and Weiner, S. (2008). Phytolith-rich layers from the Late Bronze and Iron Ages at Tel Dor (Israel): Mode of formation and archaeological significance. *Journal of Archaeological Science*, 35, 57–75.

Alberts, E.E., Moldenhauer, W.C. and Foster, G.R. (1980). Soil aggregates and primary particles transported in rill and interrill flow. *Soil Science Society American Journal*, 44, 590–595.

Aldeias, V. and Bicho, N. (2016). Embedded behavior: Human activities and the construction of the mesolithic shellmound of Cabeco da Amoreira, Muge, Portugal. *Geoarchaeology: An International Journal*, 31, 530–549.

Aldeias, V., Goldberg, P., Sandgathe, D., Berna, F., Dibble, H.L., Mcpherron, S.P., Turq, A. and Rezek, Z. (2012). Evidence for Neandertal use of fire at Roc de Marsal (France). *Journal of Archaeological Science*, 39, 2414–2423.

Aldeias, V., Goldberg, P., Dibble, H.L. and El-Hajraoui, M. (2014a). Deciphering site formation processes through soil micromorphology at Contrebandiers Cave, Morocco. *Journal of Human Evolution*, 69, 8–30.

Aldeias, V., Goldberg, P., Sandgathe, D., Dibble, H.L., Mcpherron, S., Chiotti, L., Guérin, G., Bruxelles, L., Sinet-Mathiot, V. and Turq, A. (2014b). Insights into site formation processes of the Middle and Upper Paleolithic layers in the western section of La Ferrassie (Dordogne). In: *4th Annual Meeting of European Society for the study of Human Evolution (ESHE),* Florence, 18–20 September, 2014.

Aldeias, V., Dibble, H.L., Sandgathe, D., Goldberg, P. and McPherron, S.J.P. (2016). How heat alters underlying deposits and implications for archaeological fire features: A controlled experiment. *Journal of Archaeological Science*, 67, 64–79.

Allué, E., Cabanes, D., Solé, A. and Sala, R. (2012). *Hearth functioning and forest resource exploitation based on the archeobotanical assemblage from Level J. High Resolution Archaeology and Neanderthal Behavior.* Springer.

Alperson-Afil, N., Richter, D. and Goren-Inbar, N., (2007). Phantom hearths and the use of fire at Gesher Benot Ya'akov, Israel. *PaleoAnthropology*, 1–15.

Anderson, D.G., Cornelison, J.E. and Sherwood, S.C. (2013). Archeological Investigations at Shiloh Indian Mounds National Historic Landmark (40HR7): 1999–2004. National Park Service, Southeast Archeological Center.

Angelucci, D.E. and Zilhão, J. (2009). Stratigraphy and Formation Processes of the Upper Pleistocene Deposit at Gruta da Oliveira, Almonda Karstic System, Torres Novas, Portugal. *Geoarchaeology: An International Journal*, 24, 277–310.

Angelucci, D.E., Boschian, G., Fontanals, M., Pedrotti, A. and Vergès, J.M. (2009). Shepherds and karst: The use of caves and rock-shelters in the Mediterranean region during the Neolithic. *World Archaeology*, 41, 191–214.

Angelucci, D.E., Anesin, D., Susini, D., Villaverde, V., Zapata, J. and Zilhão, J. (2013). Formation processes at a high resolution Middle Paleolithic site: Cueva Antón (Murcia, Spain). *Quaternary International*, 315, 24–41.

Arpin, T.L., Mallol, C., and Goldberg, P. (2002). A new method of analyzing and documenting micromorphological thin sections using flatbed scanners: Applications in geoarchaeological studies. *Geoarchaeology: An International Journal*, 17, 305–313.

Arroyo, A.B.M., Ruiz, M.D.L., Bernabeu, G.V., Román, R.S., Morales, M.R.G. and Straus, L.G. (2008). Archaeological implications of human-derived manganese coatings: A study of blackened bones in El Mirón Cave, Cantabrian Spain. *Journal of Archaeological Science*, 35, 801–813.

Artioli, G. (2010). *Scientific Methods and Cultural Heritage: An introduction to the application of materials science to archaeometry and conservation science.* Oxford: Oxford University Press.

Bailey, G. (2007). Time perspectives, palimpsests and the archaeology of time. *Journal of Anthropological Archaeology*, 26, 198–223.

Bailey, R. (2017). Layers and gaps: The nature of the stratigraphical record. *Geology Today*, 33, 24–31.

Bailey, G. and Galanidou, N. (2009). Caves palimpsests and dwelling spaces: Examples from the Upper Palaeolithic of south-east Europe. *World Archaeology*, 41, 215–241.

Bal, L. (1973). *Micromorphological Analysis of Soils – Lower Levels in the Organization of Organic Soil Materials.* Soil Survey Paper 6. Wageningen: Soil Survey Institute.

Balbo, A.L., Madella, M., Vila, A. and Estévez, J. (2010). Micromorphological perspectives on the stratigraphical excavation of shell middens: A first approximation from the ethnohistorical site Tunel VII, Tierra del Fuego (Argentina). *Journal of Archaeological Science*, 37, 1252–1259.

Balbo, A.L., Iriarte, E., Arranz, A., Zapata, L., Lancelotti, C., Madella, M., Teira, L., Jiménez, M., Braemer, F. and Ibáñez, J.J. (2012). Squaring circle. Social and environmental implications of pre-pottery neolithic building technology at Tell Qarassa (South Syria). *PLoS ONE*, 7, e42109.

Balek, C.L. (2002). Buried artifacts in stable upland sites and the role of bioturbation: A review. *Geoarchaeology: An International Journal*, 17, 41–51.

Balme, J. and Paterson, A. (2014). *Archaeology in Progress: A student guide to archaeological analyses*, 2nd edn. Chichester: Wiley-Blackwell.

Banerjea, R.Y., Bell, M., Matthews, W. and Brown, A. (2015b). Applications of micromorphology to understanding activity areas and site formation processes in experimental hut floors. *Archaeological and Anthropological Sciences*, 7, 89–112.

Banerjea, R.Y., Fullford, M., Bell, M., Clarke, A. and Matthews, W. (2015a). Using experimental archaeology and micromorphology to reconstruct timber-framed buildings from Roman Silchester: A new approach. *Antiquity*, 89, 1174–1188.

Baranski, M.Z., Garcia-Suarez A., Klimowicz, A., Love S. and Pawlowska, K., (2015). The architecture of neolithic catalhoyuk as a process. In: A. Hodder and A. Marciniak (Eds.) *Assembling Catalhoyuk*. European Association of Archaeologists, Maney Publishing, pp. 111–126.

Bargalló, A., Gabucio, M.J. and Rivals, F. (2016). Puzzling out a palimpsest: Testing an interdisciplinary study in level O of Abric Romaní. *Quaternary International*, 417, 51–65.

Barham, A.J. (1995). Methodological approaches to archaeological context recording: X-radiography as an example of a supportive recording, assessment and interpretive technique. In: T. Barham and R.I. Macphail (Eds.) *Archaeological Sediments and Soils, Analysis, Interpretation and Management*. London: University of London, Institute of Archaeology Publications, pp. 145–182.

Barker, P. (1993). *Techniques of Archaelogical Excavation*, 3rd edn. London: Routledge.

Bar-Matthews, M. and Ayalon, A. (2011). Mid-Holocene climate variations revealed by high-resolution speleothem records from Soreq Cave, Israel and their correlation with cultural changes. *Holocene*, 21, 163–171.

Bar-Matthews, M., Marean, C.W., Jacobs, Z., Karkanas, P., Fisher, E.C., Herries, A.I.R., Brown, K., Williams, H.M., Bernatchez, J., Ayalon, A. and Nilssen, P.J. (2010). A high resolution and continuous isotopic speleothem record of paleoclimate and paleoenvironment from 90 to 53 ka from Pinnacle Point on the south coast of South Africa. *Quaternary Science Reviews*, 29, 2131–2145.

Baruch, U., Werker, E. and Bar-Yosef, O. (1992). Charred wood remains from Kebara cave, Israel: Preliminary results. *Actualités Botaniques*, 2, 531–538.

Bar-Yosef, O. and Alon, D. (1988). Nahal Hemar cave. The excavations. *Atiqot*, 18, 1–30.

Bar-Yosef, O., Vandermeersch, B., Arensburg, B., Belfer-Cohen, A., Goldberg, P., Laville, H., Meignen, L., Rak, Y., Speth, J.D., Tchernov, E., Tillier, A.M. and Weiner, S. (1992). The excavations in Kebara cave, Mt Carmel. *Current Anthropology*, 33, 497–550.

Bateman, M.D., Carr, A.S., Dunajko, A.C., Holmes, P.J., Roberts, D.L., McLaren, S.J., Bryant, R.G., Marker, M.E. and Murray-Wallace, C.V. (2011). The evolution of coastal barrier systems: A case study of the Middle-Late Pleistocene Wilderness barriers, South Africa. *Quaternary Science Reviews*, 30, 63–81.

Bathurst, R.R., Zori, D. and Byock, J. (2010). Diatoms as bioindicators of site use: Locating turf structures from the Viking Age. *Journal of Archaeological Science*, 37, 2920–2928.

Baykara, İ., Mentzer, S.M., Stiner, M.C., Asmerom, Y., Güleç, E.S. and Kuhn, S.L. (2015). The Middle Paleolithic occupations of Üçağızlı II Cave (Hatay, Turkey): Geoarcheological and archeological perspectives. *Journal of Archaeological Science: Reports*, 4, 409–426.

Beckman, G.G. and Smith, K.J. (1974). Micromorphological changes in surface soils following wetting, drying and trampling. In: Rutherford, G.K. (Ed.) *Soil Microscopy*. Kingston, Ontario: The Limestone Press, pp. 832–845.

Behrensmeyer, A.K., Gordon, K.D. and Yanagi, G.T. (1986). Trampling as a cause of bone surface damage and pseudo-cut marks. *Nature*, 319, 768–771.

Bell, M., Caseldine, A. and Neumann, H. (2000). *Prehistoric Intertidal Archaeology in the Welsh Severn Estuary*. York: Council for British Archaeology.

Benito-Calvo, A., Martinez-Moreno, J., Mora, R., Roy, M. and Roda, X. (2011). Trampling experiments at Cova Gran de Santa Linya, Pre-Pyrenees, Spain: Their relevance for archaeological fabrics of the Upper Middle Paleolithic assemblages. *Journal of Archaeological Science*, 38, 3652–3661.

Benvenuti, M. (2003). Facies analysis and tectonic significance of lacustrine fan-deltaic successions in the Pliocene–Pleistocene Mugello Basin, Central Italy. *Sedimentary Geology*, 157, 197–234.

Bergadà, M.M., Villaverde, V. and Román, D. (2013). Microstratigraphy of the Magdalenian sequence at Cendres Cave (Teulada-Moraira, Alicante, Spain): Formation and diagenesis. *Quaternary International*, 315, 56–75.

Berggren, Å. (2009). The relevance of stratigraphy. *Archaeological Dialogues*, 16, 22–25.

Berna, F. (2017). Geo-ethnoarchaeological study of the traditional Tswana dung floor from the Moffat Mission Church, Kuruman, North Cape Province, South Africa. *Archaeological and Anthropological Sciences*, 9, 1–9.

Berna, F. and Goldberg, P. (2008). Assessing Paleolithic pyrotechnology and associated hominin behavior in Israel. *Israel Journal of Earth Sciences*, 56, 107–121.

Berna, F., Matthews, A. and Weiner, S. (2004). Solubilities of bone mineral from archaeological sites: The recrystallization window. *Journal of Archaeological Science*, 31, 867–882.

Berna, F., Behar, A., Shahack-Gross, R., Berg, J., Boaretto, E., Gilboa, A., Sharon, I., Shalev, S., Shilstein, S., Yahalom-Mack, N., Zorn, J.R. and Weiner, S. (2007). Sediments exposed to high temperatures: Reconstructing pyrotechnological processes in Late Bronze and Iron Age Strata at Tel Dor (Israel). *Journal of Archaeological Science*, 34, 358–373.

Berna, F., Goldberg, P., Horwitz, L.K., Brink, J., Holt, S., Bamford, M. and Chazan, M. (2012). Microstratigraphic evidence of in situ fire in the Acheulean strata of Wonderwerk Cave, Northern Cape province, South Africa. *Proceedings of the National Academy of Sciences*, 109, 7593–7594.

Bernatchez, J.A. (2010). Taphonomic implications of orientation of plotted finds from Pinnacle Point 13B (Mossel Bay, Western Cape Province, South Africa). *Journal of Human Evolution*, 59, 274–288.

Bertran, P. (1993). Deformation-induced microstructures in soils affected by mass movements. *Earth Surface Processes and Landforms*, 18, 645–660.

Bertran, P. (1994). Dégradation des niveaux d'occupation paléolithiques en contexte périglaciaire: Exemples et implications archéologiques. *Paleo*, 6, 285–302.

Bertran, P. and Raynal, J.P. (1991). Apport de la micromorphologie a l'étude archéologique du village mıdııval de Saint-Victor de Massiac (Cantal, France). *Revue Archéologique du Centre de la France*, 30, 135–170.

Bertran, P. and Texier, J.P. (1999). Facies and microfacies of slope deposits. *Catena*, 35, 99–121.

Bertran, P., Francou, B. and Texier, J.P. (1995). Stratified slope deposits: The stone-banked sheets and lobes model. In: Slaymaker, O. (Ed.) *Steepland Geomorphology*. John Wiley & Sons, pp. 147–169.

Bertran, P., Hıtu, B., Texier, J.P. and Van Steijn, H. (1997). Fabric characteristics of subaerial slope deposits. *Sedimentology*, 44, 1–16.

Bertran, P., Caner, L., Langohr, R., Lemée, L. and d'Errico, F. (2008). Continental palaeoenvironments during MIS 2 and 3 in southwestern France: The La Ferrassie rockshelter record. *Quaternary Science Reviews*, 27, 2048–2063.

Bertran, P., Klaric, L., Lenoble, A., Masson, B. and Vallin, L. (2010). The impact of periglacial processes on Palaeolithic sites: The case of sorted patterned grounds. *Quaternary International*, 214, 17–29.

Bertran, P., Lenoble, A., Todisco, D., Desrosiers, P.M. and Sorensen, M. (2012). Particle size distribution of lithic assemblages and taphonomy of Palaeolithic sites. *Journal of Archaeological Science*, 39, 3148–3166.

Bertran, P., Beauval, C., Boulogne, S., Brenet, M., Costamagno, S., Feuillet, T., Laroulandie, V., Lenoble, A., Malaurent, P. and Mallye, J.B. (2015). Experimental archaeology in a mid-latitude periglacial context: Insight into site formation and taphonomic processes. *Journal of Archaeological Science*, 57, 283–301.

Bettinger, R.L. (1994). Prehistory of the Crooked Creek area. In: Hall, C.A. Jr and Widawski, B. (Eds.) *The Crooked Creek Guidebook*. Los Angeles: The Regents of the University of California, pp. 123–132.

Bettis III, E.A. and Mandel, R.D. (2002). The effects of temporal and spatial patterns of Holocene erosion and alluviation on the archaeological record of the Central and Eastern Great Plains, USA. *Geoarchaeology: An International Journal*, 17, 141–154.

Binford, L.R. (1978). Dimensional analysis of behavior and site structure: Learning from an eskimo hunting stand. *American Antiquity*, 43, 330–361.

Binford, L.R. (1981). Behavioral archaeology and the 'Pompeii Premise'. *Journal of Anthropological Research*, 37, 195–208.

Binford, L.R. (1987). Researching ambiguity: Frame of reference and site structure. In: Kent, S. (Ed.) *Methods and Theory for Activity Area Research*. New York: Columbia University Press, pp. 449–512.

Binford, L.R. (1996). Hearth and home: The spatial analysis of ethnographically documented rock shelter occupations as a template for distinguishing between human and hominid use of sheltered space. In: Conard, N.J. and Wendorf, F. (Eds.) *Middle Paleolithic and Middle Stone Age Settlement Systems*. Forlì, Italy: ABACO Edizioni, pp. 229–239.

Binford, L.R. and Ho, C.K. (1985). Taphonomy at a distance: Zhoukoudian, 'The cave home of Beijing man?' *Current Anthropology*, 26, 413–429.

Birkeland, P.W. (1999). *Soils and Geomorphology*, 3rd edn. New York: Oxford University Press.

Black, S. and Thoms, A. (2014). Hunter-gatherer earth ovens in the archaeological record: Fundamental concepts. *American Antiquity*, 79, 204–226.

Blair, T.C. and McPherson, J.C. (1994). Alluvial fans and their natural distinction from rivers based on morphology, hydraulic processes, sedimentary process, and facies assemblages. *Journal of Sedimentary Research*, A64, 450–489.

Blair, T.C. and McPherson, J.G. (2007). Quaternary sedimentology of the Rose Creek fan delta, Walker Lake, Nevada, USA, and implications to fan-delta facies models. *Sedimentology*, 54, 1–37.

Blikra, R.H. and Nemec, W. (1998). Postglacial colluvium in western Norway: Depositional processes, facies and palaeoclimatic record. *Sedimentology*, 45, 909–959.

Bloom, A.L. (1998). *Geomorphology: A Systematic Analysis of Late Cenozoic Landforms*, 3rd edn. New Jersey: Prentice Hall.

Boaz, N.T. and Behrensmeyer, A.K. (1976). Hominid taphonomy: Transport of human skeletal parts in an artificial fluviatile environment. *American Journal of Physical Anthropology*, 45, 53–60.

Boaz, N.T., Ciochon, R.L., Xu, Q. and Liu, J. (2000). Large mammalian carnivores as a taphonomic factor in the bone accumulation at Zhoukoudian. *Acta Anthropologica Sinica*, 19, 224–234.

Bocco, G. (1991). Gully erosion: Processes and models. *Progress in Physical Geography*, 15, 392–406.

Bocek, B. (1986). Rodent ecology and burrowing behavior: Predicted effects on archaeological site formation. *American Antiquity*, 51, 589–603.

Boggs, S. Jr. (2006). *Principles of Sedimentology and Stratigraphy*, 4th edn. New Jersey: Prentice Hall.

Boivin, N. (2000). Life rhythms and floor sequences: Excavating time in rural Rajasthan and Neolithic Çatalhöyük. *World Archaeology*, 31, 367–88.

Boivin, N. (2005). Comments on A. Jones, 'Archaeometry and materiality: Materials-based analysis in theory and practice. *Archaeometry*, 46(3), 327–338, 2004, and reply. Comments I: Post-textual archaeology and archaeological science. *Archaeometry*, 47, 175–179.

Bordes, F. (1975). Sur la notion de sol d'habitat en préhistoire Paléolithique. Bulletin de la Soc. *Préhistorique Française*, 72, 139–144.

Bordes, F. and Lafille, J. (1962). Decouverte d'un squelette d'enfant mousterien dans le gisement du Roc de Marsal, commune de Campagne-du-Bugue (Dordogne). *Compte Rendus Hebdomadaires des Seances de l'Academie des Sciences*, 254, 714–715.

Bosch, R.F. and White, W.B. (2004). Lithofacies and transport of clastic sediment in karstic aquifers. In: Sasowsky, I.D. and Mylroie, J. (Eds.) *Studies of Cave Sediments – Physical and Chemical Records of Paleoclimate*. New York: Springer, pp. 1–22.

Boschian, G. (1997). Sedimentology and soil micromorphology of the Late Pleistocene and Early Holocene deposits of Grotta dell'Edera (Trieste Karst, northeastern Italy). *Geoarchaeology: An International Journal*, 12, 227–249.

Boschian, G. and Miracle, P.T. (2007). Shepherds and caves in the karst of Istria (Croatia). *Atti della Società Toscana de scienze naturali, Memorie, Serie*, 112, 173–180.

Boschian, G. and Montagniari-Kokelj, E. (2000). Prehistoric shepherds and caves in the Trieste Karst (northeastern Italy). *Geoarchaeology: An International Journal*, 15, 331–371.

Bowler, J.M., Johnston, H., Olley, J.M., Prescott, J.R., Roberts, R.G., Shawcross, W. and Spooner, N.A. (2003). New ages for human occupation and climatic change at Lake Mungo. *Australia Nature (London)*, 421, 837–840.

Boynton, R.S. (1980). *Chemistry and Technology of Lime and Limestone*. New York: John Wiley & Sons.

Brammer, H. (1971). Coatings in Seasonally Flooded Soils. *Geoderma*, 6, 5–16.

Bray, W. and Trump, D. (1982). *Hearth. The Penguin Dictionary of Archaeology*. London: Penguin, pp. 108.

Brochier, J.E. (2002). Les sédiments anthropiques. Méthodes d'étude et perspectives. In: Miskovsky J.C. (Ed.) *Géologie de la Préhistoire: Méthodes, techniques, applications*. Paris: Géopré éditions, pp. 453–477.

Brochier, J.E. and Thinon, M. (2003). Calcite crystals, starch grains aggregates or... POCC? Comment on 'calcite crystals inside archaeological plant tissues'. *Journal of Archaeological Science*, 30, 1211–1214.

Brochier, J.E., Villa, P., Giacomarra, M. and Tagliacozzo, A. (1992). Shepherds and sediments: Geo-ethnoarchaeology of pastoral sites. *Journal of Anthropological Archaeology*, 11, 47–102.

Brooks A. and Yellen, J. (1987). The preservation of activity areas in the archaeological record: Ethnoarchaeological and archaeological work in northwest Ngamiland, Botswana. In: Kent, S. (Ed.) *Methods and Theory for Activity Area Research*. New York: Columbia University Press, pp. 63–106.

Brown, A.G. (1997). *Alluvial Geoarchaeology. Floodplain archaeology and environmental change*. Cambridge: Cambridge University Press.

Brown III, M.R. and Harris, E.C. (1993). Interfaces in archaeological stratigraphy. In: Harris, E.C., Brown III, M.R. and G.J. Brown (Eds.) *Practices of Archaeological Stratigraphy*. London: Academic Press, pp. 7–20.

Bruxelles, L., Clarke, R.J., Maire, R., Ortega, R. and Stratford, D. (2014). Stratigraphic analysis of the Sterkfontein StW 573 Australopithecus skeleton and implications for its age. *Journal of Human Evolution*, 70, 36–48.

Bryant V.M. and Dean C.W. (2006). Archaeological coprolite science: The legacy of Eric O. Callen (1912–1970). *Paelaeogeography, Palaeoclimatology, Palaeoecology*, 237, 51–66.

Bull, P.A. and Goldberg, P. (1985). Scanning electron microscope analysis of sediments from Tabun cave, Mount Carmel, Israel. *Journal of Archaeological Science*, 12, 177–185.

Bull, I.D., Betancourt, P.P. and Evershed, R.P. (1999) Chemical evidence supporting the existence of a structured agricultural manuring regime on Pseira Island, Crete during the Minoan Age. *Aegaeum*, 20, 69–75.

Bull, I.D., Elhmmali, M.M., Perret, P., Matthews, W., Roberts, D.J. and Evershed, R.P. (2005) Biomarker evidence of faecal deposition in archaeological sediments at Çatalhöyük. In: Hodder, I. (Ed.) *Inhabiting Chatalhoyuk: Reports from the 1995–99 Seasons*. Cambridge: Mcdonald Institute for Archaeological Research and British Institute of Archaeology at Ankara, pp. 415–420.

Bullard, R.G. (1970). Geological Studies in Field Archaeology. *The Biblical Archaeologist*, 33, 98–132.

Bullock, P., Fedoroff, N., Jongerius, A., Stoops, G. and Tursina, T. (1985). *Handbook for Soil Thin Section Description*. Wolverhampton: Waine Research Publishers.

Bunte, K. and Abt, S.R. (2001). *Sampling Surface and Subsurface Particle-Size Distributions in Wadable Gravel- and Cobble-Bed Streams for Analyses in Sediment Transport, Hydraulics, and Streambed Monitoring*. General Technical Report RMRS-GTR-74. Fort Collins, CO: US Department of Agriculture, Forest Service, Rocky Mountain Research Station.

Buol, S.W., Southard R.J., Graham, R.C. and McDaniel, P.A. (2003). *Soils Genesis and Classification*, 5th edn. Ames, Iowa: Iowa State Press.

Burke, H. and Smith, C. (2004). *The Archaeologist's Field Handbook*. Crows Nest: Allen and Unwin.

Burns, K. and Gabet, E.J. (2015). The effective viscosity of slurries laden with vegetative ash. *Catena*, 135, 350–357.

Burroni, D., Donahue, R.E., Pollard, A.M. and Mussi, M. (2002). The surface alteration features of flint artefacts as a record of environmental processes. *Journal of Archaeological Science*, 29, 1277–1287.

Butzer, K.W. (1973). Spring sediments from the Acheulian site at Amanzi (Uitenhage District, South Africa). *Quaternaria*, 17, 299–319.

Butzer, K.W. (1976). *Geomorphology from the Earth*. New York: Harper & Row.

Butzer, K.W. (1978). Sediment stratigraphy of the Middle Stone Age sequence at Klasies River Mouth, Tsitsikama Coast, South Africa. *The South African Archaeological Bulletin*, 33, 141–151.

Butzer, K.W. (1981). Cave sediments, Upper Pleistocene stratigraphy and Mousterian facies in Cantabrian Spain. *Journal of Archaeological Science*, 8, 133–183.

Butzer, K.W. (1982). *Archaeology as Human Ecology: Method and Theory for a Contextual Approach*. Cambridge: Cambridge University Press.

Butzer, K.W. (2008). Challenges for a cross-disciplinary geoarchaeology: The intersection between environmental history and geomorphology. *Geomorphology*, 101, 402–411.

Butzer, K.W. (2009). Evolution of an interdisciplinary enterprise: The *Journal of Archaeological Science* at 35 years. *Journal of Archaeological Science*, 36, 1842–1846.

Butzer, K.W. and Hansen, C.L. (1968). *Desert and River in Nubia: Geomorphology and prehistoric environments at the Aswan Reservoir*. Madison: University of Wisconsin Press.

Cabanes, D., Mallol, C., Exposito, I. and Baena, J. (2010). Phytolith evidence for hearths and beds in the late Mousterian occupations of Esquilleu cave (Cantabria, Spain). *Journal of Archaeological Science*, 37, 2947–2957.

Cahen, D. and Moeyersons, J. (1977). Subsurface movements of stone artifacts and their implications for the prehistory of central Africa. *Nature*, 266, 812–815.

Camarós, E., Cueto, M., Teira, L., Münzel, S.C., Plassard, F., Arias, P. and Rivals, F. (2017). Bears in the scene: Pleistocene complex interactions with implications concerning the study of Neanderthal behavior. *Quaternary International*, 435A, 237–246.

Campbell, C.V. (1967). Lamina, laminaset, bed and bedset. *Sedimentology*, 8, 7–26.

Cant, D.J. (1982). Fluvial facies model. In: Scholle, P.A. and Spearing, D. (Eds.) *Sandstone Depositional Environments*. Tulsa, Oklahoma: American Association of Petroleum Geologists, Memoir 31, pp. 115–138.

Canti, M.G. (1997). An investigation of microscopic calcareous spherulites from herbivore dungs. *Journal of Archaeological Science*, 24, 219–231.

Canti, M.G. (1998). The micromorphological identification of faecal spherulites from archaeological and modern materials. *Journal of Archaeological Science*, 25, 435–444.

Canti, M.G. (1999). The production and preservation of faecal spherulites: Animals, environment and taphonomy. *Journal of Archaeological Science*, 26, 251–258.

Canti, M.G. (2003a). Earthworm activity and archaeological stratigraphy: A review of products and processes. *Journal of Archaeological Science*, 30, 135–148.

Canti, M.G. (2003b). Aspects of the chemical and microscopic characteristics of plant ashes found in archaeological soils. *Catena*, 54, 339–361.

Canti, M.G. (2009). Geoarchaeological studies associated with remedial measures at Silbury Hill, Wiltshire, UK. *Catena*, 78, 301–309.

Canti, M.G. and Linford, N. (2000). The effects of fire on archaeological soils and sediments: Temperature and colour relationships. *Proceedings of the Prehistoric Society*, 66, 385–395.

Canti, M.G. and Piearce, T.G. (2003). Morphology and dynamics of calcium carbonate granules produced by different earthworm species. *Pedobiologia*, 47, 511–521.

Canuto, M.A., Charton, J.P. and Bell, E.E. (2010). Let no space go to waste: Comparing the uses of space between two Late Classic centers in the El Paraíso valley, Copan, Honduras. *Journal of Archaeological Science*, 37, 30–41.

Capitan, L. and Peyrony, D. (1909). Deux squelettes humains au milieu de foyers de l'époque moustérienne. *Comptes rendus des séances de l'Académie des Inscriptions et Belles-Lettres*, 53, 797–806.

Carpentier, F. and Vandermeulen, B. (2016). High-resolution photography for soil micromorphology slide documentation. *Geoarchaeology: An International Journal*, 31, 603–607.

Carran, D., Hughes, J., Leslie, A. and Kennedy, C. (2011). A short history of the use of lime as a building material beyond Europe and North America. *International Journal of Architectural Heritage*, 6, 117–146.

Carrancho, Á., Villalaín, J.J., Vallverdú, J. and Carbonell, E. (2016). Is it possible to identify temporal differences among combustion features in Middle Palaeolithic palimpsests? The archaeomagnetic evidence: A case study from level O at the Abric Romaní rock-shelter (Capellades, Spain). *Quaternary International*, 417, 39–50.

Chame M. (2003). Terrestrial mammal feces: A morphometric summary and description. *Memoir, Institute Oswaldo Cruz, Rio de Janeiro*, 98 (Suppl. I), 71–94.

Chazan, M., Ron, H., Matmon, A., Porat, N., Goldberg, P., Yates, R., Avery, M., Sumner, A. and Horwitz, L.K. (2008). Radiometric dating of the Earlier Stone Age sequence in Excavation I at Wonderwerk Cave, South Africa: Preliminary results. *Journal of Human Evolution*, 55, 1–11.

Clauss, M., Lechner-Doll, M. and Streich, W.J. (2002). Faecal particle size distribution in captive wild ruminants: An approach to the browser/grazer dichotomy from the other end. *Oecologia*, 131, 343–349.

Coard, R. (1999). One bone, two bones, wet bones, dry bones: Transport potentials under experimental conditions. *Journal of Archaeological Science*, 26, 1369–1375.

Cobo-Sánchez, L, Aramendi, J. and Domínguez-Rodrigo, M. (2014). Orientation patterns of wildebeest bones on the lake Masek floodplain (Serengeti, Tanzania) and their relevance to interpret anisotropy in the Olduvai lacustrine floodplain. *Quaternary International*, 322–323, 277–284.

Cohen, D.J., Bar-Yosef, O., Wu, X., Patania, I. and Goldberg, P. (2017). The emergence of pottery in China: Recent dating of two early pottery cave sites in south China. *Quaternary International*, 441, 36–48.

Cohen-Ofri, I., Weiner, L., Boaretto, E., Mintz, G. and Weiner, S. (2006). Modern and fossil charcoal: Aspects of structure and diagenesis. *Journal of Archaeological Science*, 33, 428–439.

Collcutt, S.N. (1987). Archaeostratigraphy: A geoarchaeologist's viewpoint. *Stratigraphica Archaeologica*, 2, 11–18.

Collinson, J.D. (1996). Alluvial sediments. In: Reading, H.G. (Ed.) *Sedimentary Environments: Processes, Facies and Stratigraphy*, 3rd edn. New York: Wiley, pp. 37–82.

Collinson, J.D. and Thompson, D.B. (1989). *Sedimentary Structures*, 2nd edn. London: Unwin Hyman.

Cook, S.R., Clarke, A.S. and Fulford, M.G. (2005). Soil geochemistry and detection of early Roman precious metal and copper alloy working at the Roman town of Calleva Atrebatum (Silchester, Hampshire, UK). *Journal of Archaeological Science*, 32, 805–812.

Cornwall, I.W. (1958). *Soils for the Archaeologists*. London: Phoenix House Ltd.

Costa, J.E. (1983). Paleohydraulic reconstruction of flash-flood peak from boulder deposits in the Colorado Front Range. *Geological Society of American Bulletin*, 94, 985–1004.

Costa, J.E. (1988). Rheologic, geomorphic, and sedimentologic differentiation of water floods, hyperconcentrated flows, and debris flows, In: Baker, V.R., Kochel, R.C. and Patton, P.C. (Eds.) *Flood Geomorphology*. New York: John Wiley & Sons, pp. 113–122.

Couchoud, I. (2003). Processus géologiques de formation du site moustérien du Roc de Marsal (Dordogne, France). *Paleo*, 15, 51–68.

Coudé-Gaussen, G. and Rognon, P. (1988). Caractérisation sédimentologique et conditions paléoclimatiques de la mise en place de loess au nord du Sahara à partir de l'exemple du Sud-Tunisien. *Bulletin de la Société Géologique de France*, IV (6), 1081–1090.

Courty, M.A. (2001). Microfacies analysis assisting archaeological stratigraphy. In: Goldberg, P., Holliday, V.T. and Ferring, C.R. (Eds.) *Earth Sciences and Archaeology*. New York: Kluwer, pp. 205–239.

Courty, M.A., Goldberg, P. and Macphail, R.I. (1989). *Soils and Micromorphology in Archaeology*. Cambridge: Cambridge University Press.

Courty, M.A., Macphail, R.I. and Wattez, J. (1991). Soil micromorphological indicators of pastoralism with special reference to Arene Candide, Fianle Ligure, Italy. *Rivista di Studi Liguri*, A. LVII, 127–150.

Courty, M.A., Goldberg, P. and Macphail, R.I. (1994a). Ancient people – lifestyles and cultural patterns. In: Etchevers, J.D. (Ed.) *Transactions of the 15th World Congress of Soil Science, International Society of Soil Science, Mexico, Volume 6a*. Acapulco: International Society of Soil Science, pp. 250–269.

Courty, M.A., Marlin, C., Dever, L., Temblay, P. and Vachier, P. (1994b). The Properties, Genesis and Environmental Significance of Calcitic Pendents from the High Arctic (Spitsbergen). *Geoderma*, 61, 71–102.

Courty, M.A., Carbonell, E., Poch, J.V. and Banerjee, R. (2012). Microstratigraphic and multi-analytical evidence for advanced Neanderthal pyrotechnology at Abric Romani (Capellades, Spain). *Quaternary International*, 247, 294–313.

Coussot, P. and Meunier, M. (1996). Recognition, classification and mechanical description of debris flows. *Earth-Science Reviews*, 40, 209–227.

Cremaschi, M., Fedoroff, N., Guerreschi, A., Huxtable, J., Colombi, N., Castellettill. L. and Masperoll, A. (1990). Sedimentary and pedological processes in the Upper Pleistocene loess of northern Italy. The Bagaggera Sequence. *Quaternary International*, 5, 23–38.

Cremeens, D.L. (2005). Micromorphology of Cotiga Mound, West Virginia. *Geoarchaeology: An International Journal*, 20, 581–597.

Cremeens, D.L., Landers, D.B. and Frankenberg, S.R. (1997). Geomorphic setting and stratigraphy of Cotiga Mound, Mingo County, West Virginia. *Geoarchaeology: An International Journal*, 12, 459–477.

Crombé, P., Langohr, R. and Louwagie, G. (2015). Mesolithic hearth-pits: Fact or fantasy? A reassessment based on the evidence from the sites of Doel and Verrebroek (Belgium). *Journal of Archaeological Science*, 61, 158–171.

Crowther, J., Macphail, R.I. and Cruise, G.M. (1996). Short-term burial change in a humic rendzina, Overton Down Experimental Earthwork, Wiltshire, England. *Geoarchaeology: An International Journal*, 11, 95–117.

Cullen, D.J. (1988). Mineralogy of nitrogenous guano on the Bounty Islands, SW Pacific Ocean. *Sedimentology*, 35, 421–428.

Dan, J. and Yaalon, D.H. (1982). Evolution of Reg Soils in Southern Israel and Sinai. *Geoderma*, 28, 173–202.

Dart, A. (1986). Sediment accumulation along Hohokam canals. *Kiva*, 51, 63–84.

Darwin, C. (1881). *The Formation of Vegetable Mould, through the Action of Worms, with Observations on their Habits*. London: John Murray.

Davidson, D.A., Wilson, C.A., Lemos, I.S. and Theocharopoulos, S.P. (2010). Tell formation processes as indicated from geoarchaeological and geochemical investigations at Xeropolis, Euboea, Greece. *Journal of Archaeological Science*, 37, 1564–1571.

Deal, K. (2012). *Fire effects on flaked stone, ground stone, and other stone artifacts*. Chapter 4, USDA Forest Service General Technical Report, RMRS-GTR-42, Volume 3, pp. 97–111.

Dekker, L.W. and De Weerd, M.D. (1973). The value of soil survey for archaeology. *Geoderma*, 10, 169–178.

Dibble, H.L., Chase, P.G., Mcpherron, S.P. and Tuffreau, A. (1997). Testing the reality of a 'living floor' with archaeological data. *American Antiquity*, 62, 629–651.

Dibble, H.L., Berna, F., Goldberg, P., McPherron, S.P., Mentzer, S., Niven, L., Richter, D., Théry-Parisot, I., Sandgathe, D. and Turq, A. (2009). A Preliminary Report on Pech de l'Azé IV, Layer 8 (Middle Paleolithic, France). *Paleoanthropology*, 182–219.

Dibble, H.L., Aldeias, V., Goldberg, P., Mcpherron, S.P., Sandgathe, D. and Steele, T.E. (2015). A critical look at evidence from La Chapelle-aux-Saints supporting an intentional Neandertal burial. *Journal of Archaeological Science*, 53, 649–657.

Dirks, P.H.G.M., Kibii, J.M., Kuhn, B.F., Steininger, C., Churchill, S.E., Kramers, J.D., Pickering, R., Farber, D.L., Mériaux, A.S., Herries, A.I.R., King, G.C.P. and Berger, L.R. (2010). Geological setting and age of *Australopithecus sediba* from Southern Africa. *Science*, 328, 205–208.

Dodonov, A.E. (1995). Geoarchaeology of Palaeolithic sites in loesses of Tajikistan (Central Asia). In: Johnson, E. (Ed.) *Ancient Peoples and Landscapes*. Lubbock: Museum of Texas Tech University, pp. 127–136.

Domínguez-Rodrigo, M.D., Uribelarrea, U., Santonja, M., Bunn, H.T., García-Pérez, A., Pérez-González, A, Panera J., Rubio-Jara, S., Mabulla, A., Baquedano, E., Yravedra, J. and Diez-Martín, F. (2014). Autochthonous anisotropy of archaeological materials by the action of water: Experimental and archaeological reassessment of the orientation patterns at the Olduvai sites. *Journal of Archaeological Science*, 41, 44–68.

Donahue, J. and Adovasio, J.M. (1990). Evolution of sandstone rockshelters in eastern North America; A geoarchaeological perspective. In: Lasca N.P. and Donahue J. (Eds.) *Archaeological Geology of North America, Centennial, Vol*ume 4. Boulder: Geological Society of America, pp. 231–251.

Draut, A.E., Rubin, D.M., Dierker, J.L., Fairley, H.C., Griffiths, R.E., Jr., J.E.H., Hunter, R.E., Kohl, K., Leap, L.M., Nials, F.L., Topping, D.J. and Yeatts, M. (2008). Application of sedimentary-structure interpretation to geoarchaeological investigations in the Colorado River Corridor, Grand Canyon, Arizona, USA. *Geomorphology*, 101, 497–509.

Durand, N., Monger, H.C. and Canti, M.G. (2010). Calcium carbonate features. In: Stoops, G., Marcelino, V. and Mees, F. (Eds.) *Interpretation of Micromorphological Features of Soils and Regoliths*. Amsterdam: Elsevier, pp. 149–194.

Duwe, S., Eiselt, B.S., Darling, J.A., Willis, M.D. and Walker, C. (2016). The pueblo decomposition model: A method for quantifying architectural rubble to estimate population size. *Journal of Archaeological Science*, 65, 20–31.

Dykes, A. (2007). Mass movements in cave sediments: Investigation of a ~40,000-year-old guano mudflow inside the entrance of the Great Cave of Niah, Sarawak, Borneo. *Landslides*, 4, 279–290.

Eren, M.I., Durant, A., Neudorf, C., Haslam, M., Shipton, C., Bora, J., Korisettar, R. and Petraglia, M. (2010). Experimental examination of animal trampling effects on artifact movement in dry and water saturated substrates: A test case from South India. *Journal of Archaeological Science*, 37, 3010–3021.

Evstatiev, D., Gergova, D. and Rizzo, V. (2006). Geoarchaeological characteristics of the Thracian Tumuli in Bulgaria. *Helis*, 4, 156–168.

Fanning, P. and Holdaway, S. (2001). Stone artifact scatters in western NSW, Australia: Geomorphic controls on artifact size and distribution. *Geoarcheology: An International Journal*, 16, 667–686.

Farrand, W.R. (1975). Sediment analysis of a prehistoric rockshelter: The Abri Pataud. *Quaternary Research*, 5, 1–26.

Farrand, W.R. (1984). Stratigraphic classification: Living within the law. *Quarterly Reviews of Archaeology*, 5, 1–5.

Feder, K.L. (1997). Data preservation: Recording and collecting. In: Hester, T.R., Shafer, H.J. and Feder K.L. (Eds.) *Field Methods in Archaeology*, 7th edn. Mountain View, CA: Mayfield Publishing Co, pp. 113–142.

Fedoroff, N., Courty, M.A. and Thompson, M.L. (1990). Micromorphological evidence of palaeoenvironmental change in Pleistocene and Holocene paleosols. In: Douglas, L.A. (Ed.) *Soil Micromorphology: A Basic and Applied Science. Developments in Soil Science, Volume 19*. Amsterdam: Elsevier, pp. 653–666.

Fernández Rodriguez, C., Rego, P.R. and Cortizas, A.M. (1995). Characterization and depositional evolution of hyaena (*Crocuta crocuta*) coprolites from La Valifia Cave (northwest Spain). *Journal of Archaeological Science*, 22, 597–607.

Ferring, C.R. (1986). Rates of fluvial sedimentation: Implications for archaeological variability. *Geoarchaeology: An International Journal*, 1, 259–274.

Ferring, R., Oms, O., Agustí, J., Berna, F., Nioradze, M., Shelia, T., Tappen, M., Vekua, A., Zhvania, D. and Lordkipanidze, D. (2011). Earliest human occupations at Dmanisi (Georgian Caucasus) dated to 1.85–1.78 Ma. *Proceedings of the National Academy of Sciences*, 108, 10432–10436.

Field, J. and Banning, E.B. (1998). Hillslope processes and archaeology in Wadi Ziqlab, Jordan. *Geoarchaeology: An International Journal*, 13, 595–616.

Fitzpatrick, E.A.F. (1993). *Soil Microscopy and Micromorphology*. New York, Chichester: John Wiley.

Folk, R.L. (1974). *Petrology of Sedimentary Rocks*. Austin, Texas: Hemphill Publishing Company.

Folk, R.L. and Hoops, G.K. (1982). An early Iron-Age layer of glass made from plants at Tel Yin'am, Israel. *Journal of Field Archaeology*, 9, 455–466.

Ford, D. (1997). Dating and paleo-environmental studies of speleothems. In: Hill, C. and Forti, P. (Eds.) *Cave Minerals of the World*, 2nd edn. Huntsville: National Speleological Society, pp. 271–284.

Ford, D. and Williams, P. (2007). *Karstic Geomorphology and Hydrology*, revised edition. Chichester: John Wiley.

Forte, M., Delli Unto, N., Johnsson, K. and Lercari, N. (2015). Interpretation process at Catalhoyuk using 3D. In: Hodder I. and Marciniak A. (Eds.) *Assembling Catalhoyuk. Themes in Contemporary Archaeology 1, EAA*. Leeds: Maney Publishing, pp. 43–58.

Fox, C.A. and Tarnokai, C. (1990). The micromorphology of a sedimentary peat doposit from the Pacific temperate wetland region of Canada. In: Douglas, L.A. (Ed.) *Soil Micromorphology: A Basic and Applied Science. Developments in Soil Science, Volume 19.* Amsterdam: Elsevier, pp. 311–319.

Fraysse, F., Pokrovsky, O.S., Schott, J. and Meunier, J.D. (2009). Surface chemistry and reactivity of plant phytoliths in aqueous solutions. *Chemical Geology*, 258, 197–206.

Frederick, C.D., Bateman, M.D. and Rogers, R. (2002). Evidence for eolian deposition in the Sandy Uplands of East Texas and the implications for archaeological site integrity. *Geoarchaeology: An International Journal*, 17, 191–217.

French, C. (2003). *Geoarchaeology in Action: Studies in Soil Micromorphology and Landscape Evolution.* London: Routledge.

Friesem, D., Boaretto, E., Eliyahu-Behar, A. and Shahack-Gross, R. (2011). Degradation of mud brick houses in an arid environment: A geoarchaeological model. *Journal of Archaeological Science*, 38, 1135–1147.

Friesem, D. Tsartsidou, G., Karkanas, P. and Shahack-Gross, R. (2014a). Where are the roofs? A geo-ethnoarchaeological study of mud structures and their collapse processes, focusing on the identification of roofs. *Archaeological and Anthropological Sciences*, 6, 73–92.

Friesem, D., Karkanas, P. Tsartsidou, G., and Shahack-Gross, R. (2014b). Sedimentary processes involved in mud brick degradation in temperate environments: A micromorphological approach in an ethnoarchaeological context in northern Greece. *Journal of Archeological Science*, 41, 556–567.

Frumkin, A. (Ed.) (2013). *Karst Geomorphology, Volume 6.* Amsterdam: Elsevier.

Fuchs, M., Kandel, A.W., Conard, N.J., Walker, S.J. and Felix-Henningsen, P. (2008). Geoarchaeological and chronostratigraphical investigations of open-air sites in the Geelbek Dunes, South Africa. *Geoarchaeology: An International Journal*, 23, 425–449.

Galloway, W.E. and Hobday, D.K. (1996). Terrigenous clastic depositional systems: Applications to fossil fuel and groundwater resources. Berlin: Springer Verlang.

Gardner, R.A.M. (1981). Reddening of dune sands – evidence from southeast India. *Earth Surface Processes*, 6, 459–468.

Garfinkel, Y. (1993). The Yarmukian Culture in Israel. *Paléorient*, 19, 115–134.

Garfinkel, Y., Ben-Shlomo, D. and Marom, N. (2012). Sha'ar Hagolan: A majorpottery Neolithic settlement and arstistic center in Jordan valley. *Eurasian Prehistory*, 8, 97–143.

Gargett, R.H. (1989). Grave Shortcomings. The Evidence for Neandertal Burial. *Current Anthropology*, 30, 157–190.

Gargett, R.H. (1999). Middle Palaeolithic burial is not a dead issue: The view from Qafzeh, Saint-Césaire, Kebara, Amud, and Dederiyeh. *Journal of Human Evolution*, 37, 27–90.

Gargett, R.H., Bricker, H.M., Clark, G., Lindly, J., Farizy, C., Masset, C., Frayer, D.W., Montet-White, A., Gamble, C. and Gilman, A. (1989). Grave Shortcomings: The Evidence for Neandertal Burial [and Comments and Reply]. *Current Anthropology*, 30, 157–190.

Gé, T., Courty, M.A., Matthews, W. and Wattez, J. (1993). Sedimentary formation processes of occupation surfaces. In: Goldberg, P., Nash, D.T. and Petraglia, M.D. (Eds.) *Formation Processes in Archaeological Context. Monographs in World Archaeology 17.* Madison: Prehistory Press, pp. 149–163.

Gebhardt, A. and Langohr, R. (1999). Micromorphological study of construction materials and living floors in the Medieval Motte of Werken (West Flandres, Belgium). *Geoarchaeology: An International Journal*, 14, 595–620.

Gifford, D.P. (1978). Ethnoarchaological observations of natural processes affecting cultural materials. In: Gould, R.A. (Ed.) *Explorations in Ethnoarchaeology.* Albuquerque: University of New Mexico Press, pp. 77–101.

Gifford, D.P. and Behrensmeyer, A.K. (1977). Observed formation and burial of a recent human occupation site in Kenya. *Quaternary Research*, 8, 245–266.

Gifford-Gonzalez, D.P., Damrosch, D.B., Damrosch, D.R., Pryor. J. and Thunen, R.L. (1985). The third dimension in site structure: An experiment in trampling and vertical dispersal. *American Antiquity*, 50, 803–818.

Gile, L.H., Peterson, F.F. and Grossman, R.B. (1966). Morphological and genetic sequences of carbonate accumulation in desert soils. *Soil Science*, 101, 347–361.

Gillieson, D. (2009). *Caves: Processes, Development and Management.* Oxford: John Wiley & Sons.

Gladfelter, B.G. (2001). Archaeological sediments in humid alluvial environments. In: Stein, J.K. and Farrand, W.R. (Eds.) *Sediments in Archaeological Context.* Salt Lake City: University of Utah Press, pp. 93–125.

Gleeson, P. and Grosso, G. (1976). Ozette site. In: Croes, D. (Ed.) *The Excavation of Water Saturated Archaeological Sites (Wet Sites) on the Northwest Coast of North America.* Ottawa: National Museum of Man, pp. 13–44.

Glob, P.V. (1965). *The Bog People: Iron-Age Man Preserved.* New York: New York Review of Books.

Glover, I.C. (1979). The effects of sink action on archaeological deposits in caves: An Indonesian example. *World Archaeology*, 10, 302–317.

Goldberg, P. (1973). *Sedimentology, Stratigraphy and Paleoclimatology of et-Tabun Cave, Mount Carmel, Israel.* Ph.D. Doctoral dissertation, The University of Michigan.

Goldberg, P. (1977). Late quaternary stratigraphy of Gebel Maghara. In: Bar-Yosef, O. and Phillips, J.L. (Eds.) *Prehistoric Investigations in Gebel Maghara, Northern Sinai, Qedem. Monographs of the Institute of Archaeology 7.* Jerusalem, pp. 11–31.

Goldberg, P. (1978). Granulométrie des sédiments de la grotte de Taboun, Mont-Carmel, Israël. *Géologie Méditerranéenne*, 4, 371–383.

Goldberg, P. (1979a). Micromorphology of Pech-de-l'Azé II sediments. *Journal of Archaeological Science*, 6, 17–47.

Goldberg, P. (1979b). Geology of the late Bronze Age mudbrick from Tel Lachish. Tel Aviv. *Journal of the Tel Aviv Institute of Archaeology*, 6, 60–71.

Goldberg, P. (1983). Geology of sites Boker and Boker Tachtit and their surroundings. In: Marks, A.E. (Ed.) *Prehistory and Paleoenvironments in the Central Negev, Israel, Volume III.* Dallas: Southern Methodist University Press, pp. 39–62.

Goldberg, P. (1987). The geology and stratigraphy of Shiqmim. In: Levy, T.E. (Ed.) *Shiqmim 1: Studies Concerning Chalcolithic Societies in the Northern Negev Desert, Israel (1982–1984).* Oxford: British Archaeological Reports, pp. 35–43 and 435–444.

Goldberg, P. (2000a). Micromorphology and site formation at Die Kelders Cave I, South Africa. *Journal of Human Evolution*, 38, 43–90.

Goldberg, P. (2000b). Micromorphological aspects of site formation at Keatley Creek. In: Hayden, B. (Ed.) *The Ancient Past of Keatley Creek.* Burnaby, BC: Archaeology Press, Simon Fraser University, pp. 79–95.

Goldberg, P. (2001). Some micromorphological aspects of prehistoric cave deposits. *Cahiers D'archéologie du CELAT, Série Archéométrie*, 10, 161–175.

Goldberg, P. (2003). Some observations on Middle and Upper Palaeolithic ashy cave and rockshelter deposits in the Near East. In: Goring-Morris, A.N. and Belfer-Cohen, A. (Eds.) *More than Meets the Eye: Studies on Upper Palaeolithic Diversity in the Near East.* Oxford: Oxbow Books, pp. 19–32.

Goldberg, P. (2008). Raising the Bar. In: Sullivan, III A.P. (Ed.) *Archaeological Concepts for the Study of the Cultural Past.* Salt Lake City: University of Utah Press, pp. 24–39.

Goldberg, P. and Bar-Yosef, O. (1998). Site formation processes in Kebara and Hayonim caves and their significance in Levantine prehistoric caves. In: Akazawa, E.A. (Ed.) *Neandertals and Modern Humans in Western Asia.* New York: Plenum, pp. 107–125.

Goldberg, P. and Berna F. (2010). Micromorphology and context. *Quaternary International*, 214, 56–62.

Goldberg, P. and Guy, J. (1996). Micromorphological observations of selected rock ovens, Wilson-Leonard Site, Central Texas. In: Castelletti, L. and Cremaschi, M. (Eds.) *XIII International Congress of Prehistoric and Protohistoric Sciences: Colloquium VI Micromorphology of Deposits of Anthropogenic Origin.* Forlí, Italia: ABACO, pp. 115–122.

Goldberg, P. and Holliday, V.T. (1998). Geology and stratigraphy of the Wilson–Leonard Site. In: Wilson Leonard, A.N. (Ed.) *11,000-year Archeological Record of Hunter-Gatherers in Central Texas.* Studies in Archaeology, Volume 31. Austin: Texas Archeological Research Laboratory, University of Texas at Austin, pp. 77–121.

Goldberg, P. and Macphail, R.I. (2003). Strategies and techniques in collecting micromorphology samples. *Geoarchaeology: An International Journal*, 18, 571–578.

Goldberg, P. and Macphail, R.I. (2006). *Practical and Theoretical Geoarchaeology.* Oxford: Blackwell.

Goldberg, P. and Nathan, Y. (1975). The phosphate mineralogy of et-Tabun cave, Mount Carmel, Israel. *Mineralogical Magazine*, 40, 253–258.

Goldberg, P. and Sherwood, S. (1994). Micromorphology of Dust Cave Sediments: Some preliminary results. *Journal of Alabama Archaeology*, 40, 56–64.

Goldberg, P. and Sherwood, S.C. (2006). Deciphering human prehistory through the geoarcheological study of cave sediments. *Evolutionary Anthropology*, 15, 20–36.

Goldberg, P. and Whitbread, I. (1993). Micromorphological study of a Bedouin tent floor. In: Goldberg, P., Nash, D.T., and Petraglia, M.D. (Eds.) *Formation Processes in Archaeological Context.* Monographs in World Prehistory, World Archaeology 17. Madison: Prehistory Press, pp. 165–187.

Goldberg, P., Weiner, S., Bar-Yosef, O., Xu, Q. and Liu, J. (2001). Site formation processes at Zhoukoudian, China. *Journal of Human Evolution*, 41, 483–530.

Goldberg, P., Laville, H., Meignen, L. and Bar-Yosef, O. (2007). Stratigraphy and geoarchaeological history of Kebara Cave, Mount Carmel. In: Bar-Yosef, O. and Meignen, L. (Eds.) *Kebara Cave Mt. Carmel, Israel: The Middle and Upper Paleolithic Archaeology, Part 2. American School of Prehistoric Research Bulletin 49.* Cambridge: Peabody Museum of Archaeology and Ethnology Harvard University. pp. 49–89.

Goldberg, P., Miller, C.E., Schiegl, S., Ligouis, B., Berna, F., Conard, N.J. and Wadley, L. (2009a). Bedding, hearths, and site maintenance in the Middle Stone Age of Sibudu Cave, KwaZulu-Natal, South Africa. *Archaeological and Anthropological Science*, 1, 95–122.

Goldberg, P., Berna, F. and Macphail, R.I. (2009b). Comment on DNA from Pre-Clovis Human Coprolites in Oregon, North America. *Science*, 325, 148a.

Goldberg, P., Meignen, L. and Mallol, C. (2009c). Geoarchaeology, site formation, and transitions. In: Shea, J. and Lieberman, D. (Eds.) *Transitions in Prehistory: Essays in Honor of Ofer Bar-Yosef.* Oxfordand, Oakville: American School of Prehistoric Research Publication, Oxbow/David Brown, pp. 431–443.

Goldberg, P., Dibble, H., Berna, F., Sandgathe, D., McPherron, S.J.P. and Turq, A. (2012). New evidence on Neandertal use of fire: Examples from Roc de Marsal and Pech de l'Azé IV. *Quaternary International*, 247, 325–340.

Goldberg, P., Berna, F. and Chazan, M. (2015). Deposition and Diagenesis in the Earlier Stone Age of Wonderwerk Cave, Excavation 1, South Africa. *African Archaeological Review*, 32, 613–643.

Goldberg, P., Aldeias, V., Balzeau, A., Bruxelles, L., Chiotti, L., Crevecoeur, I., Dibble, H.L., Gómez-Olivencia, A., Guérin, G., Hublin, J.J., Maureille, B., Mcpherron, S. J.P., Madelaine, S., Sandgathe, D.M., Steele, T., Talamo, S. and Turq, A. (2016). On the context of the Neanderthal skeletons at La Ferrassie: New evidence on old data. In: *6th Annual Meeting of the European Society for the Study of Human Evolution (ESHE), Madrid. Proceedings of the European Society for the Study of Human Evolution, Vol. 5.* European Society for the Study of Human Evolution, pp. 108

Goldberg, P., Miller, C.E. and Mentzer, S.M. (2017a). Recognizing fire in the Paleolithic archaeological record. *Current Anthropology*, 58, S175–S190.

Goldberg, P., Aldeias, V., Dibble, H., Mcpherron, S., Sandgathe, D. and Turq, A. (2017b). Testing the Roc de Marsal Neandertal 'burial' with geoarchaeology. *Archaeological and Anthropological Sciences*, 9, 1005–1015.

Golding, K.A., Simpson, I.A., Schofield, J.E. and Edwards, K.J. (2011). Norse–Inuit interaction and landscape change in southern Greenland? A geochronological, Pedological, and Palynological investigation. *Geoarchaeology: An International Journal*, 26, 315–345.

Goodman-Elgar, M. (2008). The devolution of mudbrick: Ethnoarchaeology of abandoned earthen dwellings in the Bolivian Andes. *Journal of Archaeological Science*, 35, 3057–3071.

Goren, Y. and Goldberg, P. (1991). Petrographic thin sections and the development of Neolithic plaster production in Northern Israel. *Journal of Field Archaeology*, 18, 131–140.

Goren, Y. and Goring-Morris, A.N. (2008). Early pyrotechnology in the Near East: Experimental lime-plaster production at the Pre-Pottery Neolithic B site of Kfar HaHoresh, Israel. *Geoarchaeology: An International Journal*, 23, 779–798.

Goren-Inbar, N., Alperson, N., Kislev, M.E., Simchoni, O., Melamed, Y., Ben-Nun, A. and Werker, E. (2004). Evidence of hominin control of fire at Gesher Benot Ya'akov, Israel. *Science*, 304, 725–727.

Goring-Morris, A.N. and Goldberg, P. (1990). Late Quaternary dunce incursion in the southern Levant: Archaeology, chronology and palaeoenvironments. *Quaternary International*, 3, 115–137.

Gose, W. (2000). Paleomagnetic studies of burned rocks. *Journal of Archaeological Science*, 27, 409–421.

Goudge, C.K. (2007). Geochemical exploration: Sample collection and survey design. *SIPES Quarterly*, 44, 25–30.

Guccione, M.J., Lafferty, R.H., III and Cummings, L.S. (1988). Environmental constraints of human settlement in an evolving Holocene alluvial system: The Lower Mississippi Valley. *Geoarchaeology: An International Journal*, 3, 65–84.

Guérin, G., Frouin, M., Talamo, S., Aldeias, V., Bruxelles, L., Chiotti, L., Dibble, H. L., Goldberg, P., Hublin, J.J., Jain, M., Lahaye, C., Madelaine, S., Maureille, B., Mcpherron, S.J.P., Mercier, N., Murray, A.S., Sandgathe, D., Steele, T.E., Thomsen, K.J. and Turq, A. (2015). A multi-method luminescence dating of the Palaeolithic sequence of La Ferrassie based on new excavations adjacent to the La Ferrassie 1 and 2 skeletons. *Journal of Archaeological Science*, 58, 147–166.

Guérin, G., Frouin, M., Tuquoi, J., Thomsen, K.J., Goldberg, P., Aldeias, V., Lahaye, C., Mercier, N., Guibert, P., Jain, M., Sandgathe, D., Mcpherron, S.P., Turq, A. and Dibble, H.L. (2017). The complementarity of luminescence dating methods illustrated on the Mousterian sequence of the Roc de Marsal: A series of reindeer-dominated, Quina Mousterian layers dated to MIS 3. *Quaternary International*, 433B, 102–115.

Gur-Arieh, S., Mintz, E., Boaretto, E. and Shahack-Gross, R. (2013). An ethnoarchaeological study of cooking installations in rural Uzbekistan: Development of a new method for identification of fuel sources. *Journal of Archaeological Science*, 40, 4331–4347.

Gur-Arieh, S., Shahack-Gross, R., Maeir, A.M., Lehmann, G., Hitchcock, L.A. and Boaretto, E. (2014). The taphonomy and preservation of wood and dung ashes found in archaeological cooking installations: Case studies from Iron Age Israel. *Journal of Archaeological Science*, 46, 50–67.

Hampton, B.A. and Horton, B.K. (2007). Sheetflow fluvial processes in a rapidly subsiding basin, Altiplano plateau, Bolivia. *Sedimentology*, 54, 1121–1147.

Händel, M., Simon, U., Einwögerer, T. and Neugebauer-Maresch, C. (2009). Loess deposits and the conservation of the archaeological record – The Krems-Wachtberg example. *Quaternary International*, 198, 46–50.

Händel, M., Einwögerer, T., Simon, U. and Neugebauer-Maresch, C. (2014). Krems-Wachtberg excavations 2005–12: Main profiles, sampling, stratigraphy, and site formation. *Quaternary International*, 351, 38–49.

Hanson, N.R. (1958). *Patterns of Discovery: An Inquiry into the Conceptual Foundations of Science.* Cambridge: Cambridge University Press.

Hanson, M. and Cain, C.R. (2007). Examining histology to identify burned bone. *Journal of Archaeological Science*, 34, 1902–1913.

Harris, E. (1989). *Principles of Archaeological Stratigraphy*, 2nd edn. London: Academic Press.

Harris, C. and Ellis, S. (1980). Micromorphology of soils in soliflucted materials, Okstindan, Northern Norway. *Geoderma*, 23, 11–29.

Harris, E.C., Brown, M.R. and Brown, G.J. (1993). *Practices of Archaeological Stratigraphy*. London, San Diego: Academic Press.

Harrison, K. (2013). The application of forensic fire investigation techniques in the archaeological record. *Journal of Archaeological Science*, 40, 955–959.

Hassan, F.A. (1985). Fluvial Systems and Geoarchaeology in Arid Lands: With Examples from North Africa, the Near East and the American Southwest. In: Stein, J.K. and Farrand, W.R. (Eds.) *Archaeological Sediments in Context*. Orono: Center for the Study of Early Man, Institute for Quaternary Studies, University of Maine at Orono, pp. 53–68.

Hayden, B. (1997). *The pithouses of Keatley Creek: Complex hunter-gatherers of the Northwest Plateau*. Fort Worth, TX: Harcourt Brace College Publishers.

Hayden, B. (2000). *The Ancient Past of Keatley Creek, Volume 1*. Burnaby, BC: Archaeology Press.

Hayden, B. (2005). The emergence of large villages and large residential corporate group structures among complex hunter-gatherers at Keatley Creek. *American Antiquity*, 70, 169–174.

Hayden, B. and Spafford, J. (1993). The Keatley Creek Site and corporate group archaeology. *BC Studies*, 99, 106–139.

Hedges, R.E.M. and Millard, A.R. (1995). Bones and groundwater: Towards the modelling of diagenetic processes. *Journal of Archaeological Science*, 22, 155–164.

Heimdahl, J., Menander, H. and Karlsson, P. (2005). A new method for urban geoarchaeological excavation, example from Norrköping, Sweden. *Norwegian Archaeological Review*, 38, 102–112.

Hein, F.J. (1984). Deep-sea and fluvial braided channel conglomerates: A comparison of two case studies. In: Koster, E.H. and Steel, R.J. (Eds.) *Sedimentology of Gravels and Conglomerates*. Canadian Society of Petroleum Geologists, Memoir 10, pp. 33–49.

Helwing, B. (2009). What's the news? Thinking about McAnany and Hodder's 'Thinking about stratigraphic sequence in social terms'. *Archaeological Dialogues*, 16, 25–31.

Henry, D. (2012). The palimpsest problem, hearth pattern analysis, and Middle Paleolithic site structure. *Quaternary International*, 247, 246–266.

Henry, D.O., Hietala, H.J., Rosen, A.M., Demidenko, Y.E., Usik, V.I. and Armagan, T.L. (2004). Human behavioral organization in the Middle Paleolithic: Were Neanderthals different? *American Anthropologist*, 106, 17–31.

Hill, C. and Forti, P. (1997). *Cave Minerals of the World*, 2nd edn. Huntsville: National Speleological Society.

Hill, A. and Walker, A. (1972). Procedures in vertebrate taphonomy; notes on a Ugandan Miocene fossil locality. *Journal of the Geological Society of London*, 128, 399–406.

Hinchliffe, S. (1999). Timing and significance of talus slope reworking, Trotternish, Skye, northwest Scotland. *The Holocene*, 9, 483–494.

Hodder, I. (1987). The meaning of discard: Ash and domestic space in Baringo. In: Kent, S. (Ed.) *Methods and Theory for Activity Area Research*. New York: Columbia University Press, pp. 424–448.

Hodder, I. (1992). *Theory and Practice in Archaeology*. London: Routledge.

Hodder, I. and Cessford, C. (2004). Daily practice and social memory at Catalhoyuk. *American Antiquity*, 69, 17–40.

Hoffecker, J.F. (1987). Upper Pleistocene loess stratigraphy and Paleolithic site chronology on the Russian Plain. *Geoarchaeology: An International Journal*, 2, 259–284.

Holliday, V.T. (1990). Pedology in archaeology. In: N.P. Lasca and J. Donahue (Eds.) *Archaeological Geology of North America*. Boulder: Geological Society of America, pp. 525–540.

Holliday, V.T. (1997) *Paleoindian Geoarchaeology of the Southern High Plains*. Texas Archaeology and Ethnohistory Series. Austin: University of Texas Press.

Holliday, V.T. (2004). *Soils and Archaeological Research*. Oxford: Oxford University Press.

Holliday, V.T., Hoffecker, J.F., Goldberg, P., Macphail, R.I., Forman, S.L., Anikovich, M. and Sinitsyn, A. (2007). Geoarchaeology of the Kostenki–Borshchevo Sites, Don River Valley, Russia. *Geoarchaeology: An International Journal*, 22, 181–228.

Holliday, V.T., Huckell, B.B., Weber, R.H., Hamilton, M.J., Reitze, W.T. and Mayer, J.H. (2009). Geoarchaeology of the Mockingbird Gap (Clovis) Site, Jornada del Muerto, New Mexico. *Geoarchaeology: An International Journal*, 24, 348–370.

Holmes, S. and Wingate, M. (2002). *Building with lime*, revised edition. Bourton on Dunsmore: Intermediate Technology Publications Ltd.

Horwitz, L.K. and Goldberg, P. (1989). A study of Pleistocene and Holocene hyaena coprolites. *Journal of Archaeological Science*, 16, 71–94.

Horwitz, L.K. and Smith, P. (1988). The effects of striped hyaena activity on human remains. *Journal of Archaeological Science*, 15, 471–481.

Hosfield, R.T. (2011). Rolling stones: Understanding river-rolled Paleolithic artifact assemblages. In: Brown, A.G., Basell, L.S. and Butzer, K.W. (Eds.) *Geoarchaeology, Climate Change, and Sustainability: Geological Society of America Special Paper 476*. Geological Society of America, pp. 37–52.

Howarth, F.G. and Hoch, H. (2012). Adaptive shifts. In: White, W.B. and Culver, D.C. (Eds.) *Encyclopedia of Caves*, 2nd edn. Amsterdam: Elsevier, pp. 9–17.

Huaqing, L. (1989). Effect of chemical adsorption of calcium oxalate monohydrate on DTA baseline. *Thermochimica Acta*, 141, 151–157.

Huckleberry, G. (1999). Assessing Hohokam canal stability through stratigraphy. *Journal of Field Archaeology*, 26, 1–18.

Huckleberry, G. (2001). Archaeological sediments in dryland alluvial environments. In: Stein, J.K. and Farrand, W.R. (Eds.) *Sediments in Archaeological Context*. Salt Lake City: University of Utah Press.

Huisman, D.J., Jongmans, A.G. and Raemaekers, D.C.M. (2009). Investigating Neolithic land use in Swifterbant (NL) using micromorphological techniques. *Catena*, 78, 185–197.

Huisman, D.J., Braadbaart, F., van Wijk, I.M. and van Os, B.J.H. (2012). Ashes to ashes, charcoal to dust: Micromorphological evidence for ash-induced disintegration of charcoal in Early Neolithic (LBK) soil features in Elsloo (The Netherlands). *Journal of Archaeological Science*, 39, 994–1004.

Huisman, D.J., Ngan-Tillard, D., Tensen, M.A., Laarman, F.J. and Raemaekers, D.C.M. (2014). A question of scales: Studying Neolithic subsistence using micro CT scanning of midden deposits. *Journal of Archaeological Science*, 49, 585–594.

Hutchinson, G.E. (1950). *The Biogeochemistry of Vertebrate Excretion*. New York: American Museum of Natural History.

Hutson, S.R., Stanton, T.W., Magnoni, A., Terry, R. and Craner, J. (2007). Beyond the buildings: Formation processes of ancient Maya houselots and methods for the study of non-architectural space. *Journal of Anthropological Archaeology*, 26, 442–473.

Iacovou, M. (2017). Laona tumulus at Palaipaphos. From recognition to method of investigation (in Greek). In: Vlachou V. and Gadolou A. (Eds.) *ΤΕΡΨΙΣ, Studies in Mediterranean Archaeology – In honour of Nota Kourou*. Brussels: CReA-Patrimoine, pp. 317–329.

Incropera, F.P., DeWitt, D.P., Bergman, T.L. and Lavine, A.S. (2007). *Fundamentals of heat and mass transfer*, 6th edn. John Wiley & Sons.

Imeson, A.C. and Kwaad, F.J.P.M. (1980). Gully types and gully prediction. *Geografisch Tijdschrift*, 14, 430–441.

Isaac, G. (1967). Towards the interpretation of occupational debris: Some experiments and observations. *Kroeber Anthropological Society Papers*, 37, 31–57.

Ismail-Meyer, K., Rentzel, P. and Wiemann, P. (2013). Neolithic Lakeshore Settlements in Switzerland: New Insights on Site Formation Processes from Micromorphology. *Geoarchaeology: An International Journal*, 28, 317–339.

Jacobs, Z., Duller, G.A.T., Wintle, A.G. and Henshilwood, C.S. (2006). Extending the chronology of deposits at Blombos Cave, South Africa, back to 140 ka using optical dating of single and multiple grains of quartz. *Journal of Human Evolution*, 51, 255–273.

Jelinek, A.J., Farrand, W.R., Haas, G., Horowitz, A. and Goldberg, P. (1973). New excavations at the Tabun Cave, Mount Carmel, Israel, 1967–1972: A preliminary report. *Paléorient*, 1/2, 151–183.

Joffe, J.S. (1949). *Pedology*, 2nd edn. New Brunswick, NJ: Pedology Publications.

John, J. and Kočar, P. (2009). Trial excavation of a talus cone at the Middle Eneolithic site of Radkovice-Osobovská skála and its archaeobotanical analysis. *Fines Transire*, 18, 209–213.

Johnson, D.L. (1990). Biomantle evolution and the redistribution of earth materials and artifacts. *Soil Science*, 149, 84–102.

Johnson, D.L. (2002). Darwin would be proud: Bioturbation, dynamic denudation, and the power of theory in science. *Geoarchaeology: An International Journal*, 17, 7–40.

Johnson, D.L. and Balek, C.L. (1991). The genesis of Quaternary landscapes with stone-lines. *Physical Geography*, 12, 385–394.

Jones, O. and Fallu, D. (2015). Between a rock and a hard place: Preservation issues of human remains in a Mycenaean Tholos. In: *Archaeological Institute of America Annual Meeting*, 8–11 January 2015.

Joukowsky, M. (1980). *A Complete Manual of Field Archaeology*. Englewood Cliffs, NJ: Prentice-Hall.

Karkanas, P. (2001). Site formation processes in Theopetra cave: A record of climatic change during the Late Pleistocene and early Holocene in Thessaly, Greece. *Geoarchaeology: An International Journal*, 16, 373–399.

Karkanas, P. (2006). Late Neolithic household activities in marginal areas: The micromorphological evidence from the Kouveleiki caves, Peloponnese, Greece. *Journal of Archaeological Science*, 33, 1628–1641.

Karkanas, P. (2007). Identification of lime plaster in prehistory using petrographic methods: A review and reconsideration of the data on the basis of experimental and case studies. *Geoarchaeology: An International Journal*, 22, 775–795.

Karkanas, P. (2010a). Preservation of anthropogenic materials under different geochemical processes: A mineralogical approach. *Quaternary International*, 210, 63–69.

Karkanas, P. (2010b). Geology, stratigraphy and site formation processes of the Upper Palaeolithic and later sequence in Klissoura Cave 1. *Eurasian Prehistory*, 7, 15–36.

Karkanas, P. (2015) Site formation processes in Palamari. In: Parlama L., Theochari, M., Romanou, C. and Bonatsos, S. (Eds.) *The Fortified Settlement in Palamari, Skyros*. Athens: TDPEAE, pp. 201–222 (in Greek with extended English abstract).

Karkanas, P. (2017). Chemical alteration. In: Allan, S.G. (Ed.) *Encyclopedia of Geoarchaeology*. Dordrecht: Springer, pp. 129–139.

Karkanas, P. (in press). Geoarcheological analysis of Mycenae. In: Maggidis, C. (Ed.) *The Lower Town of Mycenae: Geophysical Survey 2003–2012, Volume 1*. Athens: Archaeological Society.

Karkanas, P. and Efstratiou, N. (2009). Floor sequences in Neolithic Makri, Greece: Micromorphology reveals cycles of renovation. *Antiquity*, 83, 955–967.

Karkanas, P. and Goldberg, P. (2008). Micromorphology of sediments: Decipheirng archaeological context. *Israel Journal of Earth Sciences*, 56, 63–71.

Karkanas, P. and Goldberg, P. (2010a). Site formation processes at Pinnacle Point Cave 13B (Mossel Bay, Western Cape Province, South Africa): Resolving stratigraphic and depositional complexities with micromorphology. *Journal of Human Evolution*, 59, 256–273.

Karkanas P. and Goldberg P. (2010b). Phosphatic features. In: Stoops, G., Eswaran, H., Marcelino, V. and Mees, F. (Eds.) *Micromorphological features of soils and regoliths. Their relevance for pedogenic studies and classifications*. Amsterdam: Elsevier, pp. 521–541.

Karkanas, P. and Goldberg, P. (2013). Micromorphology of cave sediments. In: John, F.S. (Editor-in-chief) and Frumkin, A. (Volume Editor). *Treatise on Geomorphology, Volume 6, Karst Geomorphology*. San Diego: Academic Press, pp. 286–297.

Karkanas, P. and Goldberg, P. (2017a). Soil micromorphology. In: Gilbert, A.S. (Ed.) *Encyclopedia of Geoarchaeology*. Dordrecht: Springer, pp. 830–841.

Karkanas, P. and Goldberg, P. (2017b). Cave settings. In: Gilbert, A.S. (Ed.) *Encyclopedia of Geoarchaeology*. Dordrecht: Springer, pp 108–118.

Karkanas, P. and van de Moortel, A. (2014). Micromorphological analysis of sediments at the Bronze Age site of Mitrou, central Greece: Patterns of floor construction and maintenance. *Journal of Archaeological Science*, 43, 198–213.

Karkanas, P., Kyparissi-Apostolika, N., Bar-Yosef, O. and Weiner, S. (1999). Mineral assemblages in Theopetra, Greece: A framework for understanding diagenesis in a prehistoric cave. *Journal of Archaeological Science*, 26, 1171–1180.

Karkanas, P., Bar-Yosef, O., Goldberg, P. and Weiner, S. (2000). Diagenesis in prehistoric caves: The use of minerals that form in situ to assess the completeness of the archaeological record. *Journal of Archaeological Science*, 27, 915–929.

Karkanas, P., Rigaud, J.P., Simek, J.F., Albert, R.M. and Weiner, S. (2002). Ash bones and guano: A study of the minerals and phytoliths in the sediments of Grotte XVI, Dordogne, France. *Journal of Archaeological Science*, 29, 721–732.

Karkanas, P., Koumouzelis, M., Kozlowski, J.K., Sitlivy, V., Sobczyk, K., Berna, F. and Weiner, S. (2004). The earliest evidence for clay hearths: Aurignacian features in Klisoura Cave 1, southern Greece. *Antiquity*, 78, 513–525.

Karkanas, P., Shahack-Gross, R., Ayalon, A., Bar-Matthews, M., Barkai, R., Frumkin, A., Gopher, A. and Stiner, M. C. (2007). Evidence for habitual use of fire at the end of the Lower Paleolithic: Site-formation processes at Qesem Cave, Israel. *Journal of Human Evolution*, 53, 197–212.

Karkanas, P., Schepartz, L.A., Miller-Antonio, S., Wei, W. and Weiwen, H. (2008). Late Middle Pleistocene climate in southwestern China: Inferences from the stratigraphic record of Panxian Dadong Cave, Guizhou. *Quaternary Science Reviews*, 27, 1555–1570.

Karkanas, P., Dabney, M.K., Smith, R., Angus, K. and Wright, J.C. (2012). The goearchaeology of Mycenean chamber tombs. *Journal of Archaeological Science*, 37, 2722–2732.

Karkanas, P., Brown, K.S., Fisher, E.C., Jacobs, Z. and Marean, C.W. (2015a). Interpreting human behavior from depositional rates and combustion features through the study of sedimentary microfacies at site Pinnacle Point 5-6, South Africa. *Journal of Human Evolution*, 85, 1–21.

Karkanas, P., White, D., Lane, S.C., Stringer, C., Davies, W., Cullen, L. V., Smith, C.V., Ntinou, M., Tsartsidou, G. and Kyparissi-Apostolika, N. (2015b). Tephra chronostratigraphy and climatic events between the MIS6/5 transition and the beginning of MIS3 in Theopetra Cave, central Greece. *Quaternary Science Reviews*, 118, 170–181.

Kaufmann, C., Gutiérrez, M.A., Álvarez, M.C., González, M.E. and Massigoge, A. (2011). Fluvial dispersal potential of guanaco bones (*Lama guanicoe*) under controlled experimental conditions: The influence of age classes to the hydrodynamic behavior. *Journal of Archaeological Science*, 38, 334–344.

Keely, H.C.M., Hudson, G.E. and Evans, J. (1977). Trace element contents of human bones in various states of preservation. 1. The soil silhouette. *Journal of Archaeological Science*, 4, 19–24.

Kenyon, K.M. (Ed.) (1981). *Excavations at Jericho, Volume III*. Jerusalem: British School of Archeology.

Kidder, T.R., Arco, L.J., Ortmann, A.L., Schilling, T., Boeke, C., Bielitz, R. and Adelsberger, K.A. (2009). *Poverty Point Mound A: Final Report of the 2005 and 2006 Field Seasons* (unpublished manuscript). Baton Rouge, LA: Louisiana Division of Archaeology and the Louisiana Archaeological Survey and Antiquities Commission, http://anthropology.artsci.wustl.edu/files/anthropology/imce/PP_Md_A_report_final.pdf.

Kilfeather, A.A. and van der Meer, J.J.M. (2008). Pore size, shape and connectivity in tills and their relationship to deformation processes. *Quaternary Science Reviews*, 27, 250–266.

Kipfer, B.A. (2007). *The Archaeologist's Fieldwork Companion.* Malden, MA, Oxford: Blackwell Publications.

Kittel, P. (2014). Slope deposits as an indicator of anthropopressure in the light of research in Central Poland. *Quaternary International*, 324, 34–55.

Knapp, E.P., Terry, D.O., Harbor, D.J. and Thren, R.C. (2004). Reading Virginia's paleoclimate from geochemistry and sedimentology of clastic cave sediments. In: Sasowsky, I.D. and Mylroie, J. (Eds.) *Studies of Cave Sediments: Physical and Chemical Records of Paleoclimate.* New York: Kluwer/Plenum, pp. 95–106.

Knudson, K.J., Frink, L., Hoffman, B.W. and Price, T.D. (2004). Chemical characterization of Arctic soils: Activity area analysis in contemporary Yup'ik fish camps using ICP-AES. *Journal of Archaeological Science*, 31, 443–456.

Kooistra, M.J. and Pulleman, M.M. (2010). Features related to faunal activity. In: Stoops, G., Marcelino, V. and Mees, F. (Eds.) *Interpretation of Micromorphological Features of Soils and Regoliths.* Amsterdam: Elsevier Science, pp. 397–418.

Koromila, G., Karkanas, P., Kotzamani, G., Harris, K., Hamilakis, Y. and Kyparissi-Apostolika, N. (in press). Humans, animals, and the landscape in Neolithic Koutroulou Magoula, Central Greece: An approach through micromorphology and plant remains in dung. In A. Sarris, E. Kalogiropoulou, T. Kalayci and L. Karimali (Eds.) *Communities, Landscapes and Interaction in Neolithic Greece.* Proceedings of International Conference, Rethymno 29–30, May 2015.

Koulidou, S. (1998). *Depositional Patterns in Abandoned Modern Mud-Brick Structures.* Unpublished M.Sc. thesis. University of Sheffield.

Kourampas, N., Simpson, I.A., Perera, N., Deraniyagala, S.U. and Wijeyapala, W.H. (2009). Rockshelter sedimentation in a dynamic tropical landscape: Late Pleistocene–Early Holocene archaeological deposits in Kitulgala Beli-lena, southwestern Sri Lanka. *Geoarchaeology: An International Journal*, 24, 677–714.

Krajcarz, M.T., Cyrek, K., Krajcarz, M., Mroczek, P., Sudoł, M., Szymanek, M., Tomek, T. and Madeyska, T. (2016). Loess in a cave: Lithostratigraphic and correlative value of loess and loess-like layers in caves from the Kraków-Częstochowa Upland (Poland). *Quaternary International*, 399, 13–30.

Krasilnikov, P. and Arnold, R. (2009). French soil classification system. In: Krasilnikov, P., Marti, J.-J.I., Arnold, R., Shoba, S. (Eds.) *A Handbook of Soil Terminology, Correlation and Classification.* London: Earthscan, pp. 102–115.

Krauskopf, K.B. (1979). *Introduction to Geochemistry*, 2nd edn. New York: McGraw-Hill Co.

Krinsley, D.H. and Doornkamp, J.C. (2011). *Atlas of Quartz Sand Surface Textures.* Cambridge: Cambridge University Press.

Krumbein, W.C. and Pettijohn, F.J. (1938). *Manual of Sedimentary Petrography.* New York: Appleton Century-Crofts, Inc.

Krumbein, W.C. and Sloss L.L. (1963). *Stratigraphy and Sedimentation.* San Francisco and London: Freeman.

Kuhn, S.L., Stiner, M.C., Güleç, E., Özer, I., Yilmaz, H., Baykara, I., AçIkkol, A., Goldberg, P., Molina, K.M., Ünay, E. and Suata-Alpaslan, F. (2009). The early Upper Paleolithic occupations at ÜçagIzlI Cave (Hatay, Turkey). *Journal of Human Evolution*, 56, 87–113.

Kühn, P., Pietsch, D. and Gerlach, I. (2010). Archaeopedological analyses around a Neolithic hearth and the beginning of Sabaean irrigation in the oasis of Ma'rib (Ramlat as-Sab'atayn, Yemen). *Journal of Archaeological Science*, 37, 1305–1310.

Kuman, K., Inbar, M. and Clarke, R.J. (1999). Palaeoenvironments and cultural sequence of the Florisbad Middle Stone Age hominid site, South Africa. *Journal of Archaeological Science*, 26, 1409–1425.

Kužvart, M. and Böhmer, M. (1984). Prospecting and exploration of mineral deposits. *Development in Economic Geology 21.* Amsterdam: Elsevier.

Lachniet, M.S., Larson, G.J., Strasser, J.C., Lawson, D.E., Evenson, E.B. and Alley, R.B. (1998). Microstructures of glacigenic sediment-flow deposits, Matanuska Glacier, Alaska. In: Mickelson, D.M. and Attig, J.W. (Eds.) *Glacial Processes Past and Present.* Geological Society of America Special Paper 337. Boulder: Geological Society of America, pp. 45–57.

Lafille, J. (1961). Le gisement dit 'Roc de Marsal', commune de de Campagne du Bugue (Dordogne): Note préliminaire. *Bulletin de la Société Préhistorique Française*, 58, fasc. 11–12, 712–713.

LaMotta, V.M. and Schiffer, M.B. (1999). Formation processes of house floor assemblages. In: Allison, P.M. (Ed.) *The Archaeology of Household Activities.* London: Routledge, pp. 19–29.

Lancelotti, C. and Madella, M. (2012). The 'invisible' product: Developing markers for identifying dung in archaeological contexts. *Journal of Archaeological Science*, 39, 953–963.

Lane, C.S., Cullen, V.L., White, D., Bramham-Lawa, C.W.F. and Smith, V.C. (2014). Cryptotephra as a dating and correlation tool in archaeology. *Journal of Archaeological Science*, 42, 42–50.

Laville, H., Rigaud, J.P. and Sackett, J. (1980). *The Rockshelters of the Perigord.* New York: Academic Press.

Le Tensorer, J.M., Jagher, R., Rentzel, P., Hauck, T., Ismail-Meyer, K., Pümpin, C. and Wojtczak, D. (2007). Long-term site formation processes and the natural springs Nadaouiyeh and Hummal in the El Kowm oasis, central Syria. *Geoarchaeology: An International Journal*, 22, 621–639.

Lehmann, J., Kern, D.C., Glaser, B. and Woods, W.I. (Eds.) (2007). *Amazonian Dark Earths: Origin Properties Management.* New York: Springer Science and Business Media.

Leigh, D.S. (2001). Buried artifacts in sandy soils; techniques for evaluating pedoturbation versus sedimentation. In: Goldberg, P., Holliday, V.T. and Ferring, C.R. (Eds.) *Earth Sciences and Archaeology.* New York: Kluwer-Plenum, pp. 269–293.

Lenoble, A. and Bertran, P. (2004). Fabric of Palaeolithic levels: Methods and implications for site formation processes. *Journal of Archaeological Science*, 31, 457–469.

Lenoble, A., Bertran, P. and Lacrampe, F. (2008). Solifluction-induced modifications of archaeological levels: Simulation based on experimental data from a modern periglacial slope and application to French Palaeolithic sites. *Journal of Archaeological Science*, 35, 99–110.

Lewis, D.W. and McConchie, D. (1995). *Analytical Sedimentology.* New York: Chapman and Hall.

Li, R., Carter, J.A., Xie, S., Zou, S., Gu, Y., Zhu, J. and Xiong, B. (2010). Phytoliths and microcharcoal at Jinluojia archeological site in middle reaches of Yangtze River indicative of paleoclimate and human activity during the last 3000 years. *Journal of Archaeological Science*, 37, 124–132.

Lim, H.S., Lee, Y.I., Yi, S., Kim, C.-B., Chung, C.H., Lee, H.J. and Choi, J.H. (2007). Vertebrate burrows in late Pleistocene paleosols at Korean Palaeolithic sites and their significance as a stratigraphic marker. *Quaternary Research*, 68, 213–219.

Lima, H.N., Schaefer, C.E.R., Mello, J.W.V., Gilkes, R.J. and Ker, J.C. (2002). Pedogenesis and pre-Colombian land use of 'Terra Preta Anthrosols' ('Indian black earths') of western Amazonia. *Geoderma*, 110, 1–17.

Lindholm, R.C. (1987). *A Practical Approach to Sedimentology.* London: Allen and Unwin.

Lisá, L., Hošek, J., Bajer, A., Matys Grygar, T. and Vandenberghe, D. (2013). Geoarchaeology of Upper Palaeolithic loess sites located within a transect through Moravian valleys, Czech Republic. *Quaternary International*, 351, 25–37.

Long, A.J., Waller M.P. and Stupples, P. (2006). Driving mechanisms of coastal change: Peat compaction and the destruction of late Holocene coastal wetlands. *Marine Geology*, 225, 63–84.

Love, S. (2012). The geoarchaeology of mudbricks in architecture: A methodological study from Çatalhöyük, Turkey: *Geoarchaeology: An International Journal*, 27, 140–156.

Lowe, J.J. and Walker, M.J.C. (2015). *Reconstructing Quaternary Environments*, 3rd edn. London: Routledge.

Lucas G. (2005). *The Archaeology of Time.* London: Routledge

Lyell, C. (1863). *Geological Evidences of the Antiquity of Man.* London: J. Murray.

Maca, A.L. (2009). Remembering the basics. Social and stratigraphic debates and biases. *Archaeological Dialogues*, 16, 31–38.

Machado, J. and Pérez, L. (2016). Temporal frameworks to approach human behavior concealed in Middle Palaeolithic palimpsests: A high-resolution example from El Salt Stratigraphic Unit X (Alicante, Spain). *Quaternary International*, 417, 66–81.

Machado, J., Hernández, C.M., Mallol, C. and Galván, B. (2013). Lithic production, site formation and Middle Palaeolithic palimpsest analysis: In search of human occupation episodes at Abric del Pastor Stratigraphic Unit IV (Alicante, Spain). *Journal of Archaeological Science*, 40, 2254–2273.

Macphail, R.I. (1981). Soil and botanical studies of the 'Dark Earth'. In: Jones, M. and Dimbleby, G.W. (Eds.) *The Environment of Man: The Iron Age to the Anglo-Saxon Period.* British Series 87. Oxford: British Archaeological Reports, pp. 309–331.

Macphail, R.I. (1986). Paleosols in archaeology: Their role in understanding Flandrian pedogenesis. In: Wright, V.P. (Ed.) *Paleosols.* Oxford: Blackwell Scientific Publications, pp. 263–290.

Macphail, R.I. (1987). A review of soil science in archaeology in England. In: Keeley, H.C.M. (Ed.) *Environmental Archaeology: A Regional Review, Volume II.* Occasional Paper No. 1. London: Historic Buildings and Monuments Commission for England, pp. 332–379.

Macphail, R.I. (1994). The reworking of urban stratigraphy by human and natural processes. In: Hall, A.R. and Kenward, H.K. (Eds.) *Urban-Rural Connexions: Perspectives from Environmental Archaeology.* Volume Monograph 47. Oxford: Oxbow, pp. 13–43.

Macphail, R.I. (1999). Sediment micromorphology. In: Roberts, M.B. and Parfin, S.A. (Eds.) *Boxgrove: A Middle Pleistocene Hominid site at Eartham Quarry, Boxgrove, West Sussex.* London: English Heritage, pp. 118–148.

Macphail, R.I. (2008). Soils and Archaeology. In: Pearsall, D.M. (Ed.) *Encyclopedia of Archaeology.* New York: Academic Press, pp. 2064–2072.

Macphail, R.I. (2016). House Pits and Grubenhäuser. In: Gilbert, A.S. (Ed.) *Encyclopedia of Geoarchaeology.* Dordrecht: Springer Scientific, pp. 435–432.

Macphail, R.I. (2017). Privies and latrines. In Gilbert, A.S. (Ed.) *Encyclopedia of Geoarchaeology.* Dordrecht: Springer, pp. 682–687.

Macphail, R.I. and Crowther, J. (2007). Soil micromorphology, chemistry and magnetic susceptibility at Huizui (Yiluo region, Henan province, northern China), with special focus on a typical

Yangshao floor sequence. *Indo-Pacific Prehistory Association Bulletin*, 27, 103–113.

Macphail, R.I. and Cruise, G.M. (1998). *Report on the soil micromorphology and chemistry of the Saxon Grubenhäuser and Late Medieval dovecote at Stratton, Bedfordshire*. Bedford: Bedfordshire County Archaeological Service.

Macphail, R.I. and Cruise, G.M. (2001). The soil micromorphologist as team player: A multianalytical approach to the study of European microstratigraphy. In: Goldberg, P., Holliday, V. and Ferring, R. (Eds.) *Earth Science and Archaeology*. New York: Kluwer Academic/ Plenum Publishers, pp. 241–267.

Macphail, R.I. and Goldberg, P. (1990). The micromorphology of tree subsoil hollows: Their significance to soil science and archaeology. In: Douglas, L.A. (Ed.) *Soil-Micromorphology: A Basic and Applied Science. Developments in Soil Science* 19. Amsterdam: Elseveir, pp. 425–429.

Macphail, R.I. and Goldberg, P. (1999). The soil micromorphological investigation of Westbury Cave. In: Andrews, P., Cook, J., Currant, A. and Stringer, C. (Eds.) *Westbury Cave. The Natural History Museum Excavations 1976–1984*. Bristol: CHERUB (Centre for Human Evolutionary Research at the University of Bristol), pp. 59–86.

Macphail, R.I. and Goldberg, P. (2010). Archaeological materials. In: Stoops, G., Marcelino, V. and Mees, F. (Eds.) *Interpretation of Micromorphological Features of Soils and Regoliths*. Amsterdam: Elsevier, pp. 589–622.

Macphail, R.I. and Goldberg, P. (2012). Soil micromorphology of Gibraltar Cave coprolites. In: Barton, R.N.E., Stringer, C.B. and Finlayson, J.C. (Eds.) *Neanderthals in Context*. Oxford: Oxbow Books, pp. 240–242.

Macphail, R., Romans, J.C.C. and Robertson, L. (1987). The application of micromorphology to the understanding of Holocene soil development in the Brithish Isles: With special reference to early cultivation. In: Fedorof, N., Bresson, L.M. and Courty, M.A. (Eds.) *Micromorphologie des sols – Soil micromorphology*. Proceedings of the VIIth International Working Meeting on Soil Micromorphology, July 1985, Paris. Paris: Association Française pour l' Étude du Sol, pp. 647–656.

Macphail, R.I., Courty, M.A. and Gebhardt, A. (1990). Soil micromorphological evidence of early agriculture in north-west Europe. *World Archaeology*, 22, 53–69.

Macphail, R., Courty, M.A., Hather, J. and Wattez, J. (1997). The soil micromorphological evidence of domestic occupation and stabling activities. In: Maggi, R. (Ed.) *Arene Candide: A functional and environmental assessment of the Holocene sequences excavated by L. Bernando 'Brea (1940–1950)*. Roma: Memorie dell'Instituto Italiano di Paleontologia Umana, pp. 53–88.

Macphail, R.I., Galinié, H. and Verhaeghe, F. (2003). A future for dark earth? *Antiquity*, 77, 349–358.

Macphail, R.I., Cruise, G.M., Allen, M.J., Linderholm, J. and Reynolds, P. (2004). Archaeological soil and pollen analysis of experimental floor deposits; with special reference to Butser Ancient Farm, Hampshire, UK. *Journal of Archaeological Science*, 31, 175–191.

Macphail, R.I., Crowther, J. and Cruise, J. (2007). Micromorphology and post-Roman town research: The examples of London and Magdeburg. In: Henning, J. (Ed.) *Post-Roman towns, trade, and settlement in Europe and Byzantium, Volume 1, The Heirs of the Roman West*. Berlin: Walter de Gruyter, pp. 303–317.

Macphail, R.I., Haită, C., Bailesy, D.W., Andreescu, R. and Mirea, P. (2008). The soil micromorphology of enigmatic Early Neolithic pit-features at Măgura, southern Romania. *Studii de Preistorie*, 5, 61–77.

Macphail, R.I., Allen, M.J., Crowther, J., Cruise, G.M. and Whittaker, J.E. (2010). Marine inundation: Effects on archaeological features, materials, sediments and soils. *Quaternary International*, 214, 44–55.

Macphail, R.I., Cruise, G.M., Courty, M.A., Crowther, J. and Linderholm, J. (2016). E6 Gudbrandsdalen Valley Project (Brandrud, Fryasletta, Grytting and Øybrekka), Oppland, Norway: Soil micromorphology (with selected microchemistry, bulk soil chemistry, carbon polymer, particle size and pollen analyses). In: Gundersen, I.M. (Ed.) *Gårdogutmark i Gudbrandsdalen*. Arkeologiske undersøkelser i Fron 2011–2012. Kristiansand: Portal Forlag, pp. 304–317.

Majewsky, M. (2014). Human impact on Subatlantic slopewash processes and landform development at Lake Jasień (northern Poland). *Quaternary International*, 324, 56–66.

Malinsky-Buller, A., Hovers, E. and Marder, O. (2011). Making time: 'Living floors', 'palimpsests' and site formation processes – A perspective from the open-air Lower Paleolithic site of Revadim Quarry, Israel. *Journal of Anthropological Archaeology*, 30, 89–101.

Mallol, C. (2006). What's in a beach? Soil micromorphology of sediments from the Lower Paleolithic site of 'Ubeidiya, Israel. *Journal of Human Evolution*, 51, 185–206.

Mallol, C. and Henry, A. (2017). Ethnoarchaeology of Paleolithic fire: Methodological considerations. *Current Anthropology*, 58, S217–S229.

Mallol, C. and Hernández, C. (2016). Advances in palimpsest dissection. *Quaternary International*, 417, 1–2.

Mallol, C. and Mentzer, S. (2015). Contacts under the lens: Perspectives on the role of microstratigraphy in archaeological research. *Archaeological and Anthropological Sciences*, 1–25. doi.org/10.1007/s12520-015-0288-6.

Mallol, C., Marlowe, F.W., Wood, B.M. and Porter, C.C. (2007). Earth, wind, and fire: Ethnoarchaeological signals of Hadza fires. *Journal of Archaeological Science*, 34, 2035–2052.

Mallol, C., Mentzer, S.M. and Wrinn, P.J. (2009). A micromorphological and mineralogical study of site formation processes at the Late Pleistocene site of Obi-Rakhmat, Uzbekistan. *Geoarchaeology: An International Journal*, 24, 548–575.

Mallol, C., Cabanes, D. and Baena, J. (2010). Microstratigraphy and diagenesis at the upper Pleistocene site of Esquilleu Cave (Cantabria, Spain). *Quaternary International*, 214, 70–81.

Mallol, C., Hernández, C.M., Cabanes, D., Machado, J., Sistiaga, A., Pérez, L. and Galván, B. (2013). Human actions performed on simple combustion structures: An experimental approach to the study of Middle Palaeolithic fire. *Quaternary International*, 315, 3–15.

Mandel, R.D. (2008). Buried Paleoindian-age landscapes in stream valleys of the Central Plains, USA. *Geomorphology*, 101, 342–361.

Mandel, R. and Bettis, E.A. III (2001). Use and analysis of soils by archaeologists and geoscientists; a North American perspective. In: Goldberg, P., Holliday, V.T. and Ferring, C.R. (Eds.) *Earth Sciences and Archaeology*. New York: Kluwer Academic/Plenum Publishers, pp. 173–204.

March, R.J., Lucquin A., Joly, D., Ferreri, J.C. and Muhieddine, M. (2014). Processes of formation and alteration of archaeological fire structures: Complexity viewed in the light of experimental approaches. *Journal of Archaeological Method and Theory*, 21, 1–45.

Marra, A.C., Villa, P., Beauval, C., Bonfiglio, L. and Goldberg, P. (2004). Same predator, variable prey: Taphonomy of two Upper Pleistocene hyena dens in Sicily and SW France. *Revue de Paleobiologie*, 23, 787–801.

Marriner, N. (2009). Currents and trends in the archaeological sciences. *Journal of Archaeological Science*, 37, 2811–2815.

Martini, I. (2011). Cave clastic sediments and implications for speleogenesis: New insights from the Mugnano Cave (Montagnola Senese, Northern Apennines, Italy). *Geomorphology*, 134, 452–460.

Massari, F. and Parea, G.C. (1998). Progradational gravel beach sequences in a moderate- to high-energy, microtidal marine environment. *Sedimentology*, 35, 881–913.

Matarazzo, T. (2015). *Micromorphological analysis of activity areas sealed by Vesuvius; Avellino Eruption: The Early Bronze Age Village of Afragola in Southern Italy*. Oxford: Archaeopress Archaeology.

Matarazzo, T., Berna, F. and Goldberg, P. (2010). Occupation surfaces sealed by the Avellino eruption of Vesuvius at the Early Bronze Age Village of Afragola in Southern Italy: A micromorphological analysis. *Geoarchaeology: An International Journal*, 25, 437–466.

Matthews, W. (1995). Micromorphological characterisation of occupation deposits and microstratigraphic sequences at Abu Salabikh, Southern Iraq. In: Barham, A.J. and Macphail, R.I. (Eds.) *Archaeological Sediments and Soils: Analysis, Interpretation and Management*. London: Institute of Archaeology, University College, pp. 41–76.

Matthews, W. (2005). Life-cycle and life-course of buildings. In: Hodder, I. (Ed.) *Çatalhöyük Perspectives. Reports from the 1995–1999 seasons by members of the Çatalhöyük team*. Cambridge: McDonald Institute Monographs and British School of Ankara, pp. 125–149.

Matthews, W. (2010). Geoarchaeology and taphonomy of plant remains and microarchaeological residues in early urban environments in the Ancient Near East. *Quaternary International*, 214, 98–113.

Matthews, W. and Postage, J.N. (1994). The imprint of living in an early Mesopotamian city: Questions and answers. In: Luff, R. and Rowley-Conway, P. (Eds.) *Whither Environmental Archaeology?* Oxbow Monograph 38. Oxford: Oxbow, pp. 171–212.

Matthews, W., French, C., Lawrence, T. and Cutler, D.F. (1996). Multiple surfaces: The micromorphology. In: Hodder, I. (Ed.) *On the Surface: Catalhoyuk 1993–95*. Cambridge: MacDonald Institute for Research and British Institute of Archaeology of Ankara, pp. 301–342.

Matthews, W., French, C.A.I., Lawrence, T., Cutler, D.F. and Jones, M.K. (1997). Microstratigraphic traces of site formation processes and human activities. *World Archaeology*, 29, 281–308.

Maureille, B., Mann, A., Beauval, C., Bordes, J.G., Bourguignon, L., Costamagno, S., Couchoud, I., Fauquignon, J., Garralda, M.D., Geigl, E.M., Grün, R., Guibert, P., Lacrampe-Cuyaubère, F., Laroulandie, V., Marquet, J.C., Meignen, L., Mussini, C., Rendu, W., Royer, A., Seguin, G., Tixier, J-P. (2010). Les Pradelles à Marillac-le-Franc (Charente). Fouilles 2001–2007: Nouveaux résultats et synthèse. In: Buisson-Catil, J. and Primault, J. (Eds.) *Préhistoire entre Vienne et Charente. Hommes et Sociétés du Paléolithique, Volume mém. XXVIII*. Chauvigny: Association des Publications Chauvinoises, pp. 145–162.

Maureille, B., Holliday, T., Royer, A., Pelletier, M., Couture-Veschambre, C., Discamps, E., Gómez-Olivencia, A., Lahaye, C., Le Gueut, E. and Lacrampe-Cuyaubère, F. (2015). New data on the possible Neandertal burial at Regourdou (Montignac-sur-Vézère, Dordogne, France). *XXXVIème rencontres internationales d'Archéologie et d'Histoire d'Antibes*. Editions APDCA, pp. 175–191.

Mazza, P.P.A. and Ventra, D. (2011). Pleistocene debris-flow deposition of the hippopotamus-bearing Collecurti bonebed (Macerata, Central Italy): Taphonomic and paleoenvironmental analysis. *Palaeogeography, Palaeoclimatology, Palaeoecology*, 310, 296–314.

McAnany, P.A. and Hodder, I. (2009a). Thinking about stratigraphic sequence in social terms. *Archaeological Dialogues*, 16, 1–22.

McAnany, P.A. and Hodder, I. (2009b). Thinking about archaeological excavation in reflexive terms. *Archaeological Dialogues*, 16, 41–49.

McBrearty, S. (1990). Consider the humble termite: Termites as agents of post-depositional disturbance at African archaeological sites. *Journal of Archaeological Science*, 17, 111–143.

McCubbin, D.G. (1982). Barrier-Island and Strand Plain Facies. In: Scholle, P.A. and Spearing, D. (Eds.) *Sandstone Depositional Environments*. AAPG Memoir 3, pp. 247–280.

McDowell, R.W. and Stewart, I. (2005). Phosphorus in fresh and dry dung of grazing dairy cattle, deer, and sheep: Sequential fraction and phosphorus-31 nuclear magnetic resonance analyses. *Journal of Environmental Quality*, 34, 598–607.

McGowan, G. and Prangnell, J. (2006). The significance of vivianite in archaeological settings. *Geoarchaeology: An International Journal*, 21, 93–111.

McIntosh, R.J. (1974). Archaeology and mud wall decay in a West African village. *World Archaeology*, 6, 154–171.

McIntosh, R.J. (1977). The excavation of mud structures: An experiment from West Africa. *World Archaeology*, 9, 185–199.

McPherron, S. (2005). Artifact orientations and site formation processes from total station proveniences. *Journal of Archaeological Science*, 32, 1003–1014.

Mees, F., Verschuren, D., Nijs, R. and Dumont, H. (1991). Holocene evolution of the crater lake at Malha, Northwest Sudan. *Journal of Paleolimnology*, 5, 227–253.

Meignen, L., Bar-Yosef, O., Goldberg, P. and Weiner, S. (2001) Le feu au Paléolithique moyen: Recherches sur les structures de combustion et le statut des foyers. *L'exemple du Proche-Orient. Paléorient*, 26, 9–22.

Meignen, L., Goldberg, P. and Bar-Yosef, O. (2007). The hearths at Kebara Cave and their role in site formation processes. In: Bar-Yosef, O. and Meignen, L. (Eds.) *Kebara Cave Mt. Carmel, Israel. The Middle and Upper Paleolithic Archaeology. Part I*. American School of Prehistoric Research Bulletin 49, Peabody Museum of Archaeology and Ethnology, Harvard University, pp. 91–122.

Meignen, L., Goldberg, P., Albert, R.M. and Bar-Yosef, O. (2009). Structures de combustion, choix des combustibles et degré de mobilité des groupes dans le Paléolithique moyen du Proche-Orient: Exemples des grottes de Kébara et d'Hayonim (Israël). In: Théry-Parisot, I., Costamagno, S. and Henry, A. (Eds.) *Gestion des combustibles au Paléolithique et Mésolithique: Nouveaux outils, nouvelles interprétations/Fuel management during the Paleolithic and Mesolithic period: New tools, new interpretations*. BAR International Series 1914. Oxford: Archeopress, pp. 111–118.

Mentzer, S.M. (2011). *Macro- and Micro-Scale Geoarchaeology of Üçağızlı Caves I and II, Hatay, Turkey*. Ph.D. Thesis, University of Arizona.

Mentzer, S.M. (2014). Microarchaeological approaches to the identification and interpretation of combustion features in prehistoric archaeological sites. *Journal of Archaeological Method and Theory*, 21, 616–688.

Mentzer, S.M. (2017b). Hearths and combustion features. In: Gilbert, A.S. (Ed.) *Encyclopedia of Geoarchaeology*. Dordrecht: Springer, pp. 411–425.

Mentzer, S.M. (2017a). Rockshelter settings. In: Gilbert, A.S. (Ed.) *Encyclopedia of Geoarchaeology*. Dordrecht: Springer, pp. 725–741.

Mentzer, S.M. and Quade, J. (2013). Compositional and isotopic analytical methods in archaeological micromorphology. *Geoarchaeology: An International Journal*, 28, 87–97.

Mentzer, S. Romano, D.G. and Voyatzis, M.E. (2017). Micromorphological contributions to the study of the ritual behavior at the ash altar to Zeus on Mt. Lykaion, Greece. *Archaeological and Anthropological Sciences*, 9, 1017–1043.

Menzies, J. and Zaniewski, K. (2003). Microstructures within a modern debris flow deposit derived from Quaternary glacial diamicton – a comparative micromorphological study. *Sedimentary Geology*, 157, 31–48.

Miall, A.D. (2016). *Stratigraphy: A modern synthesis*. Dordrecht: Springer.

Middleton, G.V. (1973). Johannes Walther's Law of the Correlation of Facies. *Geological Society of America Bulletin*, 84, 979–988.

Middleton, W.D. (2004). Identifying chemical activity residues on prehistoric house floors: A methodology and rationale for multi-elemental characterization of a mild acid extract of anthropogenic sediments. *Archaeometry*, 46, 47–65.

Middleton, W.D. and Price, T.D. (1996). Identification of activity areas by multi-element characterization of sediments from modern and archaeological house floors using inductively coupled plasma-atomic emission spectroscopy. *Journal of Archaeological Science*, 23, 673–687.

Miedema, R., Jongmans, A.G. and Slager, S. (1974), Micromorphological observations on pyrite and its oxidation products in four Holocene alluvial soils in the Netherlands. In: Rutherford, G.K. (Ed.) *Soil Microscopy*. Kingston, Ontario: The Limestone Press, pp. 772–794.

Miedema, R., Bal, L. and Pons, L.J. (1976). Morphological and physio-chemical aspects of three soils developed in peat in The Netherlands and their classification. *Netherlands Journal of Agricultural Science*, 24, 247–265.

Mikkelsen, J.H., Langohr, R. and Macphail, R.I. (2007). Soilscape and land-use evolution related to drift sand movements since the Bronze Age in Eastern Jutland, Denmark. *Geoarchaeology: An International Journal*, 22, 155–180.

Milek, K.B. (2012). Floor formation processes and the interpretation of site activity areas: An ethnoarchaeological study of turf buildings at Thverá, northeast Iceland. *Journal of Anthropological Archaeology*, 31, 119–137.

Milek, K.B. and French, C.A.I. (2007). Soils and sediments in the settlement and harbour at Kaupang. In: Skre, D. (Ed.) *Kaupang in Skiringssal*. Aarhus: Aarhus University Press, pp. 321–360.

Milek, K.B. and Roberts, H.M. (2013). Integrated geoarchaeological methods for the determination of site activity areas: A study of a Viking Age house in Reykjavik, Iceland. *Journal of Archaeological Science*, 40, 1845–1865.

Miller, N.F. (1984). The use of dung as fuel: An ethnographic example and an archaeological application. *Paléorient*, 10, 71–79.

Miller, C.E. (2015). *A tale of two Swabian caves: Geoarchaeological investigatios at Hohle Fels and Geißenklösterle*. Tübingen: Kerns Verlag.

Miller, C.E. and Sievers, C. (2012). An experimental micromorphological investigation of bedding construction in the Middle Stone Age of Sibudu. *South Africa. Journal of Archaeological Science*, 39, 3039–3051.

Miller, C.E., Conard, N.J., Goldberg, P. and Berna, F. (2010). *Dumping, sweeping and trampling: Experimental micromorphological analysis of anthropogenically modified combustion features*. Palethnologie: Revue Francophone en Préhistoire, http://www.palethnologie.org.

Miller, C.E., Goldberg, P. and Berna, F. (2013). Geoarchaeological investigations at Diepkloof Rock Shelter, Western Cape, South Africa. *Journal of Archaeological Science*, 40, 3432–3452.

Mills, B.J. (2009). From the ground up. Depositional history, memory and materiality. *Archaeological Dialogues*, 16, 38–40.

Moeyersons, J. (1978). The behavior of stones and stone implements, buried in consolidating and creeping Kalahari Sands. *Earth Surface Processes*, 3, 115–128.

MoLAS, (1994). *Archaeological Site Manua*, 3rd edn. London: Museum of London Archaeological Service.

Moore, A.M.T. (2000). The building and layout of Abu Hureyra 2. In: Moore, A.M.T., Hillman, G.C. and Legge, A.J. (Eds.) *Village on the Euphrates*. Cambridge: Oxford University Press, pp. 261–275.

Moorey, P.R.S. (1994). *Ancient Mesopotamian Materials and Industries: The archaeological evidence*. Oxford: Clarendon Press.

Morin, E. (2006). Beyond stratigraphic noise: Unraveling the evolution of stratified assemblages in faunalturbated sites. *Geoarchaeology: An International Journal*, 21, 541–565.

Morley, M.W., Goldberg, P., Sutikna, T., Tocheri, M.W., Prinsloo, L.C., Jatmiko, Saptomo, E.W., Wasisto, S. and Roberts, R.G. (2017). Initial micromorphological results from Liang Bua, Flores (Indonesia): Site formation processes and hominin activities at the type locality of Homo floresiensis. *Journal of Archaeological Science*, 77, 125–142.

Movius, H.L. (1966). The hearths of the Upper Perigordian and Aurignacian horizons at the Abri Pataud, Les Eyzies (Dordogne), and their possible significance. *American Anthropologist*, 68, 296–325.

Mücher, H.J. and De Ploey, J. (1977). Experimental and micromorphological investigation of erosion and redeposition of loess by water. *Earth Surface Processes*, 2, 117–124.

Mücher, H.J., Ploey, J.D. and Savat, J. (1981). Response of loess materials to simulated translocation by water: Micromorphological observations. *Earth Surface Processes and Landforms*, 6, 331–336.

Mücher, H.J., van Steijn, H. and Kwaad, F. (2010). Colluvial and mass wasting deposits. In: Stoops, G., Marcelino, V. and Mees, F. (Eds.) *Interpretation of Micromorphological Features of Soils and Regoliths*. Amsterdam: Elsevier, pp. 37–48.

Münzel, S.C., Stiller, M., Hofreiter, M., Mittnik, A., Conard, N.J. and Bocherens, H. (2011). Pleistocene bears in the Swabian Jura (Germany): Genetic replacement, ecological displacement, extinctions and survival. *Quaternary International*, 245, 225–237.

Mysterud, I. (1983). Characteristics of summer beds of European brown bears in Norway. *Bears: Their Biology and Management*, 5, 208–222.

Nadel, D., Weiss, E., Simchoni, O., Tsatskin, A., Danin, A. and Kislev, M. (2004). Stone Age hut in Israel yields world's oldest evidence of bedding. *Proceedings of the National Academy of Sciences*, 101, 6821–6826.

Nadon, G.C. and Issler, D.R. (1997). The compaction of floodplain sediments: Timing, magnitude and implications. *Geoscience Canada*, 24, 37–43.

Nemec, W. and Kazanci, N. (1999). Quaternary colluvium in west-central Anatolia: Sedimentary facies and palaeoclimatic significance. *Sedimentology*, 46, 139–170.

Nemec, W. and Steel, R.J. (1984). Alluvial and coastal conglomerates: Their significant features and some comments on gravelly mass-flow deposits. In: Koster, E.H. and Steel, R.J. (Eds.) *Sedimentology of Gravels and*

Conglomerates. Canadian Society of Petroleum Geologists Memoir 10, pp. 1–31.

Newell, R. (1987). Reconstruction of the partitioning and utilization of outside space in a late prehistoric/early historic Inupiat village. In: Kent, S. (Ed.) *Methods and Theory for Activity Area Research.* New York: Columbia University Press, pp. 107–235.

Nicholson, R.A. (1993). A morphological investigation of burnt animal bone and an evaluation of its utility in archaeology. *Journal of Archaeological Science*, 20, 411–428.

Nicholson, P. (2010). Kilns and firing structures. In: Wendrich, W. (Ed.) *UCLA Encyclopedia of Egyptology 1.* Los Angeles: UCLA, http://digital2.library.ucla.edu/viewItem.do?ark=21198/zz0025sr24.

Nicosia, C. and Devos, Y. (2014). Urban dark earth. In: Smith, C. (Ed.) *Encyclopedia of Global Archaeology.* Dordrecht: Springer, pp. 7532–7540.

Nicosia, C., Langohr, R., Mees, F., Arnoldus-Huyzendveld, A., Bruttini, J. and Cantini, F. (2012). Medieval Dark Earth in an active alluvial setting from the Uffizi gallery complex in Florence, Italy. *Geoarchaeology: An International Journal*, 27, 105–122.

Nielsen, A.E. (1991). Trampling the archaeological record: An experimental study. *American Antiquity*, 56, 483–503.

Nodarou, E., Frederick, C. and Hein, A. (2008). Another (mud)brick in the wall: Scientific analysis of Bronze Age earthen construction materials from East Crete. *Journal of Archaeological Science*, 35, 2997–3015.

Ntinou, M. and Kyparissi-Apostolika, N. (2016). Local vegetation dynamics and human habitation from the last interglacial to the early Holocene at Theopetra cave, central Greece: The evidence from wood charcoal analysis. *Vegetation History and Archaeobotany*, 25, 191–206.

O'Brien, N.R. (1987). The effects of bioturbation on the fabric of shale. *Journal of Sedimentary Petrology*, 57, 449–445.

O'Brien, M.J. and Lyman, R.L. (2002). *Seriation, Stratigraphy, and Index Fossils. The Backbone of Archaeological Dating.* New York: Kluwer Academic Publishers.

O'Connor, S., Barham, A., Aplin, K. and Maloney, T. (2017). Cave stratigraphies and cave breccias: Implications for sediment accumulation and removal models and interpreting the record of human occupation. *Journal of Archaeological Science*, 77, 143–159.

Olszewski, D.I., Dibble, H.L., Schurmans, U.A., Mcpherron, S.P. and Smith, J.R. (2005). High desert paleolithic survey at Abydos, Egypt. *Journal of Field Archaeology*, 30, 283–303.

Ortmann, A.L. and Kidder, T.R. (2013). Building Mound A at Poverty Point, Louisiana: Monumental public architecture, ritual practice, and implications for hunter-gatherer complexity. *Geoarchaeology: An International Journal*, 28, 66–86.

Oswald, D. (1987). The organization of space in residential buildings: A cross-cultural perspective. In: Kent, S. (Ed.) *Methods and Theory for Activity Area Research.* New York: Columbia University Press, pp. 295–344.

Overstreet, D.F. and Kolb, M.F. (2002). Geoarchaeological contexts for Late Pleistocene archaeological sites with human-modified woolly mammoth remains in southeastern Wisconsin, USA. *Geoarchaeology: An International Journal*, 18, 91–114.

Pagliai, M. and Stoops, G. (2010). Physical and biological surface crusts and seals. In: Stoops, G., Marcelino, V. and Mees, F. (Eds.) *Interpretation of Micromorphological Features of Soils and Regoliths.* Amsterdam: Elsevier, pp. 419–440.

Papadopoulos, J.K., Bejko, L. and Morris, S.P. (2008). Reconstructing the prehistoric burial tumulus of Lofkend in Albania. *Antiquity*, 82, 686–701.

Parkinson, W.A. Gyucha, A., Karkanas, P., Papadopoulos, N., Tsartsidou, G., Sarris, A., Duffy, R.R. and Yerkes, R. W. (in press). A landscape of tells: Geophysics and microstratigraphy at two Neolithic tell sites on the Great Hungarian Plain. *Journal of Archaeological Science: Reports*, doi.org/10.1016/j.jasrep.2017.07.002.

Parnell, J.J., Terry, R.E. and Nelson, Z. (2002). Soil chemical analysis applied as an interpretive tool for ancient human activities in Piedras Negras, Guatemala. *Journal of Archaeological Science*, 29, 379–404.

Paul, M.A. and Barras, B.F. (1998). A geotechnical correction for post-depositional sediment compression: Examples from the Forth valley, Scotland. *Journal of Quaternary Science*, 13, 171–176.

Pavlopoulos, K., Karkanas, P., Triantaphyllou, M., Karymbalis, E., Tsourou, Th., and Palyvos, N. (2006). Palaeoenvironmental evolution of the coastal plain of Marathon, Greece, during the Late Holocene: Deposition environment, climate and sea-level changes. *Journal of Coastal Research*, 22, 424–438.

Pécsi, M. (1990). Loess is not just the accumulation of dust. *Quaternary International*, 7–8, 1–21.

Perry, C. and Taylor, K. (Eds.) (2007). *Environmental Sedimentology.* Oxford: John Wiley & Sons.

Petraglia, M.D. (1993). The genesis and alteration of archaeological patterns at the Abri Dufaure, an Upper Paleolithic rockshelter and slope site in southwestern France. In: Goldberg, P., Nash, D.T. and Petraglia, M.D. (Eds.) *Formation Processes in Archaeological Context.* Madison: Prehistory Press WI, pp. 97–112.

Petraglia, M.D. and Potts, R. (1994). Water flow and the formation of early Pleistocene artifact sites in Olduvai Gorge, Tanzania. *Journal of Anthropological Archaeology*, 13, 228–254.

Peyrony, D. (1934). *La Ferrassie. Moustérien, Périgordien, Aurignacien.* Paris: Editions Leroux.

Phillips, E. (2006). Micromorphology of a debris flow deposit: Evidence of basal shearing, hydrogracturing, liquefaction and rotational deformation during emplacement. *Quaternary Science Reviews*, 25, 720–738.

Picard, M.D. and High, L.R. (1973). Sedimentary structures of ephemeral streams. *Developments in Sedimentology 17*. Amsterdam: Elsevier.

Pickering, R. and Kramers, J.D. (2010). Re-appraisal of the stratigraphy and determination of new U-Pb dates for the Sterkfontein hominin site, South Africa. *Journal of Human Evolution*, 59, 70–86.

Pierson, T.C. and Scott, K.M. (1985). Downstream dilution of a lahar: Transition from debris flow to hyperconcentrated streamflow. *Water Resource Research*, 21, 1511–1524.

Pietsch, D. (2013). Krotovinas – soil archives of steppe landscape history. *Catena*, 104, 257–264.

Pietsch, D., Schenk, K., Japp, S. and Schnelle, M. (2013). Standardised recording of sediments in the excavation of the Sabaean town of Sirwah, Yemen. *Journal of Archaeological Science*, 40, 2430–2445.

Pietsch, D., Kühn, P., Lisitsyn, S., Markova, A. and Sinitsyn, A. (2014). Krotovinas, pedogenic processes and stratigraphic ambiguities of the Upper Palaeolithic sites Kostenki and Borshchevo (Russia). *Quaternary International*, 324, 172–179.

Piperno, D.R. (1988). *Phytolith Analysis: An Archaeological and Geological Perspective.* San Diego: Academic Press.

Plint, G.A. (2014). Mud dispersal across a Cretaceous prodelta: Storm-generated, wave-enhanced sediment gravity flows inferred from mudstone microtexture and microfacies. *Sedimentology*, 61, 609–647.

Poduska, K.M., Regev, L., Boaretto, E., Addadi, L., Weiner, S., Kronik, L. and Curtarolo, S. (2011). Decoupling local disorder and optical effects in infrared spectra: Differentiating between calcites with different origins. *Advanced Materials*, 23, 550–554.

Poinar, H.N, Hofreiter, M., Spaulding, W.G., Martin, P.S., Stankiewicz, B.A., Bland, H., Evershed, R.P., Possnert, G. and Pääbo. S. (1998). Molecular coproscopy: Dung and diet of the extinct ground sloth *Nothrotheriops shastensis*. *Science*, 281, 402–406.

Polo-Díaz, A., Benito-Calvo, A., Martínez-Moreno, J. and Mora Torcal, R. (2016). Formation processes and stratigraphic integrity of the Middle-to-Upper Palaeolithic sequence at Cova Gran de Santa Linya (southeastern Prepyrenees of Lleida, Iberian Peninsula). *Quaternary International*, 417, 16–38.

Pope, M. and Roberts, M. (2005). Observations on the relationship between Palaeolithic individuals and artefact scatters at the Middle Pleistocene site of Boxgrove, UK. In: Gamble, C. and Porr, M. (Eds.) *The Hominid Individual in Context. Archaeological Investigations of Lower and Middle Palaeolithic Landscapes. Locales and Artefacts.* London and New York: Routledge, pp. 81–97.

Pope, R.J.J., Wilkinson, K.N. and Millington, A.C. (2003). Human and climatic impact on Late Quaternary deposition in the Sparta basin piedmont: Evidence from alluvial fan systems. *Geoarchaeology: An International Journal*, 18, 685–724.

Pope, R., Wilkinson, K., Skourtsos, E., Triantaphyllou, M. and Ferrier, G. (2008). Clarifying stages of alluvial fan evolution along the Sfakian piedmont, southern Crete: New evidence from analysis of post-incisive soils and OSL dating. *Geomorphology*, 94, 206–225.

Postma, G. (1986). Classification for sediment gravity-flow deposits based on flow conditions during sedimentation. *Geology*, 14, 291–294.

Prentiss, W.C., Lenert, M., Foor, T.A., Goodale, N.B. and Schlegel, T. (2003). Calibrated radiocarbon dating at Keatley Creek: The chronology of occupation at a complex hunter-gatherer village. *American Antiquity*, 68, 719–735.

Prentiss, W.C., Lenert, M., Foor, T.A. and Goodale, N.B. (2005). The emergence of complex hunter-gatherers on the Canadian Plateau: A response to Hayden. *American Antiquity*, 70, 175–180.

Prentiss, A.M., Lyons, N., Harris, L.E., Burns, M.R. and Godin, T.M. (2007). The emergence of status inequality in intermediate scale societies: A demographic and socio-economic history of the Keatley Creek site, British Columbia. *Journal of Anthropological Archaeology*, 26, 299–327.

Purdue, L., Miles, W., Woodson, K., Darling, A. and Berger, J.F. (2010). Micromorphological study of irrigation canal sediments: Landscape evolution and hydraulic management in the middle Gila River valley (Phoenix Basin, Arizona) during the Hohokam occupation. *Quaternary International*, 216, 129–144.

Pye, K. (1995). The nature, origin and accumulation of loess. *Quaernary Science Reviews*, 14, 653–667.

Pye, K. and Tsoar, H. (2009). *Aeolian Sand and Sand Dunes.* Berlin: Springer.

Quattropani, L., Charlet, L., De Lumley, H. and Menu, M. (1999). Early Palaeolithic bone diagenesis in the Arago Cave at Tautavel, France. *Mineralogical Magazine*, 63, 801–812.

Rapp, G. Jr. and Hill, C.L. (2006). *Geoarchaeology. The Earth-Science Approach to Archaeological Interpretation*, 2nd edn. New Haven: Yale University Press.

Reading, H.G. (Ed.) (2009). *Sedimentary Environments: Processes, Facies and Stratigraphy*, 3rd edn. Oxford: John Wiley & Sons.

Reading, H.G. and Collinson, J.D. (1996). Clastic coasts. In: Reading, H.G. (Ed.) *Sedimentary Environments:*

Processes, Facies and Stratigraphy, 3rd edn. New York: Wiley, pp. 154–228.

Reddy, S.N. (1998). Fueling the hearths in India: The role of dung in paleoethnobotanical interpretation. *Paleorient*, 24, 61–69.

Regev, L., Zukerman, A., Hitchcock, L., Maeir, A., Weiner, S. and Boaretto, E. (2010). Iron Age hydraulic plaster from Tell es-Safi/Gath, Israel. *Journal of Archaeological Science*, 37, 3000–3009.

Regev, L., Cabanes, D., Homsher, R., Kleiman, A., Weiner, S., Finkelstein, I. and Shahack-Gross, R. (2015). Geoarchaeological investigation in a domestic Iron Age quarter, Tel Megiddo, Israel. *Bulletin of the American Schools of Oriental Research*, 374, 135–157.

Reineck, H.E. and Singh, I.B. (1980). *Depositional Sedimentary Environments: With Reference to Terrigenous Clastics*. Berlin: Springer-Verlag.

Reinhard, K.J. and Bryant, V.M. (2002). *Coprolite analysis: A biological perspective on archaeology*. Lincoln: School of Natural Resources, University of Nebraska, pp. 245–287.

Rellini, I., Firpo, M., Martino, G., Riel-Salvatore, J. and Maggi, R. (2013). Climate and environmental changes recognized by micromorphology in Palaeolithic deposits at Arene Candide (Liguria, Italy). *Quaternary International*, 315, 42–55.

Rendu, W., Beauval, C., Crevecoeur, I., Bayle, P., Balzeau, A., Bismuth, T., Bourguignon, L., Delfour, G., Faivre, J.P. and Lacrampe-Cuyaubère, F. (2014). Evidence supporting an intentional Neandertal burial at La Chapelle-aux-Saints. *Proceedings of the National Academy of Sciences*, 111, 81–86.

Rendu, W., Beauval, C., Crevecoeur, I., Bayle, P., Balzeau, A., Bismuth, T., Bourguignon, L., Delfour, G., Faivre, J.-P., Lacrampe-Cuyaubère, F., Muth, X., Pasty, S., Semal, P., Tavormina, C., Todisco, D., Turq, A. and Maureille, B. (2016). Let the dead speak…comments on Dibble et al.'s reply to 'Evidence supporting an intentional burial at La Chapelle-aux-Saints'. *Journal of Archaeological Science*, 69, 12–20.

Renfrew, C. (1976). Archaeology and the earth sciences. In: Davidson, D.A. and Shackley, M.L. (Eds.) *Geoarchaeology: Earth Science and the Past*. London: Duckworth, pp. 1–5.

Renfrew, C. and Bahn, P. (1991). *Archaeology: Theories, Methods and Practice*. London: Thames and Hudson.

Renfrew C. and Bahn, P. (2005). *Archaeology, the Key Concepts*. London: Routledge.

Rentzel, P. (1998). Ausgewähite Grubenstrukturen aus spätlatènezeitlichen Fundstelle Basel-Gasfabrik: Geoarchäologische interpretation der Grubenfüllungen. *Jahresbericht der Archäologischen Bodenforschung des Kantons Basel-Stadt 1995*. Basel.

Rentzel, P. and Narten, G.B. (2000). Zur Entstehung von Gehniveaus in sandig-lehmigen Ablagerungen – Experimente und archäologische Befunde (Activity surfaces in sandy-loamy deposits – Experiments and archaeological examples), Jahresbericht 1999. *Arcäeologische Bodnforschung des Kantons Basel-Stadt*. Basel, pp. 102–127.

Retallack, G.J. (2001). *Soils of the Past. An Introduction to Paleopedology*, 2nd edn. Oxford: Blackwell Science.

Reynard, J.P. (2014). Trampling in coastal sites: An experimental study on the effects of shell on bone in coastal sediment. *Quaternary International*, 330, 156–170.

Richter, D., Alperson-Afil, N. and Goren-Inbar, N. (2011). Employing TL methods for the verification of macroscopically determined heat alteration of flint artefacts from Palaeolithic contexts. *Archaeometry*, 53, 842–857.

Rick, J.W. (1976). Downslope movement and archaeological intrasite spatial analysis. *American Antiquity*, 41, 133–144.

Rick, T.C. (2002). Eolian processes, ground cover, and the archaeology of coastal dunes; a taphonomic case study from San Miguel Island, California, USA. *Geoarchaeology: An International Journal*, 17, 811–833.

Rigaud, J.P., Simek, J.F. and Gé, T. (1995). Mousterian fires from Grotte XVI (Dordogne, France). *Antiquity*, 69, 902–912.

Ritter, D.F., Kochel, C.R. and Miller, J.R. (2011). *Process Geomorphology*, 5th edn. Waveland Press.

Robin, G. (2010). Spatial Structures and symbolic systems in Irish and British passage tombs: The organization of architectural elements, parietal carved signs and funerary deposits. *Cambridge Archaeological Journal*, 20, 373–418.

Roebroeks, W. and Villa, P. (2011). On the earliest evidence for habitual use of fire in Europe. *Proceedings of the National Academy of Sciences of the United States of America*, 108, 5209–5214.

Rognon, P., Coudé-Gaussen, G., Fedoroff, N. and Goldberg, P. (1987). Micromorphology of loess in the northern Negev (Israel). In: Fedorof, N., Bresson, L.M. and Courty, M.A. (Eds.) *Micromorphologie des sols – Soil micromorphology*. Proceedings of the VIIth International Working Meeting on Soil Micromorphology, July 1985, Paris. Paris: Association Française pour l' Étude du Sol, pp. 631–638.

Rolland, N. (2000). Cave occupation, fire-making, hominid/carnivore coevolution and Middle Pleistocene emergence of home-base settlement systems. *Acta Anthropologica Sinica*, Supplement to Volume 19, 209–217.

Romano, D.G., Voyatzis, M.E., with appendixes by Sarris, A., Davis, G.H., Mentzer, M., Margaritis, E. and

Starkovich, B.M. (2014). Mt. Lykaion Excavation and Survey Project, Part 1: The Upper Sanctuary. *Hesperia*, 83, 569–652.

Rosen, A.M. (1986). *Cities of Clay: The Geoarcheology of Tells*. Chicago: University of Chicago Press.

Rosen, A.M. (1989). Ancient town and city sites: A view from the microscope. *American Antiquity*, 54, 564–578.

Rosen, A.M. (2003). Middle Palaeolithic plant exploitation: The microbotanical evidence. In: Henry, D.O. (Ed.) *Neanderthals in the Levant: Behavioural Organization and the Beginnings of Human Modernity*. London: Continuum, pp. 156–171.

Roskams, S. (2000). *Interpreting stratigraphy. Site evaluation, recording precedures and stratigraphic analysis*. BAR International Series 910.

Roskams, S. (2001). *Excavation*. Cambridge: Cambridge University Press.

Roskin, J., Katra, I. and Blumberg, D.G. (2013). Late Holocene dune mobilizations in the northwestern Negev dunefield, Israel: A response to combined anthropogenic activity and short-term intensified windiness. *Quaternary International*, 303, 10–23.

Rowley-Conwy, P. (1994). Dung, dirt and deposits: Site formation under conditions of near-perfect preservation at Qasr Ibrim, Egyptian Nubia. In: Luff, R. and Rowley-Conway, P. (Eds.) *Whither Environmental Archaeology?* Oxbow Monograph 38. Oxford: Oxbow, pp. 25–32.

Rudwick, M.J.S. (2008). *Words before Adam. The reconstruction of geohistory in the age of reform*. Chicago: University of Chicago Press.

Sala, I.L. (1986). Use wear and post-depositional surface modification: A word of caution. *Journal of Archaeological Science*, 13, 229–244.

Sánchez Goñi, M.F., Eynaud, F., Turon, J.L. and Shackleton, N.J. (1999). High resolution palynological record off the Iberian margin: Direct land-sea correlation for the Last Interglacial complex. *Earth and Planetary Science Letters*, 171, 123–137.

Sanders, D., Ostermann, M. and Kramers, J. (2009). Quaternary carbonate-rocky talus slope successions (Eastern Alps, Austria): Sedimentary facies and facies architecture. *Facies*, 55, 345–373.

Sandgathe, D.M. (2017). Identifying and describing pattern and process in the evolution of hominin use of fire. *Current Anthropology*, 58, S360–S370.

Sandgathe, D. and Berna, F. (2017). Fire and the genus homo: An introduction to supplement 16. *Current Anthropology*, 58, S165–S174.

Sandgathe, D.M., Dibble, H.L., Goldberg, P., Mcpherron, S.P., Turq, A., Niven, L. and Hodgkins, J. (2011a). On the role of fire in Neandertal adaptations in Western Europe: Evidence from Pech de l'Azé and Roc de Marsal, France. *PaleoAnthropology*, 2011, 216–242.

Sandgathe, D.M., Dibble, H.L., Goldberg, P. and Mcpherron, S.P. (2011b). The Roc de Marsal Neandertal child: A reassessment of its status as a deliberate burial. *Journal of Human Evolution*, 61, 243–253.

Sandgathe, D.M., Dibble, H.L., Goldberg, P., Mcpherron, S.P., Turq, A., Niven, L. and Hodgkins, J. (2011c). Timing of the appearance of habitual fire use. *Proceedings of the National Academy of Sciences*, 108, E298–E298.

Sañudo, P., Blasco, R. and Fernández Peris, J. (2016). Site formation dynamics and human occupations at Bolomor Cave (Valencia, Spain): An archaeostratigraphic analysis of levels I to XII (100–200 ka). *Quaternary International*, 417, 94–104.

Sanz, M., Daura, J., Égüez, N. and Cabanes, D. (2015). On the track of anthropogenic activity in carnivore dens: Altered combustion structures in Cova del Gegant (NE Iberian Peninsula). *Quaternary International*, 437B, 102–114.

Sawyer, A.H. (2016). *Site formation processes at three Viking Age farm middens in Skagafjörður, Iceland*. MA Thesis, University of Boston.

Schaetzl, R.J. and Anderson, S. (2005). *Soils: Genesis and Geomorphology*. Cambridge: Cambridge University Press.

Schick, K.D. (1984). *Processes of Palaeolithic Site Formation: An Experimental Study*. Ph.D. Thesis. University of California, Berkeley.

Schick, K.D. (1987). Experimentally-derived criteria for assessing hydrologic disturbance of archaeological sites. In: Nash, D.T. and Petraglia, M.D. (Eds.) *Natural Formation Processes and the Archaeological Record*. BAR International Series 352. Oxford: British Archaeological Reports, pp. 86–107.

Schick, K. (1992). Geoarchaeological analysis of an Acheulean site at Kalambo falls, Zambia. *Geoarchaeology: An International Journal*, 7, 1–26.

Schieber, J., Southard, J. and Thaisen, K. (2007). Accretion of mudstone beds from migrating floccule ripples. *Science*, 318, 1760–1763.

Schiegl, S., Lev-Yadun, S., Bar-Yosef, O., El Goresy, A. and Weiner, S. (1994). Siliceous aggregates from prehistoric wood ash: A major component of sediments in Kebara and Hayonim Caves (Israel). *Israel Journal of Earth Sciences*, 43, 267–278.

Schiegl, S., Goldberg, P., Bar-Yosef, O. and Weiner, S. (1996). Ash deposits in Hayonim and Kebara Caves, Israel: Macroscopic, microscopic and mineralogical observations, and their archaeological implications. *Journal of Archaeological Science*, 23, 763–781.

Schiegl, S., Goldberg, P., Pfretzschner, H.U. and Conard, N.J. (2003). Paleolithic burnt bone horizons from the Swabian Jura: Distinguishing between in situ fire places and dumping areas. *Geoarchaeology: An International Journal*, 18, 541–565.

Schiffer, M.B. (1972). Archaeological context and systemic context. *American Antiquity*, 37, 156–165.

Schiffer, M.B. (1976). *Behavioral Archeology*. New York: Academic Press.

Schiffer, M.B. (1983). Toward the identification of formation processes. *American Antiquity*, 48, 675–706.

Schiffer, M.B. (1985). Is there a 'Pompeii premise' in archaeology? *Journal of Anthropological Archaeology*, 41, 18–41.

Schiffer, M.B. (1987). *Formation Processes of the Archaeological Record*. Salt Lake City: University of Utah Press.

Schofield, D. and Hall, D.M. (1985). A method to measure the susceptibility of pasture soils to poaching by cattle. *Soil Use and Management*, 1, 134–138.

Scholle, P.A. and Spearing, D. (1982). *Sandstone Depositional Environments*. Tulsa, Oklahoma: American Association of Petroleum Geologists, Memoir 31.

Sedov, S.N., Khokhlova, O.S., Sinitsyn, A.A., Korkka, M.A., Rusakov, A.V., Ortega, B., Solleiroa, E., Rozanova, M.S., Kuznetsova, A.M. and Kazdymh, A.A. (2010). Late Pleistocene paleosol sequences as an instrument for the local paleographic reconstruction of the Kostenki 14 key section (Woronesh Oblast) as an example. *Eurasian Soil Science*, 43, 876–892.

Selby, M.J. (1993). *Hillslope Materials and Processes*, 2nd edn. Oxford: Oxford University Press.

Sergant, J., Crombé, P. and Perdaen, Y. (2006). The 'invisible' hearths: A contribution to the discernment of Mesolithic non-structured surface hearths. *Journal of Archaeological Science*, 33, 999–1007.

Seymour, D. and Schiffer, M. (1987). A preliminary analysis of pithouse assemblages from Snaketown, Arizona. In: Kent, S. (Ed.) *Methods and Theory for Activity Area Research*. New York: Columbia University Press, pp. 549–603.

Shackley, M.L. (1974). Stream abrasion of flint implements. *Nature*, 248, 501–502.

Shackley, M.L. (1978). The behaviour of artifacts as sedimentary particles in a fluviatile environment. *Archaeometry*, 20, 55–61.

Shahack-Gross, R. (2011). Herbivorous livestock dung: Formation, taphonomy, methods for identification, and archaeological significance. *Journal of Archaeological Science*, 38, 205–218.

Shahack-Gross, R. (2017). Archaeological formation theory and geoarchaeology: State-of-the-art in 2016. *Journal of Archaeological Science*, 79, 36–43.

Shahack-Gross, R. and Ayalon, A. (2013). Stable carbon and oxygen isotopic compositions of wood ash: An experimental study with archaeological implications. *Journal of Archaeological Science*, 40, 570–578.

Shahack-Gross, R. and Finkelstein, I. (2008). Subsistence practices in an arid environment: A geoarchaeological investigation in an Iron Age site, the Negev Highlands, Israel. *Journal of Archaeological Science*, 35, 965–982.

Shahack-Gross, R., Bar-Yosef, O. and Weiner, S. (1997). Black-coloured bones in Haynom cave, Israel: Differentiating between burning and oxide staining. *Journal of Archaeological Science*, 24, 439–446.

Shahack-Gross, R., Marshall, F. and Weiner, S. (2003). Geo-archaeology of pastoral sites: The identification of livestock enclosures in abandoned Maasai settlements. *Journal of Archaeological Science*, 30, 439–459.

Shahack-Gross, R., Berna, F., Karkanas, P. and Weiner, S. (2004a). Bat guano and preservation of archaeological remains in cave sites. *Journal of Archaeological Science*, 31, 1259–1272.

Shahack-Gross, R., Marshall, F., Ryan, K., and Weiner, S. (2004b). Reconstruction of spatial organization in abandoned Maasai settlements: Implications for site structure in the pastoral Neolithic of East Africa. *Journal of Archaeological Science*, 31, 1395–1411.

Shahack-Gross, R., Albert, R.M., Gilboa, A., Nagar-Hilman, O., Sharon, I. and Weiner, S. (2005). Geoarchaeology in an urban context: The uses of space in a Phoenician monumental building at Tel Dor (Israel). *Journal of Archaeological Science*, 32, 1417–1431.

Shahack-Gross, R., Simons, A. and Ambrose, S.H. (2008a). Identification of pastoral sites using stable nitrogen and carbon isotopes from bulk sediment samples: A case study in modern and archaeological pastoral settlements in Kenya. *Journal of Archaeological Science*, 35, 983–990.

Shahack-Gross, R., Ayalon, A., Goldberg, P., Goren, Y., Ofek, B., Rabinovich, R. and Hovers, E. (2008b). Formation processes of cemented features in Karstic Cave sites revealed using stable oxygen and carbon isotopic analyses: A case study at Middle Paleolithic Amud Cave, Israel. *Geoarchaeology: An International Journal*, 23, 43–62.

Shahack-Gross, R., Berna. F., Karkanas, P., Lemorini, C., Gopher, A. and Barkai, R. (2014). Evidence for the repeated use of a central hearth at Middle Pleistocene (300 ky ago) Qesem Cave, Israel. *Journal of Archeological Science*, 44, 12–21.

Shaw, J.W. (1977). New evidence for Aegean roof construction from Bronze Age Thera. *American Journal of Archaeology*, 81, 229–233.

Shaw, J.W. (2009). *Minoan Architecture: Materials and Techniques*. Bottega D'Erasmo: Centro di Archaeologia Cretese – Universitá de Catania.

Sheldon, N.D. and Retallack, G.J. (2001). Eequation for compaction due to burial. *Geology*, 29, 247–250.

Sheppard, P.J. and Pavlish, L.A. (1992). Weathering of archaeological cherts: A case study from the Solomon Islands. *Geoarchaeology: An International Journal*, 7, 41–53.

Sherwood, S.C. (2001). *The Geoarchaeology of Dust Cave: A Late Paleoindian Through Middle Archaic Site in the Middle Tennessee River Valley*. Doctoral Dissertation. University of Tennessee, Knoxville, TN.

Sherwood, S.C. (2008). Increasing the resolution of cave archaeology: Micromorphology and the classification of burned deposits at Dust Cave. In: Dye, D. (Ed.) *Cave Archaeology in the Eastern Woodlands: Papers in honor of Patty Jo Watson*. Knoxville: University of Tennessee Press, pp. 27–47.

Sherwood, S.C. and Chapman, J. (2005). The identification and potential significance of early Holocene prepared clay surfaces: Examples from Dust Cave and Icehouse Bottom. *Southeastern Archaeology*, 24, 70–82.

Sherwood, S.C. and Kidder, T.R. (2011). The DaVincis of dirt: Geoarchaeological perspectives on Native American mound building in the Mississippi River basin. *Journal of Anthropological Archaeology*, 30, 69–87.

Sherwood, S.C., Driskell, B.N., Randall, A.R. and Meeks, S.C. (2004). Chronology and stratigraphy at Dust Cave. *Alabama American Antiquity*, 69, 533–554.

Shillito, L.M. (2011). *Daily activities, diet and resource use at Neolithic Çatalhöyük*. BAR International Series 2232.

Shillito L.M. (2015). Middens and other trash deposits. In: Metheny, K.B. and Beaudry, M.C. (Eds.) *The Archaeology of Food: An Encyclopedia*. Rowman and Littlefield Publishers, pp. 316–318.

Shillito, L.M. and Matthews, W. (2013). Geoarchaeological investigations of midden-formation processes in the Early to Late Ceramic Neolithic levels at Çatalhöyük, Turkey ca. 8550–8370 cal BP. *Geoarchaeology: An International Journal*, 28, 25–49.

Shillito, L.M., Almond, M.J., Wicks, K., Marshall, L.J.R., Matthews, W. (2009). The use of FT-IR as a screening technique for organic residue analysis of archaeological samples. *Spectrochimica Acta – Part A: Molecular and Biomolecular Spectroscopy*, 72, 120–125.

Shillito, L.M., Matthews, W., Bull, I.D. and Almond, M.J. (2011a). The microstratigraphy of middens: Capturing daily routine in rubbish at Neolithic Çatalhöyük, Turkey. *Antiquity*, 85, 1024–1038.

Shillito, L.M., Bull, I.D., Matthews, W., Almond, M.J., Williams, J. and Evershed, R.P. (2011b). Biomolecular and micromorphological analysis of suspected faecal deposits at Neolithic Çatalhöyük, Turkey. *Journal of Archaeological Science*, 38, 1869–1877.

Shipman, P., Foster, G. and Schoeninger, M. (1984). Burnt bones and teeth: An experimental study of color, morphology, crystal structure, and shrinkage. *Journal of Archaeological Science*, 11, 307–325.

Shunk, A.J., Driese, S.G. and Clark, M.G. (2006). Latest Miocene to earliest Pliocene sedimentation and climate record derived from paleosinkhole fill deposits, Gray Fossil Site, northeastern Tennessee, USA. *Palaeogeography, Palaeoclimatology, Palaeoecology*, 231, 265–278.

Sigurdsson, H., Cashdollar, S. and Sparks, S.R.J. (1982). The eruption of Vesuvius in AD 79: Reconstruction from historical and volcanological evidence. *American Journal of Archaeology*, 86, 39–51.

Sillar, B. (2000). Dung by preference: The choice of fuel as an example of how Andean pottery production is embedded within wider technical, social, and economic practices. *Archaeometry*, 42, 43–60.

Simon, U., Händel, M., Einwögerer, T. and Neugebauer-Maresch, C. (2014). The archaeological record of the Gravettian open air site Krems-Wachtberg. *Quaternary International*, 351, 5–13.

Simpson, I.A. and Barett, J.H. (1996). Interpretation of midden formation processes at Robert's Haven, Caithness, Scotland using thin section micromorphology. *Journal of Archaeological Science*, 4, 543–556.

Simpson, I.A., Vésteinsson, O., Adderley, W.P. and McGovern, T.H. (2003). Fuel resource utilisation in landscapes of settlement. *Journal of Archaeological Science*, 30, 1401–1420.

Singer, R. and Wymer, J. (1982). *The Middle Stone Age at Klasies River Mouth in South Africa*. Chicago: University of Chicago Press.

Skaarup, J. and Grøn, O. (2004). *Møllegabet II. A submerged Mesolithic settlement in southern Denmark*. Oxford: Archaeopress.

Smith, G.A. (1986). Coarse-grained nonmarine volcanoclastic sediment: Terminology and depositional process. *Geological Society of America Bulletin*, 97, 1–10.

Snoeck, C., Lee-Thorp, J.A. and Schulting, R.J. (2014). From bone to ash: Compositional and structural changes in burned modern and archaeological bone. *Palaeogeography, Palaeoclimatology, Palaeoecology*, 416, 55–68.

Soil Survey Staff (1999). *Soil Taxonomy: A basic system of soil classification for making and interpreting soil surveys*, 2nd edn, Handbook No. 436. Washington DC: US Department of Agriculture/Natural Resources Conservation Service, pp. 869.

Sorensen, A.C. (2017). On the relationship between climate and Neandertal fire use during the Last Glacial in south-west France. *Quaternary International*, 436A, 114–128.

Soressi, M., McPherron, S.P., Lenoire, M., Dogandzic, T., Goldberg, P., Jacobs, Z., Maigrot, Y., Martisius, N.L., Miller, C.E., Rendu, W., Richards, M., Skinner, M.M., Steele, T.E., Talamo, S. and Texier, J.P. (2013). Neandertals made the first specialized bone tools in Europe. *Proceedings of the National Academy of Sciences*, 110, 14186–14190.

Speth, J.D. (2006). Housekeeping, Neanderthal-style: Hearth placement and midden formation in Kebara cave (Israel). In: Hovers, E. and Kuhn, S. (Eds.) *Transitions before the Transition: Evolution and stability in the Middle Paleolithic and Middle StoneAge*. New York, Boston, Dordrecht, London, Moscow: Springer, pp. 171–188.

Springer, G. (2005). Clastic sediments in caves. In: White, W.B. and Culver, D.C. (Eds.) *Encyclopedia of Caves*. Amsterdam: Elsevier, pp. 134–140.

Stahlschmidt, M.C., Miller, C.E., Ligouis, B., Goldberg, P., Berna, F., Urban, B. and Conard, N.J. (2015). The depositional environments of Schöningen 13 II-4 and their archaeological implications. *Journal of Human Evolution*, 89, 71–91.

Stein, J.K. (1983). Earthworm activity: A source of potential disturbance of archaeological sediments. *American Antiquity*, 48, 277–289.

Stein, J.K. (1987). Deposits for archaeologists. *Advances in Archaeological Method and Theory*, 11, 337–395.

Stein, J.K. (1990). Archaeological stratigraphy. In: Lasca, N.P. and Donahue, J. (Eds.) *Archaeological Geology of North America, Centennial Special Volume 4*. Boulder: Geological Society of America, pp. 513–523.

Stein, J.K. (2001). A review of site formation processes and their relevance to geoarchaeology. In: Goldberg, P., Holliday, V.T. and Ferring, C.R. (Eds.) *Earth Sciences and Archaeology*. New York: Kluwer Academic/Plenum Publishers, pp. 37–51.

Stephens, M., Rose, J., Gilbertson, D. and Canti, M.G. (2005). Micromorphology of cave sediments in the humid tropics: Niah Cave, Sarawak. *Asian Perspectives*, 44, 42–55.

Sternberg, R. and Lass, E. (2007). An archaeomagnetic study of two hearths from Kebara Cave, Israel. In: BAR-Yosef, O. and Meignen, L. (Eds.) *Kebara Cave, Mt Carmel, Israel: The Middle and Upper Paleolithic Archaeology, Part I*. American School of Prehistoric Research Bulletin 49, Peabody Museum of Archaeology and Ethnology, Harvard University, pp. 123–130.

Stiner, M.C., Kuhn, S.L., Weiner, S. and Bar-Yosef, O. (1995). Differential burning, recrystalization, and fragmentation of archaeological bone. *Journal of Archaeological Science*, 22, 223–237.

Stiner, M.C., Arsebük, G. and Howell, F.C. (1996). Cave bears and Paleolithic artifacts in Yarimburgaz Cave, Turkey: Dissecting a palimpsest. *Geoarchaeology: An International Journal*, 11, 279–327.

Stiner, M., Buitenhuis, H., Duru G., Kuhn, S.L., Mentzer, S.M., Munro, N.D, Pöllath, N., Quade, J., Tsartsidou, G. and Özbaşaranc, M. (2014). A forager–herder trade-off, from broad-spectrum hunting to sheep management at

Aşıklı Höyük, Turkey. *Proceedings of the National Academy of Sciences*, 111, 8404–8409.

Stolt, M.H. and Lindbo, D.L. (2010). Soil organic matter. In: Stoops, G., Marcelino, V. and Mees, F. (Eds.) *Interpretation of Micromorphological Features of Soils and Regoliths*. Amsterdam: Elsevier, pp. 369–396.

Stoops, G. (2003). *Guidelines for Analysis and Description of Soil and Regolith Thin Sections*. Madison, Wisconsin: Soil Science of America Inc.

Stow, D.A.V. (2010). *Sedimentary Rocks in the Field: A Colour Guide*. London: Manson Publishing Ltd.

Straus, L.G. (1979). Caves: A palaeoanthropological resource. *World Archaeology*, 10, 331–339.

Sullivan, M.E. and Sassoon, M. (1987). Prehistoric occupation of Loloata Island, Papua New Guinea. *Australian Archaeology*, 24, 1–9.

Surovell, T.A. and Stiner, M.C. (2001). Standardizing infra-red measures of bone mineral crystallinity: An experimental approach. *Journal of Archaeological Science*, 28, 633–642.

Syvitski, J.P. (2007). *Principles, Methods and Application of Particle Size Analysis*. Cambridge: Cambridge University Press.

Taglioretti, V., Sardella, N.H. and Fugassa, M.H. (2014). Morphometric analysis of modern faeces as a tool to identify artiodactyls' coprolites. *Quaternary International*, 352, 64–67.

Tani, M. (1995). Beyond the identification of formation processes: Behavioral inference based on traces left by cultural formation processes. *Journal of Archaeological Method and Theory*, 2, 231–252.

Tankard, A.J. (1976). The stratigraphy of a coastal cave and its palaeoclimatic significance. *Palaeoecology of Africa*, 9, 151–159.

Tankard, A.J. and Schweitzer, F.R. (1976). Textural analysis of cave sediments: Die Kelders, Cape Province, South Africa. In: Davidson, D.A. and Shackley, M.L. (Eds.) *Geoarchaeology: Earth Science and the Past*. London: Duckworth, pp. 289–316.

Tarplee, M.F.V., van der Meer, J.J.M. and Davis, G.R. (2011). The 3D microscopic 'signature' of strain within glacial sediments revealed using X-ray computed microtomography. *Quaternary Science Reviews*, 30, 3501–3532.

Terhorst, B., Kühn, P., Damm, B., Hambach, U., Meyer-Heintze, S. and Sedov, S. (2014). Paleoenvironmental fluctuations as recorded in the loess-paleosol sequence of the Upper Paleolithic site Krems-Wachtberg. *Quaternary International*, 351, 67–82.

Terry, R.E., Fernández, F.G., Parnell, J.J. and Inomata, T. (2004). The story in the floors: Chemical signatures of ancient and modern Maya activities at Aguateca, Guatemala. *Journal of Archaeological Science*, 31, 1237–1250.

Texier, J.P. (2009). *Histoire géologique de sites préhistoriques classiques du Périgord: Une vision actualisée: la Micoque, la grotte Vaufrey, le Pech de l'Azé I et II, la Ferrassie, l'abri Castanet, le Flageolet, Laugerie Haute.* Paris: Édition du Comité des travaux historiques et scientifiques.

Texier, J.P. and Meireles, J. (2003). Relict mountain slope deposits of northern Portugal: Facies, sedimentogenesis and environmental implications. *Journal of Quaternary Science*, 18, 133–150.

Théry-Parisot, I. and Costamagno, S. (2005). Propriétés combustibles des ossements: Données expérimentales et réflexions archéologiques sur leur emploi dans les sites paléolithiques. *Gallia Préhistoire*, 47, 235–254.

Thompson, T.J.U., Gauthier, M. and Islam, M. (2009). The application of a new method of Fourier transform infrared spectroscopy to the analysis of burned bone. *Journal of Archaeological Science*, 36, 910–914.

Thompson, C.E.L., Ball, S., Thompson, T.J.U. and Gowland, R. (2011). The abrasion of modern and archaeological bones by mobile sediments: The importance of transport modes. *Journal of Archaeological Science*, 38, 784–793.

Thoms, A.V. (2007). Fire-cracked rock features on sandy landforms in the Northern Rocky Mountains: Toward establishing reliable frames of reference for assessing site integrity. *Geoarchaeology: An International Journal*, 22, 477–510.

Thoms, A.V. (2008). The fire stones carry: Ethnographic records and archaeological expectations for hot-rock cookery in western North America. *Journal of Anthropological Archaeology*, 27, 443–460.

Thoms, A.V. (2009). Rocks of ages: Propagation of hot-rock cookery in western North America. *Journal of Archaeological Science*, 36, 573–591.

Thoms, A.V. (2015). Earth-oven cookery and targeted features: Variation in form and function. In: Thoms, A.V., Kibler, K. and Boyd, D. (Eds.) *Earth Ovens and Prehistoric Plant Use: Archaeological and Archaeobotanical Studies at Burned Rock Features on the Fort Hood Military Reservation in Central Texas.* Research Report No. 65. Texas: US Army, Fort Hood, pp. 199–215.

Todisco, D. and Bhiry, N. (2008a). Micromorphology of periglacial sediments from the Tayara site, Qikirtaq Island, Nunavik (Canada). *Catena*, 76, 1–21.

Todisco, D. and Bhiry, N. (2008b). Palaeoeskimo site burial by solifluction: Periglacial geoarchaeology of the Tayara site (KbFk-7), Qikirtaq Island, Nunavik (Canada). *Geoarchaeology: An International Journal*, 23, 177–211.

Toker, N.Y, Onar, V., Belli, O., Ak, S., Alpak, H. and Konyar, E. (2005). Preliminary results of the analysis of coprolite material of a dog unearthed from the Van-Yoncatepe necropolis in eastern Anatolia. *Turkish Journal of Veterinary and Animal Sciences*, 29, 759–765.

Toots, H. (1965). Orientation and distribution of fossils as environmental indicators. In: Proceedings of the Nineteenth Field Conference. Casper: Wyoming Geological Association, pp. 219–229.

Tourloukis, V., Karkanas, P. and Wallinga, J. (2015). Revisiting Kokkinopilos: Middle Pleistocene radiometric dates for stratified archaeological remains in Greece. *Journal of Archaeological Science*, 57, 355–369.

Trigger, B.G. (2006). *A History of Archaeological Thought*, 2nd edn. New York: Cambridge University Press.

Tsartsidou, G., Lev-Yadun, S., Efstratiou, N. and Weiner, S. (2008). Ethnoarchaeological study of phytolith assemblages from an agro-pastoral village in Northern Greece (Sarakini): Development and application of a Phytolith Difference Index. *Journal of Archaeological Science*, 35, 600–613.

Tsartsidou, G., Lev-Yadun, S., Efstratiou, N. and Weiner, S. (2009). Use of space in a Neolithic village in Greece (Makri): Phytolith analysis and comparison of phytolith assemblages from an ethnographic setting in the same area. *Journal of Archaeological Science*, 36, 2342–2352.

Turk, J. and Turk, M. (2010). Paleotemperature record in late Pleistocene clastic sediments at Divje babe 1 cave (Slovenia). *Journal of Archaeological Science*, 37, 3269–3280.

Turner, R., Roberts, N., Eastwood, W., Jenkins, E. and Rosen, R. (2010). Fire, climate and the origins of agriculture: Micro-charcoal records of biomass burning during the last glacial–interglacial transition in Southwest Asia. *Journal of Quaternary Science*, 25, 371–386.

Turq, A., Dibble, H., Faivre, J.P., Goldberg, P., Mcpherron, S.J.P. and Sandgathe, D. (2008). *Le Moustérien du Périgord Noir: Quoi de neuf?* Mémoire de la Société Préhistorique Française, XLVII, 83–93.

Turq, A., Dibble, H.L., Goldberg, P., Mcpherron, S.P., Sandgathe, D., Jones, H., Maddison, K., Maureille, B., Mentzer, S., Rink, J. and Steenhuyse, A. (2011). Les Fouilles Récentes du Pech de l'Azé IV (Dordogne). *Gallia Préhistoire*, 53, 1–58.

Valladas, H., Joron, J.L., Valladas, G., Arensburg, B., Bar-Yosef, O., Belfer-Cohen, A., Goldberg, P., Laville, H., Meignen, L., Rak, Y., Tchernov, E., Tillier, A.M. and Vandermeersch, B. (1987). Thermoluminescence dates for the Neanderthal burial site at Kebara in Israel. *Nature*, 330, 159–160.

Vallverdú-Poch, J. and Courty, M.A. (2012). Microstratigraphic analysis of level J deposits: A dual Paleoenvironmental-Paleoethnographic contribution to Paleolithic archeology at the Abric Romaní. In: Carbonell-i-Roura, E. (Ed.) *High Resolution Archaeology and Neanderthal Behavior.* Springer, pp. 77–134.

Vallverdú, J., Allué, E., Bischoff, J.L., Cáceres, I., Carbonell, E., Cebrià, A., García-Antón, D., Huguet, R., Ibáñez, N., Martínez, K., Pastó, I., Rosell, J., Saladié, P. and Vaquero, M. (2005). Short human occupations in the Middle

Palaeolithic level i of the Abric Romaní rock-shelter (Capellades, Barcelona, Spain). *Journal of Human Evolution*, 48, 157–174.

Vallverdú, J., Vaquero, M., Cáceres, I., Allué, E., Rosell, J., Saladié, P., Chacón, G., Ollé, A., Canals, A., Sala, R., Courty, M.A. and Carbonell, E. (2010). Sleeping activity area within the site structure of archaic human groups: Evidence from Abric Romaní level N combustion activity areas. *Current Anthropology*, 51, 137–145.

Vallverdú, J., Alonso, S., Bargalló, A., Bartrolí, R., Campeny, G., Carrancho, Á., Expósito, I., Fontanals, M., Gabucio, J. and Gómez, B. (2012). Combustion structures of archaeological level O and Mousterian activity areas with use of fire at the Abric Romaní rockshelter (NE Iberian Peninsula). *Quaternary International*, 247, 313–324.

van Keuren, S. and Roos, C.I. (2013). Geoarchaeological evidence for ritual closure of a kiva at Fourmile Ruin, Arizona. *Journal of Archaeological Science*, 40, 615–625.

van Riper, A.B. (1993). *Men among the Mammoths: Victorian Science and the Discovery of Prehistory*. Chicago: University of Chicago Press.

van Steijn, H. and Coutard, J.P. (1989). Laboratory experiments with small debris flwos: Physical properties related to sedimentary characteristics. *Earth Surface Processes and Landforms*, 14, 587–596.

van Steijn, H., Bertran, P., Francou, B., Hıtu, B. and Texier, J.P. (1995). Models for the genetic and environmental interpretation of stratified slope deposits: Review. *Permafrost and Periglacial Processes*, 6, 125–146.

van Vliet-Lanoë, B. (1985). Frost effects in soils. In: Boardman, J. (Ed.) *Soils and Quaternary Landscape Evolution*. Chichester: John Wiley & Sons Ltd, pp. 117–158.

van Vliet-Lanoe, B. (1986). Micromorphology. In: *Callow, P.* and Cornford, J.M. (Eds.) La Cotte de St. Brelade. Norwich: Geo Books, pp. 1961–1978.

van Vliet-Lanoë, B. (1998). Frost and soils: Implications for paleosols, paleoclimates and stratigraphy. *Catena*, 34, 157–183.

van Vliet-Lanoë, B. (2010). Frost action. In: Stoops, G., Marcelino, V. and Mees, F. (Eds.) *Interpretation of Micromorphological Features of Soils and Regoliths*. Amsterdam: Elsevier, pp. 81–108.

Van Vliet-Lanoë, B., Coutard, J.P. and Pissart, A. (1984). Structures caused by repeated freezing and thawing in various loamy sediments: A comparison of active, fossil and experimental data. *Earth Surface Processes and Landforms*, 9, 553–565.

Vaquero, M., Vallverdú, J., Rosell, J., Pasto, I. and Allue, E. (2001). Neandertal behavior at the Middle Palaeolithic site of Abric Romani, Capellades, Spain. *Journal of Field Archaeology*, 28, 93–114.

Velichko, A.A., Pisareva, V.V., Sedov, S.N., Sinitsyn, A.A. and Timireva, S.N. (2009). Paleogeography of Kostenki-14 (MARKINA GORA). *Archaeology Ethnology and Anthropology of Eurasia*, 37, (4), 35–50.

Ventra, D., Díaz, G.C. and Boer, P.L.D. (2013). Colluvial sedimentation in a hyperarid setting (Atacama Desert, northern Chile): Geomorphic controls and stratigraphic facies variability. *Sedimentology*, 60, 1257–1290.

Villa, P. (1976). Sols et niveaux d'habitat du paléolithique inférieur en Europe et au Proche Orient. *Quaternaria*, 19, 107–134.

Villa, P. and Courtin, J. (1983). The interpretation of stratified sites: A view from underground. *Journal of Archaeological Science*, 10, 267–281.

Villa, P. and Soressi, M. (2000). Stone tools in carnivore sites: The case of Bois Roche. *Journal of Anthropological Research*, 56, 187–215.

Villa, P., Castel, J.C., Beauval, C., Bourdillat, V. and Goldberg, P. (2004). Human and carnivore sites in the European Middle and Upper Paleolithic: Similarities and differences in bone modification and fragmentation. *Revue de Paléobiologie*, 23, 705–730.

Villagran, X.S. (2014). A redefinition of waste: Deconstructing shell and fish mound formation among coastal groups of southern Brazil. *Journal of Anthropological Archaeology*, 36, 211–227.

Villagran, X.S., Giannini, P.C.F. and DeBlasis, P. (2009). Archaeofacies analysis: Using depositional attributes to identify anthropic processes of deposition in a monumental shell mound of Santa Catarina State, southern Brazil. *Geoarchaeology: An International Journal*, 24, 311–335.

Villagran, X.S., Balbo, A.L., Madella, M., Vila, A. and Estevez, J. (2011). Experimental micromorphology in Tierra del Fuego (Argentina): Building a reference collection for the study of shell middens in cold climates. *Journal of Archaeological Science*, 38, 588–604.

Villagran, X.S., Schaefer, C.E.G.R. and Ligouis, B. (2013). Living in the cold: Geoarchaeology of sealing sites from Byers Peninsula (Livingston Island, Antarctica). *Quaternary International*, 315, 184–199.

Villeneuve, S., Hayden, B., Goldberg, P., Cross, G. and Sisk, M. (2011). *Investigating processes of aggregated village formation on the western Canadian Plateau*. Sacramento: Society for American Archaeology Annual Meeting.

Voorhies, M. (1969). Taphonomy and population dynamics of an Early Pliocene vertebrate fauna, Knox County, Nebraska. Special Paper No. 1. In: *Contributions to Geology*. Laramie: University of Wyoming, pp. 1–69.

Wadley, L. (2006). Partners in grime: Results of multi-disciplinary archaeology at Sibudu Cave. *Southern African Humanities*, 18, 315–341.

Wadley, L., Sievers, C., Bamford, M., Goldberg, P., Berna, F. and Miller, C. (2011). Middle Stone Age bedding construction and settlement patterns at Sibudu, South Africa. *Science*, 334, 1388–1391.

Walker, W.H. (2002). Stratigraphy and practical reason. *American Anthropologist*, 104, 159–177.

Walther, J. (1894). *Einleitung in die Geologie als historische Wissenschaft.* Jena: Verlag von Gustav Fischer.

Warburton, D.A. (2003). *Archaeological Stratigraphy. A Near Eastern Approach.* Neuchâtel: Researchers and Publication.

Waters, M.R. (1992). *Principles of Geoarchaeology: A North American Perspective.* Tucson: University of Arizona Press.

Wattez, J. (1988). Contribution à la connaissance des foyers préhistoriques par l'étude des cendres. *Bulletin de la Société Préhistorique Française,* 85, 352–366.

Wattez, J. (1992). Dynamique de formation des structures de combustion de la fin du Paléolithique au Néolithique moyen. Approche méthodologique et implications culturelles. Ph.D. Thesis, Université Paris I.

Wattez, J. and Courty, M.A. (1987). Morphology of ash of some plant materials. In: Fedorof, N., Bresson, L.M. and Courty, M.A. (Eds.) *Micromorphologie des Sols – Soil Micromorphology.* Proceedings of the VIIth International Working Meeting on Soil Micromorphology, July 1985, Paris. Paris: Association Française pour l' Étude du Sol, pp. 677–683.

Weaver, T. and Dale, D. (1978). Trampling effects of hikers, motorcycles and horses in meadows and forests. *Journal of Applied Ecology,* 15, 451–457.

Weiner, S. (2010). *Microarcheology. Beyond the visible archaeological record.* Cambridge: Cambridge University Press.

Weiner, S., Goldberg, P. and Bar-Yosef, O. (1993). Bone preservation in Kebara Cave, Israel using on-site Fourier transform infrared spectrometry. *Journal of Archaeological Science,* 20, 613–627.

Weiner, S., Schiegl, S., Goldberg, P. and Bar-Yosef, O. (1995). Mineral assemblages in Kebara and Hayonim, Israel: Excavation strategies, bone preservation and wood ash remnants. *Israel Journal of Chemistry,* 35, 143–154.

Weiner, S., Xu, Q., Goldberg, P., Liu, J. and Bar-Yosef, O. (1998). Evidence for the use of fire at Zhoukoudian, China. *Science,* 281, 251–253.

Weiner, S., Goldberg, P. and Bar-Yosef, O. (2002). Three-dimensional distribution of minerals in the sediments of Hayonim cave, Israel: Diagenetic processes and archaeological implications. *Journal of Archaeological Science,* 29, 1289–1308.

Weiner, S., Berna, F., Cohen-Ofri, I., Shahack-Gross, R., Maria Albert, R., Karkanas, P., Meignen, L. and Bar-Yosef, O. (2007). Mineral distribution in Kebara Cave: Diagenesis and its effects on the archaeological record. In: Bar-Yosef, O. and Meignen, L. (Eds.) *Kebara Cave Mt Carmel, Israel. The Middle and Upper Paleolithic Archaeology. Part I.* American School of Prehistoric Research Bulletin 49, Peabody Museum of Archaeology and Ethnology, Harvard University, pp. 131–146.

Wells, S.G. and Harvey, A.M. (1987). Sedimentologic and geomorphic variations in storm-generated alluvial fans, Howgill Fells, northwest England. *Geological Society of America Bulletin,* 98, 182–198.

Wendorf, F. and Schild, R. (1980). *Prehistory of the Eastern Sahara.* New York: Academic Press.

Wentworth, C.K. (1922). A scale of grade and class terms for clastic sediments. *Journal of Geology,* 30, 377–392.

Westaway, K.E., Sutikna, T., Saptomo, W.E., Jatmiko, Morwood, M.J., Roberts, R.G. and Hobbs, D.R. (2009a). Reconstructing the geomorphic history of Liang Bua, Flores, Indonesia: A stratigraphic interpretation of the occupational environment. *Journal of Human Evolution,* 57, 465–483.

Westaway, K.E., Roberts, R.G., Sutikna, T., Morwood, M.J., Drysdale, R., Zhao, J. and Chivas, A.R. (2009b). The evolving landscape and climate of western Flores: An environmental context for the archaeological site of Liang Bua. *Journal of Human Evolution,* 57, 450–464.

White, W.B. (1988). *Geomorphology and Hydrology of Karst Terrains.* New York: Oxford University Press.

White, W.B. and Culver, D.C. (Eds.) (2012). *Encyclopedia of Caves,* 2nd edn. San Diego: Academic Press.

White, R., Mensan, R., Clark, A.E., Tartar, E., Marquer, L., Bourrillon, R. Goldberg, P., Chiotti, L., Cretin, C., Rendu, W. and Pike-Tay, A. (2017). Heat and light technologies in the Vézère Aurignacian: Some glimpses of fire-feature organization and structure. *Current Anthropology,* 58, S288–S302.

Więckowski, W., Cohen, S., Mienis, H.K. and Horwitz, L.K. (2013). The excavation and analysis of porcupine dens and burrowing on ancient and recent faunal and human remains at Tel Zahara (Israel). *Bioarchaeology of the Near East,* 7, 3–20.

Wilkinson, A. and Bunting, B.T. (1975). Overland transport by rill water in a periglacial environment. *Geografiska Annaler,* 57A, 105–116.

Wilson, D.C. (1994). Identification and assessment of secondary refuse aggregates. *Journal of Archaeological Method and Theory,* 1, 41–68.

Wilson, C.A., Davidson, D.A. and Cresser, M.S. (2008). Multi-element soil analysis: An assessment of its potential as an aid to archaeological interpretation. *Journal of Archaeological Science,* 35, 412–424.

Wilson, C.A., Davidson, D.A. and Cresser, M.S. (2009). An evaluation of the site specificity of soil elemental signatures for identifying and interpreting former functional areas. *Journal of Archaeological Science,* 36, 2327–2334.

Wood, R.W. and Johnson, D.L. (1978). A survey of disturbance processes in archeological site formation. In: Schiffer, M.B. (Ed.) *Advances in Archaeological Method and Theory.* New York: Academic Press, pp. 315–381.

Woodward, J.C. and Goldberg, P. (2001). The sedimentary records in Mediterranean rockshelters and caves:

Archives of environmental change. *Geoarchaeology: An International Journal*, 16, 327–354.

Woodward, J.C., Hamlin, R.H.B., Macklin, M.G., Karkanas, P. and Kotjabopoulou, E. (2001). Quantitative sourcing of slackwater deposits at Boila Rockshelter: A record of Late glacial flooding and Palaeolithic settlement in the Pindus Mountains, northwest Greece. *Geoarchaeology: An International Journal*, 16, 501–536.

Wrangham, R. (2009). *Catching Fire: How Cooking Made Us Human*. New York: Basic Books.

Wright, V.P. and Tucker, M.E. (1991). Calcretes: An introduction. In: Wright, V.P. and Tucker M.E. (Eds.) *Calcretes*. Reprint series 2 volume of the International Association of Sedimentologists. Oxford: Blackwell, pp. 1–22.

Wright, J., Triantaphyllou, S., Dabney, M.K., Karkanas, P., Kotzamani, G. and Livarda, A. (2008). Nemea valley archaeological project, excavations at Barnavos, final report. *Hesperia*, 77, 605–672.

Wu, X., Zhang, C., Goldberg, P., Cohen, D., Pan, Y., Arpin, T. and Bar-Yosef, O. (2012). Early Pottery at 20,000 years ago in Xianrendong Cave, China. *Science*, 336, 1696–1700.

Zecchin, M., Nalin, R. and Roda, C. (2004). Raised Pleistocene marine terraces of the Crotone peninsula (Calabria, southern Italy): Facies analysis and organization of their deposits. *Sedimentary Geology*, 172, 165–185.

Zerboni, A. (2011). Micromorphology reveals in situ Mesolithic living floors and archaeological features in multiphase sites in central Sudan. *Geoarchaeology: An International Journal*, 26, 365–391.

Zilhão, J., d'Errico, F., Bordes, J.G., Lenoble, A., Texier, J.P. and Rigaud, J.P. (2006). Analysis of Aurignacian interstratification at the Châltelperronian-type site and implications for the behavioral modernity of Neandertals. *Proceedings of the National Academy of Sciences*, 103, 1243–12648.

Žychowski, J. (2011). Geological aspects of decomposition of corpses in mass graves from WW1 and 2, located in SE Poland. *Environmental Earth Sciences*, 64, 437–448.

Index

Reconstructing Archaeological Sites: Understanding the Geoarchaeological Matrix, First Edition. Panagiotis (Takis) Karkanas and Paul Goldberg.
© 2019 John Wiley & Sons Ltd. Published 2019 by John Wiley & Sons Ltd.

Sites and Place Names